Proceedings of the Symposium F
ELECTROMAGNETIC MATERIALS

International Conference on Materials for
Advanced Technologies

Proceedings of the Symposium F
ELECTROMAGNETIC MATERIALS

SUNTEC, Singapore   7 – 12 December 2003

edited by

Lim Hock, Ong Chong Kim, Serguei Matitsine and **Gan Yeow Beng**

National University of Singapore

NEW JERSEY · LONDON · SINGAPORE · SHANGHAI · HONG KONG · TAIPEI · BANGALORE

Published by

World Scientific Publishing Co. Pte. Ltd.

5 Toh Tuck Link, Singapore 596224

USA office: Suite 202, 1060 Main Street, River Edge, NJ 07661

UK office: 57 Shelton Street, Covent Garden, London WC2H 9HE

British Library Cataloguing-in-Publication Data
A catalogue record for this book is available from the British Library.

ELECTROMAGNETIC MATERIALS
Proceedings of the Symposium F, ICMAT 2003

ISBN 981-238-372-7 (pbk)

This proceedings volume is sponsored by ICMAT 2003 Conference Organiser, Materials Research Society (Singapore).

Printed in Singapore by World Scientific Printers (S) Pte Ltd

FOREWORD

International Conference on Materials for Advanced Technologies (ICMAT) is a biannual conference jointly launched by the National University of Singapore and the Materials Research Society (Singapore) in 2001. The first conference held in July 2001 was a great success, attracting 1,400 participants from 42 countries. The current ICMAT 2003 is the second in the series. In early 2003, the world was surprised by an outbreak of the Severe Acute Respiratory Syndrome (SARS), and the Organizing Committee took the decision to postpone ICMAT 2003 by about six months. Despite this disruption, support remains strong, and ICMAT 2003 will open on 8 December 2003 with a full programme of 16 Symposia, 8 Plenary Lectures, and 2 Public Lectures by Nobel Laureates. There will also be an exhibition, technical visits and social programmes.

This Symposium F is dedicated to the studies of materials that exhibit electromagnetic effects—theoretical understanding of their properties, characterization and measurement techniques, design and fabrication methods, applications. Despite the long history of this line of study, the complex phenomena of materials-electromagnetic field interactions still provide fertile ground for basic research and technology innovations. Materials with unique properties are synthesized to meet the ever more demanding specifications of modern technologies, and scientists and engineers are becoming very skillful and precise in the control of their performance. Radical ideas occasionally ignite heated debates. An example is the "meta-materials" with negative permittivity and permeability. Their intriguing properties have led scientists to re-examination some basic issues of electromagnetic wave propagation, and to explore potential ground-breaking applications. We are pleased that this Symposium will provide a forum for the presentation of a good collection of papers on this interesting topic by leading experts, including Prof V.G. Veselago, whose seminal paper of 1968 is responsible for starting this line of investigation.

On behalf of the Organizing Committee of this Symposium, I wish to thank the ICMAT 2003 Organizing Committee for setting up the grand stage on which this Symposium has the privilege to play its supporting role in a niche area. The financial supports of DSTA and DSO are gratefully acknowledged. World Scientific Publishing Co. Pte. Ltd., in its usual professional manner, has published with great efficiency, and at a cost within our budget, this handsome volume of the proceedings ready for our participants at the opening of ICMAT 2003.

I wish all participants an exciting and fruitful conference, and our guests from overseas a pleasant and enjoyable visit to Singapore.

Professor LIM Hock
Chair
Symposium F (Electromagnetic Materials)
ICMAT 2003

INVITED SPEAKERS

- B. **CHAMBERS**, The University of Sheffield, UK
- J. A. **KONG**, Massachusetts Institute of Technology, USA
- A. N. **LAGARKOV**, Institute for Theoretical and Applied Electromagnetics, Russia
- K. **LEWIS**, QinetiQ, UK
- G. D. **LI**, Institute of Physics, Chinese Academy of Science, China
- B. A. **MUNK**, The Ohio State University, USA
- C. K. **ONG**, National University of Singapore, Singapore
- V. K. **VARADAN**, Pennsylvania State University, USA
- V. G. **VESELAGO**, Moscow Institute of Physics and Technology, Russia
- X. **YAO**, Tongji University, China

ORGANISING AND SCIENTIFIC COMMITTEE

- Hardy **CHAN**, Dept of Chemistry, NUS, Singapore
- Chao-Ran **DENG**, DSO National Laboratories, Singapore
- Yeow Beng **GAN**, Temasek Laboratories, NUS, Singapore
- Kai Meng **HOCK**, Temasek Laboratories, NUS, Singapore
- Kim Seng **LEE**, Directorate of Research and Development, DSTA, Singapore
- Hock **LIM**, Temasek Laboratories, NUS, Singapore
- Serguei **MATITSINE**, Temasek Laboratories, NUS, Singapore
- Ping Ping **NEO**, Temasek Laboratories, NUS, Singapore
- Chong Kim **ONG**, Dept of Physics, NUS, Singapore

PUBLICATION AND LIAISON COMMITTEE

- Karrie **CHAN**, Temasek Laboratories, NUS, Singapore
- Yeow Beng **GAN**, Temasek Laboratories, NUS, Singapore
- Hock **LIM**, Temasek Laboratories, NUS, Singapore
- Jesslyn **LIM**, Temasek Laboratories, NUS, Singapore
- Serguei **MATITSINE**, Temasek Laboratories, NUS, Singapore
- Maryate **MUHAMAD**, Temasek Laboratories, NUS, Singapore

ACKNOWLEDGEMENT

- **ARUMUGAM** Sundaram, Temasek Laboratories, NUS, Singapore
- Kai Yew **LUM**, Temasek Laboratories, NUS, Singapore
- Anyong **QING**, Temasek Laboratories, NUS, Singapore

CONTENTS

Session F1

Opening Session

Chair: H. Lim

Microwave properties of composites and inhomogeneous structures

A. N. Lagarkov[1], K. N. Rozanov, A. P. Vinogradov, V. N. Semenenko, I. A. Ryzhikov, I. T. Iakubov,

Institute for Theoretical and Applied Electromagnetics, RAS, Moscow, Russia

and S. M. Matitsine

Temasek Laboratory, National University of Singapore, Singapore

Introduction

The solution of present-day engineering problems often calls for materials with pre-assigned frequency dispersion of permittivity $\varepsilon=\varepsilon'+i\varepsilon''$ and permeability $\mu=\mu'+i\mu''$ in the microwave frequency range. In some cases, the dispersion dependence of a fairly complex form is required, for example, that with several maxima of ε', separated by frequency domains in which $\varepsilon'<0$. Similar requirements may be placed on magnetic permeability. However, the values and frequency dependences of ε' and ε'', as well as of μ' and μ'', cannot be arbitrary, because they are related by the Kramers-Kronig relations.

This paper deals with urgent problems associated with the development of new composite materials for microwave applications. Composite fillers of different types are discussed, which enable one to obtain desired values and frequency dispersion of ε and μ, such as spherical particles, fibers, films, inclusions in the form of rings and spirals, as well as their possible combinations. Effective material parameters of composites are reviewed depending on the shape, orientation, conductivity, and volume fraction of inclusions. The restrictions on the limiting values of ε and μ are treated, as well as the form of their dispersion dependences. Problems are discussed, which are associated with the possibility of describing the microwave properties of inhomogeneous materials in terms of effective material parameters.

Dielectrics with high values of permittivity and low loss

One of the reasons for unceasing interest in the dielectric properties of metal/dielectric mixtures is that the microwave technology requires materials of high permittivity and low dielectric loss tangent. Ferroelectric ceramics are usually used as such materials in developing microwave filters, resonators, dielectric antennas, and the like. Composite materials consisting of a polymer matrix and conducting inclusions make it possible to obtain values of ε' of up to several hundred [1] and exhibit a number of other advantages such as small weight, low cost, simple manufacturing technology, and the possibility of machining. The problem is only one of the relatively high dielectric loss observed in composites.

Note that the introduction of local effective values of ε and μ is possible only in the case when, first, the wavelength significantly exceeds the characteristic scales of inhomogeneity and, second, the part of the exciting field that arose due to the wave scattering may be treated as a uniform field (for more detail, see the last section below). The first condition implies curl-less fields being treated, and the second condition is valid in the case of low concentration p of inclusions. The formula for effective permittivity ε_e of a composite may be derived within the perturbation theory,

$$\varepsilon_e = \varepsilon_h \left(1 + p\frac{4\pi i\,\sigma/\omega - \varepsilon_h}{\varepsilon_h + N(4\pi i\,\sigma/\omega - \varepsilon_h)}\right), \tag{1}$$

where σ and N denote the conductivity and the depolarization factor of the inclusions, respectively; ε_h is the matrix permittivity; and ω is the circular frequency. Note that, in accordance with Eq. (1), the effective permittivity obeys the Debye frequency dependence. In the frequency domain much below the Debye absorption peak, the dielectric loss tangent is

$$\tan\delta = (\varepsilon_h\omega)/(4\pi\sigma N). \tag{2}$$

---

[1] corresponding author, e-mail lag@eldyn.msk.ru

Equation (2) describes the loss associated with the excitation of conduction currents in conducting inclusions. One can readily derive from Eq. (1) that the use of non-spherical inclusions such as conducting fibers [1] or flake-like particles [2] produces high values of ε'; however, in so doing, $\tan\delta$ increases in accordance with Eq. (2).

As a rule, the values of $\tan\delta$ in actual composites are orders of magnitude higher than those which may be obtained from Eq. (2) using the values of σ that are characteristic of metals. One of the reasons for this may be the presence of intrinsic loss in the polymer matrix: one can readily derive that the intrinsic loss tangent of the polymer is added as a term to the right-hand part of Eq. (2). Therefore, $\tan\delta$ of the composite cannot be less than the loss tangent of the matrix. Because, for the majority of polymers, $\tan\delta \sim 10^{-2}$, while $\tan\delta \leq 10^{-3}$ is usually required for applications, this limits significantly the possibilities of using polymer-based composites as microwave materials with high values of permittivity and low loss.

As the concentration of conducting inclusions rises, $\tan\delta$ increases compared with Eq. (2) because of the interaction between inclusions. The minimal loss is attained if the inclusions are distributed uniformly. A random distribution is more frequently observed in actual composites, which leads to the emergence of conducting clusters. The most general approach to analyzing the effective properties of composites in a quasi-static approximation involves the use of the Bergmann spectral function $B(s)$ (see, for example, [3]) containing the complete information about the composite microstructure. For a two-component mixture, the spectral function is introduced by the relations

$$1-\frac{\varepsilon_e}{\varepsilon_h} = \int_0^1 \frac{B(x)}{s-x}\,dx, \qquad s = \frac{1}{1-\varepsilon_i/\varepsilon_h}. \tag{3}$$

For a pre-assigned microstructure, $B(s)$, with ε appropriately replaced by μ, describes the effective magnetic permeability as well.

For a metal/dielectric mixture in the frequency domain below the absorption peak, the real and imaginary parts of permittivity may be obtained directly from Eq. (3),

$$\varepsilon_e' = \varepsilon_h\,(1+\int_0^1 \frac{B(x)}{x}\,dx), \qquad \varepsilon_e'' = \frac{\varepsilon_h^2\,\omega}{4\pi\sigma}\int_0^1 \frac{B(x)}{x^2}\,dx. \tag{4}$$

The form of the spectral function of actual composites may be determined by the measured frequency dispersion of permittivity [4]. As a rule, the microwave permittivity of composites with conducting inclusions does not exhibit frequency dispersion. However, in view of the fact that ε_e and μ_e are described by one and the same spectral function, $B(s)$ may be recovered from the data on the frequency dependence of composites with conducting ferromagnetic inclusions. The difficulty consists in that the measurement of the magnetic permeability of fine particles presents a complex experimental problem. The permeability of particles may not be recalculated from other experimental data, for example, from the results of static measurements, either, because of its complex dependence on the crystal and magnetic structure of particles. Therefore, no independent data on this quantity are available, as a rule.

Nevertheless, this problem may still be solved [5]. The experimental data on the frequency dispersion of magnetic permeability of composites with different contents of ferromagnetic powders may be used to simultaneously derive the spectral functions $B(s)$ for all concentrations and the frequency dependence of intrinsic permeability of inclusions $\mu_i(f)$. For this purpose, Osipov et $al.$ [5] introduced parameterized function for both $B(s)$ and $\mu_i(f)$, whose parameters were found from the condition of best agreement between the predicted values of effective permeability, obtained using these functions, and the measured dispersion curves. The measurements were performed in the frequency range from 0.1 to 10 GHz for composites consisting of carbonyl iron powder in a paraffin matrix. One of the obtained spectral functions, which corresponds to the 10% iron content in the composite, is given in Fig. 1 along with the spectral function obtained for the same composite using the effective medium theory (EMT). One can see that the actual spectral function has nonzero values at lower values of the argument than the result according to the EMT. The data on the frequency dependence of intrinsic permeability of inclusions, obtained from different sets of concentrations, agree well with one another. Agreement is also obtained with the known tabular data, e.g., with Snoek's constant for iron.

It readily follows from Eq. (4) that the loss on low frequencies is approximately inversely proportional to the least value of s at which nonzero values of $B(s)$ arise. For the data in Fig. 1, this value is close to 0.1 for the EMT and to 0.01 for the recovered spectral function. Therefore, the Bergman theory predicts for this composite a ten times higher loss on low frequencies compared with the EMT. Nevertheless, this loss is still orders of magnitude lower than that observed in actual composites. In view of this, the

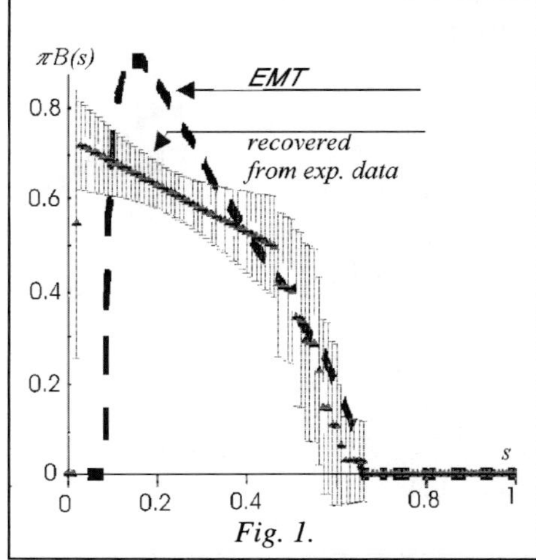

Fig. 1.

question of the nature of low-frequency dielectric loss remains open. The reasons for this loss may include the high resistance of conducting clusters, which arises due to non-ideality of electric contacts between individual inclusions, or the skin effect on the clusters [20]. Up to now, the effect of these factors has not been investigated in detail either theoretically or experimentally.

Microwave permittivity of composites filled with conducting fibers
As was noted above, composites filled with conducting fibers may exhibit high values of permittivity at a low concentration of fibers. The physical reason for this is that the polarizability of a fiber of length a and thickness b is close to that of a spherical inclusion of diameter a, but the fiber volume is $(a/b)^2$ times less. Another important property of such composites is the possibility of a high dispersion of microwave permittivity. The dispersion is associated with the dipole resonance of scattering from fibers and may be used to extend the working range or to raise the efficiency of various frequency-selective microwave materials [6]. In the presence of resonance dispersion, the real part of permittivity may be negative in some frequency range, which is of interest from the standpoint of developing left-handed materials.

The microwave dispersion of permittivity of composites filled with conducting fibers was experimentally investigated in [1]. Figures 2 and 3 reproduce some of the dispersion curves obtained in [1]. Figure 2 gives the dispersion dependence for a composite containing 0.05 vol. % of carbon fiber 10 mm long. The dispersion curve exhibits one spread absorption peak accompanied by a region of high dispersion ε'. Figure 3 gives the dispersion dependence of a composite containing fibers of different length, namely, 5 and 12 mm. This dispersion curve exhibits two distinct sharp absorption peaks, to which two regions of anomalous dispersion correspond. Therefore, by varying the conductivity and length of fibers and combining different fibers in one material, one tailor vary the form and parameters of the dielectric dispersion curve in a wide range.

An interesting feature revealed in fiber-filled composites [1] is that their permittivity is alike one predicted by the perturbation theory in accordance with Eq. (1) (see the lines in the drawings). It is surprising that the agreement is observed even in cases when the permittivity of the composite strongly differs from that of the matrix. The composite behaves as if the fibers within do not interact and the high effective permittivity is due only to the high polarizability of individual fibers. Also revealed in [1] were anomalies in the position of the dielectric absorption peak for the case when the resonant fibers are placed in a composite environment with inclusions of a different type.

Based on these data, a theory of effective permittivity of composites with conducting fibers was suggested in [7]. For testing this theory, Matitsine *et al.* [8] performed a thorough experimental

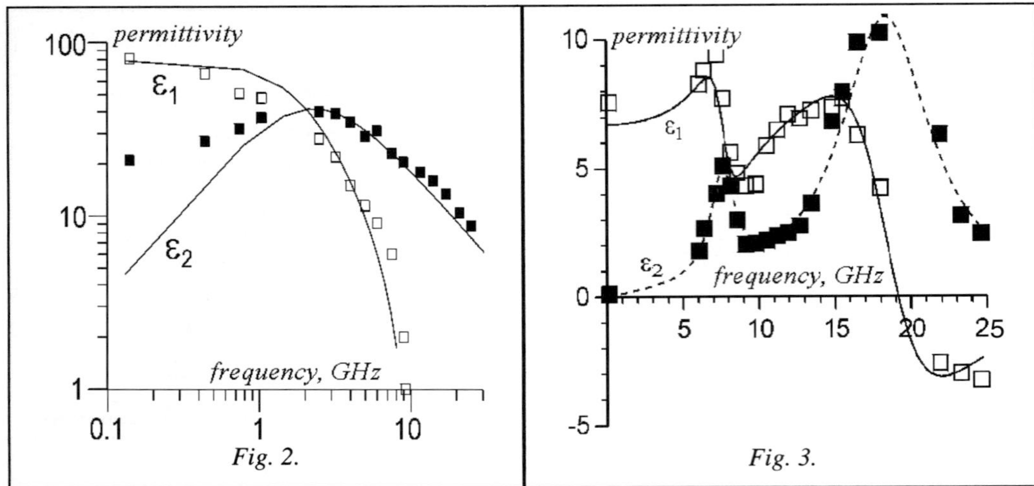

Fig. 2.

Fig. 3.

investigation of the position of resonance of long highly conducting fibers in a composite matrix. It has been demonstrated that the position of dipole resonance of fibers is fully described by the permittivity of the matrix and its anisotropy. There is no need in additional assumptions of the special properties of a fiber placed in an inhomogeneous environment.

Therefore, the experimental data of [8] indicate that the interaction between individual fibers in a composite must significantly affect its properties. The disagreement with the data of [1] may be attributed to the fact that the latter study dealt with thin sheet samples of composites, for which the interaction between fibers is indeed lower than for massive samples. The theory suggested in [7] needs to be refined to obtain a more accurate quantitative description of the properties of massive composites at high concentrations of fibers.

Ferromagnetic film-based magnetic composites
An additional restriction exists as regards attaining high values of magnetic permeability on high frequencies. This restriction is expressed by Snoek's law [9] relating the static magnetic permeability μ_s and the ferromagnetic resonance frequency f_r and valid for the majority of magnets,

$$\mu_s f_r = \gamma 4\pi M_s \, .$$
(5)

In Eq. (5), M_s is the saturation magnetization of the material, $\gamma \approx 3$ GHz/Oe, and the quantity in the right-hand part is referred to as Snoek's constant. One can approximately assume that the permeability is equal to its static value on frequencies below f_r, after which it drops abruptly. Therefore, Eq. (5) is a restriction on the high-frequency values of magnetic permeability: high values of permeability may be realized only at relatively low frequencies and, conversely, an attempt at increasing f_r brings about a decrease in μ_s. The values of μ_s and f_r strongly depend on the crystal and magnetic structure of the material, in particular, on the magnetic anisotropy field H_A; however, their product is defined only by the magnet composition, on which the value of M_s depends.

Thin magnetic films present an important case, for which an analog of Snoek's law is written as [10]

$$\mu_s f_r^{\,2} = (\gamma 4\pi M)^2$$
(6)

For the majority of practical cases, a much more lenient restriction on the value of high-frequency permeability follows from Eq. (6) than from (5), which explains the promise held by the use of thin magnetic films as high-frequency magnetic materials.

Ferromagnetic metals exhibit the highest values of M_s and, consequently, the highest possible values of $\mu_s f_r^2$. In this case, however, the possibility of high-frequency applications is further limited by the high electrical conductivity σ of these metals. First of all, the film thickness d must

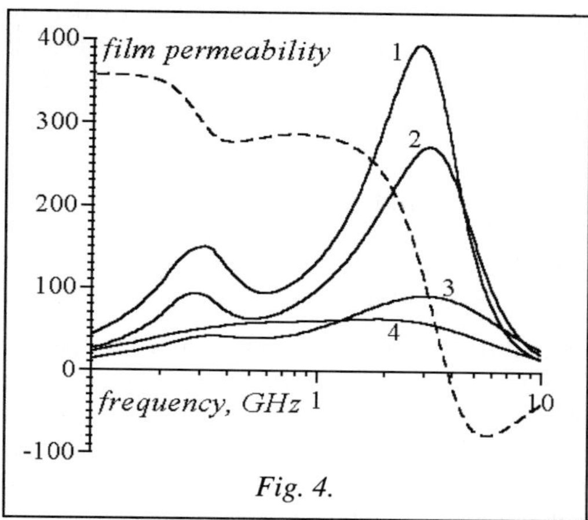

Fig. 4.

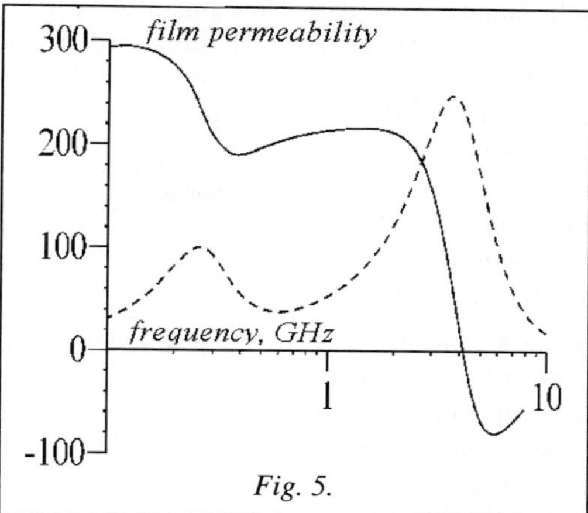

Fig. 5.

not exceed the skin depth $\delta = c/2\pi\sqrt{\sigma\mu_s f_r}$; otherwise, if the film is fairly thick, the position of the maximum of magnetic loss will be defined by the condition $\delta = d$ rather than by the ferromagnetic resonance frequency. However, in this case as well, the eddy current loss may cause the film to change from the resonance to relaxation low-frequency state if the inequality is not strong enough, $d << \delta$. Therefore, in order to produce a film with high values of μ at microwaves, the film must be made of a magnetically soft metal to eliminate the magnetization reversal loss; the metal must have high magnetic moment, exhibit the lowest possible electrical conductivity, and be rather thin. It is not only with the skin effect that the latter requirement is associated. The film must also be thin for the perpendicular anisotropy not to arise and for the magnetic moment to lie in the film plane. This condition dependent on many factors is valid if the Acher coefficient,

$$k_A = \int_0^\infty \mu''(f)f\,df \left/ \left(\frac{\pi}{2}(\gamma 4\pi M_s)^2 \right) \right. , \tag{7}$$

is unity. In Eq. (7), the integration is performed in the FMR frequency range, and the permeability corresponds to the external magnetic field applied in the film plane along the hard axis.

A change of the process conditions of film preparation makes it possible to vary its permeability within the limit established by (6). The possibilities of control are especially wide in iron films, because the properties of iron are most dependent on the deposition technology and on the presence of shallow impurities [11]. The presence of impurities (for example, nitrogen in fractions of 4–8 at.% [12, 13]) both causes a reduction of conductivity and leads to changes in the crystal structure, i.e., to changes of the anisotropy constant.

Figure 4 gives the dispersion dependences of μ' and μ'' of an iron film as functions of the film thickness. Films 0.1 to 2 μm thick were deposited on a Lavsan (Soviet equivalent of Dacron) substrate by either ion beam or magnetron sputtering. The permeability was measured at frequencies from 100 MHz to 10 GHz. The film was wound on a rod and inserted into a hollow cylinder which was an element of a standard coaxial measuring cell. The results demonstrate a wide diversity of dispersion dependences obtained under different conditions of sputtering. As a result of intense nitriding, the dispersion dependences were brought to the resonance mode; then, the FMR, to which the high-frequency peak corresponded, was shifted to a frequency close to 3 GHz. In so doing, the resistivity increased to 50 $\mu\Omega\cdot$cm. One can clearly see that the permeability continues to rise with decreasing film thickness, inasmuch as the eddy current loss decreases. In this case, the values of thickness are already lower than the value of skin depth $\delta \approx 0.8$ μm. The low-frequency peak corresponds to the domain wall motion. The form of the hysteresis loops points to the absence of a stripe domain structure. No marked perpendicular anisotropy is observed.

Bulk samples are required for a number of microwave applications. Such samples are provided by multilayer materials in which structured magnetic and dielectric layers alternate. The effective permeability of such a composite is given by Rytov's formulas. If d is the magnetic layer thickness and b is the interlayer thickness, then, on condition that $d<<\delta$, for a wave normally incident on the composite,

$$\mu_e = (d\mu_i + b)/(d + b). \tag{8}$$

The quantity b is restricted from below, because the condition of validity of the favorable expression (8) is the absence of magnetic interaction between layers. One can assume that the domain structure of a layer is closed if the interlayer thickness is not markedly less than 0.1 μm.

Figure 5 gives the dispersion dependence curves for a multilayer film consisting of four iron layers, each 0.15 μm thick, separated by layers of silicon oxide. The magnetic layers were photolithographically patterned, with the diameter of magnetic inclusion of 0.6 mm. The measured values of intrinsic permeability of the film are given in the figure. Comparison with Fig. 4 reveals that the quality of iron in the FMR region did not deteriorate, although the domain structure apparently changed, as is suggested by the variations in the low-frequency peak.

The obtained results indicate that composites of multilayer patterned magnetic films enable one to obtain dispersion dependences of the desired form and vary the shape, width, and position of resonance dependence curves.

Artificial magnets

Another possibility of producing materials exhibiting magnetic properties at microwaves is the use of artificial (i.e., not associated with the presence of a permanent magnetic moment) magnetism. The artificial magnetism is exhibited by composites based on nonmagnetic inclusions possessing pronounced resonant properties, for example, dielectric resonators with a high value of ε' ($\varepsilon' = 100...3000$) and low $\tan\delta$ or helicoid resonators of various types.

It is known [14] that a dielectric sphere has a spectrum of eigenfrequencies ω_n of TM-modes, dependent on its radius a and permittivity ε: $\omega_n/2\pi=(2n+1)c/(4a\sqrt{\varepsilon})$, $n = 1, 2, 3, ...$. In an external alternating magnetic field H, a sphere of volume V acquires the dipole magnetic moment $m=\alpha_m VH$, where its magnetic polarizability $\alpha_m=3\gamma/4\pi$, and $\gamma=(1+3\cot(ka))/(ka)-3/(ka)^2)/2$. The frequency dependence of the magnetic polarizability of a sphere is the sum of resonance functions,

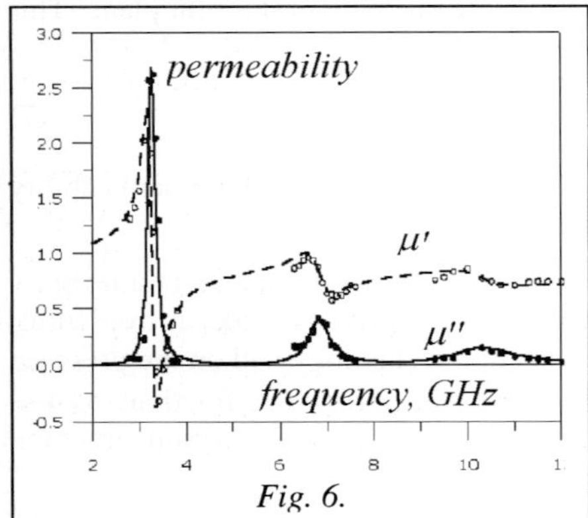

Fig. 6.

$$\alpha_m = \sum_{n=1}^{N} A_n (\beta_n/\omega_n)(\omega^2/(\omega_n^2 - \omega^2 - i\omega\beta_n)), \tag{9}$$

where A_n denotes the amplitudes, and β_n – the half-widths, of the resonance curves. The electric polarizability of a dielectric sphere with a high value of ε' is independent of frequency and equal to $3/(4\pi)$. In the case of artificial magnets, a restriction is also imposed on the magnitude of effective permeability: the magnetic efficiency defined as

$$\chi = \sum_{n=1}^{N} A_n \beta_n/\omega_n, \tag{10}$$

is rigidly restricted, $\chi < 1$. This follows from the non-negativity of the effective permeability at very high frequencies. The value of χ is reflective of the integral high-frequency magnetic loss in a composite.

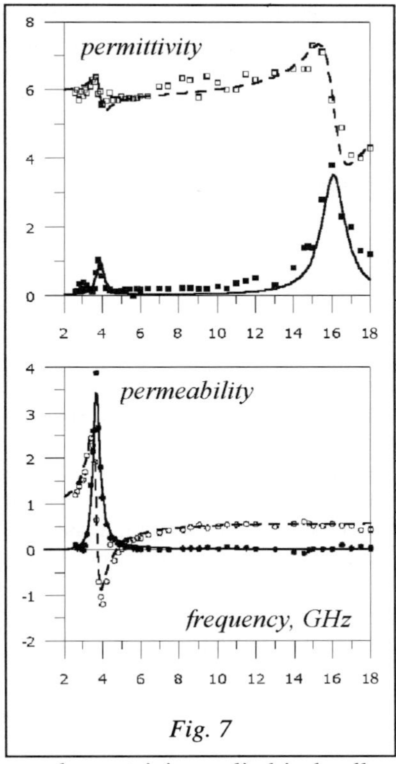

Fig. 7

Figure 6 gives the frequency dependence of effective permeability μ_e of a composite containing 20% by volume of cubes of BaSrTiO$_3$ with the permittivity $\varepsilon = 2700 + i90$. The measured values of μ_e in the vicinity of three lowest magnetic modes are fitted by the function $\mu_e = 1 + 4\pi\alpha_m$, where α_m is defined by (9); the fitting curves are indicated in the drawing by lines. The magnetic resonance amplitudes of the composite nonlinearly increase with the concentration of inclusions and are saturated at concentrations of about 80%. For the given concentration of inclusions, μ_e increases with their permittivity.

Inclusions of another type, which enable one to produce artificial magnetic properties on microwave frequencies, are conducting inclusions of various shapes. Of most interest among these inclusions are compensated spirals [15, 16] (i.e., a combination of two wire spirals, left-hand and right-hand ones, on a single axis) which have one μ_e resonance and several ε_e resonances, namely, strong high-frequency longitudinal and weak transverse resonances, see Fig. 7 [16]. For an anisotropic sample whose parameters are given in Fig. 4, $\chi = 0.4$ for the spiral diameter of 2.2 mm and the winding pitch of 0.9 mm. When the winding pitch of the spirals is reduced, the value of χ increases (for the same conductivity) and reaches a maximum at zero pitch (a particular case of broken rings or film rolls). For example, a sample containing cylindrical rolls of insulated aluminum foil has a close-to-limiting (with due regard for the fill factor) value of magnetic efficiency $\chi = 0.65$.

Isotropic composites consisting of spaced left-hand and right-hand spirals simultaneously exhibit the resonance dispersion of permittivity and permeability in one and the same frequency range. If the Q-factor of resonances is high enough, negative values of the real parts of permittivity and permeability are simultaneously reached in some frequency range in the composite. Shown in Fig. 8 is an external view of such a composite (on the left), as well as the dielectric (at the center) and magnetic (on the right) dispersion dependences. The points in the drawings indicate the measurement results, and the lines indicate their fitting by the resonance dispersion law. On above-resonance frequencies, such composites provide a practical example of left-handed materials.

Spatial dispersion in inhomogeneous materials
By virtue of the principle of causality, the frequency dependence of ε and μ implies that the system response (polarization) is delayed with respect to the exciting (local) field. This delay may be due, first, to the inertia of the current carriers leading to the frequency dependence of ε and μ of inclusions (see Eq. (1)). Second, this delay may be caused by the effect of field retardation on the inclusion scale, as in the case of spiral inclusions [17]. In the latter case, in fact, different parts of an inclusion are excited by different fields, and the overall response depends both on the inclusion scale-average field and on its derivatives. This spatial non-locality of response is referred to as spatial dispersion and, when describing harmonic fields, shows up as the dependence of ε and μ on both the frequency and the wave vector [17, 18]. If the dispersion equation $k^2 = k_0^2 \varepsilon(\omega, k)$ has only one solution for each direction, then, in describing the propagation of plane waves, everything may be reduced to the frequency dispersion $\varepsilon(\omega) = \varepsilon(\omega, k(\omega))$. However, in problems involving the diffraction on bodies whose shape varies on a scale of $2\pi/k$, one must even in this case take into account the spatial dispersion, because scattered fields contain k-harmonics whose wave vector is

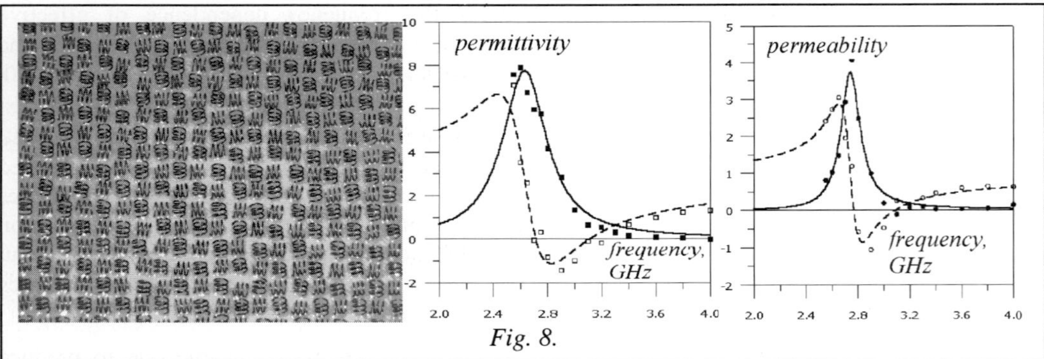

Fig. 8.

defined by the geometry of the problem rather than by the dispersion relation. In the case of increasing concentration and size of inclusions, the need to take into account the spatial dispersion arises because we can no longer treat the field from the neighboring inclusion as uniform. This effect shows up most clearly when an additional solution of the dispersion relation emerges. This means that numerous, rather than one, refracted waves are excited in the medium. From the practical standpoint, this implies the emergence of an additional channel of energy input into the medium, which may be used to improve the absorbing performance of the material. In the case of weak spatial dispersion, the excitation of a second refracted wave is possible when the quadrupole moment of inclusion becomes significant [18]. For photon crystals, the phenomenon of multiple refraction is well known and was used by us to improve the performance of radar absorbers [19].

Acknowledgments

This study was supported in part by the Russian Foundation for Basic Research, grant no. 02-02-16707.

References

[1]   A. N. Lagarkov, S. M. Matytsin, K. N. Rozanov, A. K. Sarychev, *J. Appl. Phys* **84** (7) 3806 (1998)

[2]   W. H. Emerson, *IEEE Trans. Antennas and Propagat.*, **21** (7) 484 (1973)

[3]   D. J. Bergman, *Phys. Rep. Sec. C* **43** 377, 1978

[4]   A. R. Day, A. R. Grant, A. J. Sievers, M. F. Thorpe, *Phys. Rev. Lett.*, **84** (9) 1978 (2000)

[5]   A. V. Osipov, K. N. Rozanov, N. A. Simonov, S. N. Starostenko, *J. Phys. Condens. Matter* **14** 9507 (2002)

[6]   K. N. Rozanov, S. N. Starostenko, *The European Phys. J. – Appl. Phys.* **8**, 147 (1999)

[7]   A. N. Lagarkov and A. K. Sarychev, *Phys. Rev. B* 53 (10), 6318 (1996)

[8]   S.M. Matitsine, K.M. Hock, L. Liu, Y.B. Gan, A.N. Lagarkov, K.N. Rozanov, *submitted to J. Appl. Phys.*, 2003

[9]   J. L. Snoek, *Physica*, **14**, 204 (1948)

[10]   Perrin, G., O. Acher, J. C. Peuzin, N. Vukadinovic, *J. Magn. Magn. Mater.*, **136**, 269 (1994)

[11]   S. Chikazumi, *Physics of Magnetism*, NY: Wiley (1964)

[12]   Y. M. Kim *et al.*, *IEEE Trans. Magn.*, **37**, 2288 (2001)

[13]   S. C. Byeon, F. Liu, G. J. Mankey, *IEEE Trans. Magn.*, **37**, 1770 (2001)

[14]   L. D. Landau, E. M. Lifshitz, *Electrodynamics of continuous media*, Oxford: Pergamon (1984)

[15]   A. P. Vinogradov, V. E. Romanenko, Proc. 3rd Internat. Conf. on Chiral, Bi-isotropic, and Bi-anisotropic Media, The Pennsylvania State University, USA, October 1-14, 1995, pp. 143-148

[16]   A. N. Lagarkov, V. N. Semenemko, V. A. Chistyaev, D. E. Ryabov, S. A. Tretyakov, C. R. Simovski, *Electromagnetics* **17** (3) 213 (1997)

[17]   A. P. Vinogradov, Proc. of "Bianisotropics 93", Gomel, Belarus, Ed. by A. Sihvola, S. Tretyakov, I. Semchenco: Helsinki University of Technology, Finland, 1993, pp. 22-26

[18]   A. P. Vinogradov, Electrodynamics of composites, Moscow: URSS Publishers, 2001 (*in Russian*)

[19]   A. P. Vinogradov, A. M. Karimov, A. K. Sarychev, *Sov. Phys. JETP* **67**, 2129, (1988)

[20]   A. N. Lagarkov, A. P. Vinogradov, In: Book of abstracts of the Int. Conf. ETOPIM-6, Snowbird, Utha, USA, July 15–19 2002, p. 145

An Introduction To The Phase-Switched Screen

B. Chambers and A. Tennant

Dept. of Electronic & Electrical Engineering, University of Sheffield, U.K.
e-mail: b.chambers@sheffield.ac.uk

Abstract: The Phase-Switched Screen is a novel technique for the dynamic control of radar cross-section. It relies on the modification of the signal scattered from an object in such a way as to place the signal energy outside the bandwidth of a receiving system. The paper includes a theoretical discussion of the new technique and examples of its application in practice.

1.   Introduction

Although conventional (i.e. passive) microwave absorbent materials have been widely used for many years to modify the radar cross-section (RCS) of military platforms, such materials may not have adequate performance or flexibility to satisfy future requirements. For example, a passive microwave absorber, once designed and manufactured, has fixed characteristics that are bounded by the electrical thickness at the lowest desired operating wavelength, following the so-called "Rozanov limit" [1]. Hence, if the threat for which the absorber was designed changes, then either reduced performance against the new threat must be accepted or the material must be replaced by an improved one. Active materials, however, offer the potential to overcome the Rozanov limit and to enable additional "smart" functionality such as monitoring damage, adaptive control of RCS or target appearance, Identification-Friend-or-Foe (IFF) and Absorb-While-Scan (AWS) [2]. In this paper, we introduce a particular technique for realising an active material for use at microwave frequencies and explore its potential.

Consider a semi-infinite half-plane in free-space whose surface is illuminated by a plane electromagnetic wave. Although the structure comprising the half-plane is not yet known, let the surface at any instant be able to present one of two different admittances Y_1 or Y_2 to the incident wave, the instantaneous choice of Y_1 or Y_2 being controlled by the application of a particular electrical or optical stimulus to the surface. In practice, the choice of Y_1 and Y_2 is not arbitrary and for normal incidence they should satisfy the relationship $Y_1 Y_2 = Y_0^2$, where Y_0 is the characteristic admittance of free-space. Hence, Y_2 may be written in terms of Y_1 as $Y_2 = Y_0^2/Y_1$, which is the equivalent of transforming Y_1 by a length of transmission line of length $\lambda/4$ and characteristic admittance Y_0. For the two admittance states of the surface, the incident wave experiences reflection coefficients of $+A$ and $-A$ (A is generally complex) and if these are each present for half of the time, then the surface exhibits a time-averaged reflection coefficient of zero, i.e. it appears to act as a perfect absorber at the incident frequency corresponding to the wavelength λ. This behaviour is quite different to that of a conventional microwave absorbent material since no apparent loss of energy is involved. The *modus operandi* is clarified, however, if the behaviour of the surface is examined in terms of the spectrum of the incident and reflected waves. It is clear then that by periodically changing the surface admittance, the reflected wave is being binary phase-modulated with the result that all of the energy at the incident wave frequency is redistributed into an infinite number of sidebands. If the frequency of the periodic stimulus signal that alters the surface admittance is high enough, then none of this redistributed energy falls within the pass-band of the receiving system and hence the surface mimics the behaviour of a perfect absorber, as measured by the receiver. We have called this new type of active 'absorber' the "Phase-Switched Screen" (PSS).

2. Transmission-line analysis of the single layer phase-switched screen

The transmission-line analogue of the single active layer PSS consists of a short-circuited length, d, of transmission line with characteristic admittance Y_c and propagation constant β, across whose input terminals is placed an admittance Y(t), defined as

$$Y(t) = Y_1 \qquad 0 < t < \tau$$
$$= Y_2 \qquad \tau < t < T \tag{1}$$

where τ is the 'on' time and T is the time period for one cycle of the waveform used to control the state of Y(t). Depending on the incident polarisation and angle of incidence, θ, Y_c is given by either $Y_c = Y_0 /\cos\theta$ for parallel polarisation, or $Y_c = Y_0\cos\theta$ for perpendicular polarisation. For both polarisations, β is given by $\beta = \beta_0 \cos\theta$. Then the required general relationships between Y_1 and Y_2 are given by [3]

(a) *parallel polarisation*

$$[Y_1 - j\frac{Y_0}{\cos\theta}\cot(\beta_0 d \cos\theta)][Y_2 - j\frac{Y_0}{\cos\theta}\cot(\beta_0 d \cos\theta)] = \frac{Y_0^2}{\cos^2\theta} \tag{2}$$

(b) *perpendicular polarisation*

$$[Y_1 - jY_0 \cos\theta \cot(\beta_0 d \cos\theta)][Y_2 - jY_0 \cos\theta \cot(\beta_0 d \cos\theta)] = Y_0^2 \cos^2\theta \tag{3}$$

If $\theta = 0°$ and $Y_1 = G_1$, $Y_2 = G_2$ (active layer has switchable conductance only), then (2) and (3) simplify to $G_1 G_2 = Y_0^2$. Typical reflectivity characteristics for the single active layer PSS when $G_1 = 0$ and $G_2 = \infty$ are shown in Figure 1, together with those for the normal Salisbury screen.

Reflectivity (dB)

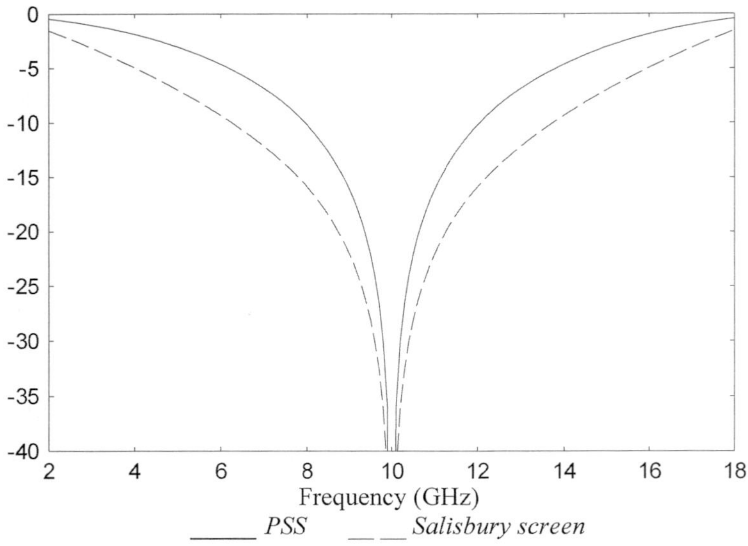

Figure 1   *Frequency responses of switched conductance PSS and Salisbury screen*
d = 7.5 mm, PSS: $G_1 = 0$, $G_2 = \infty$, Salisbury screen: R = 376.7 $\Omega/\Box$

The two curves are different due to a lack of multiple reflection phenomena in the PSS. By adjusting the 'on' time of the active layer (i.e. the ratio τ/T) we can change its effective sheet resistance and so control the depth of the PSS reflectivity null [4], as shown in Figure 2. Hence the PSS can be configured dynamically to act as a good reflector or as a variable absorber. Furthermore, by replacing the PEC backplane of the PSS by a second active layer, the structure may have these properties when viewed from either side. In the bi-directional PSS (BPSS), each active layer is switched so that at successive instants in time, $G_1 = 0$, $G_2 = \infty$; $G_1 = \infty$, $G_2 = 0$; $G_1 = 0$, $G_2 = \infty$, etc.

From Figure 1, it can be seen that the switched conductance PSS has a narrow bandwidth, but this can be broadened by making the active layer reactive. For example, if $\theta = 0°$, $Y_1 = jB_1$, $Y_2 = jB_2$, then (2) and (3) give the required relationship between B_1 and B_2 as

$$B_1 B_2 = -Y_0^2 \cos ec^2 \beta_0 d \tag{4}$$

Hence, if an active layer whose B_1 and B_2 had the form $B_1 \approx -jY_0 \cos ec\beta_0 d$ and $B_2 \approx jY_0 \cos ec\beta_0 d$ could be realised, then the bandwidth of the PSS would approach infinity. In practice, these values of B_1 and B_2 cannot be realised over a very large bandwidth, but a combination of switchable susceptance and conductance in the PSS active layer is still advantageous, as shown by the measured data in Figure 3 [5].

Although the transmission-line analysis given above enables us to examine the frequency characteristics of the PSS, it does not allow us to determine the optimum choice of switching waveform shape and frequency, particularly in the presence of pulsed incident signals; this requires a spectral analysis, which we consider next.

Reflectivity (dB)

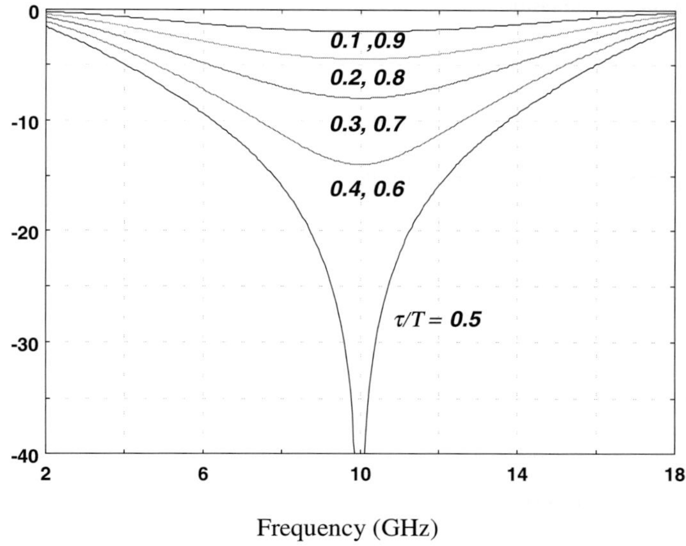

Frequency (GHz)

Figure 2   *Variation of null depth with τ/T for switched conductance PSS*

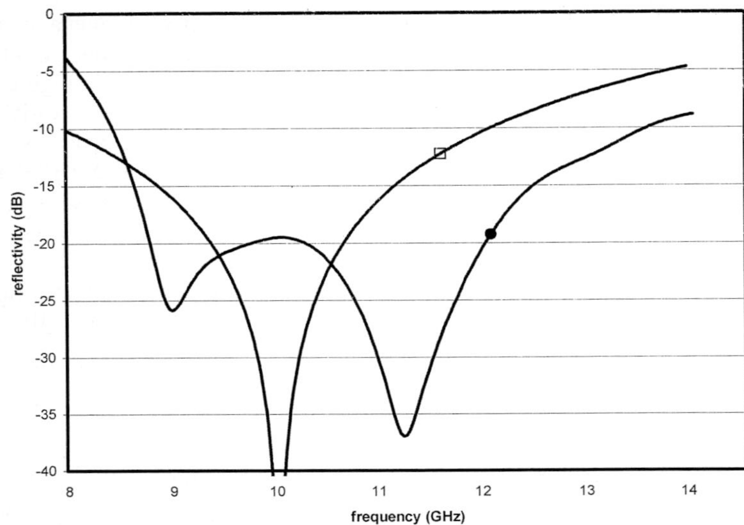

Figure 3   *Reflectivity characteristics of ideal switched conductance PSS (□)*
and experimental single layer PSS (•).

3.   Spectral analysis of the single-layer phase-switched screen

For simplicity, we will consider here only the case of a single active layer PSS operating against a periodic train of rectangular pulses of carrier frequency f_c [6], as shown in Figure 4(a). Each pulse is of length T_p and the pulse repetition period is T. After reflection from the PSS, assume that each incident carrier pulse is replaced by M cycles of a bipolar rectangular waveform, as shown in Figure 4(b). Hence, the period of this modulating waveform, 2τ, is given by

$$2\tau = \frac{T_p}{M} \tag{5}$$

and the switching frequency, f_s, is given by $f_s = \dfrac{M}{T_p}$

Since the incident pulse train is periodic, with period T, its spectrum and that of the signal reflected from the PSS may be determined using Fourier analysis. Hence, the incident and reflected signals may be written in the form

$$v(t) = \sum_{n=-\infty}^{n=+\infty} c_n e^{jn\omega_0 t} \tag{6}$$

where $\omega_0 = 2\pi/T$

The complex Fourier coefficients, c_n, are then given by

$$c_n = \frac{1}{T} \int_0^T v(t) e^{-jn\omega_0 t} dt \tag{7}$$

For the incident periodic pulse train, c_n are given by

$$c_n = \frac{-j}{2n\pi} \left(1 - e^{-j2n\pi \frac{T_p}{T}} \right) \tag{8}$$

For the PSS-modulated wave and M integer, c_n may be written in the form

$$c_n = \frac{j}{2n\pi}\,\zeta\chi\sum_{m=1}^{M} e^{-2\gamma(m-1)} \tag{9}$$

where

$$\gamma = jn\omega_0\tau\,, \qquad \zeta = 1 - e^{-\gamma}\,, \qquad \chi = \left(e^{j2\beta d} + e^{-\gamma}\right) \qquad \text{and} \qquad \beta = \frac{2\pi}{\lambda} \tag{10}$$

The more general equations for the case of non-integer M are given in [6]. The apparent PSS reflectivity at frequency f_c, for any switching frequency f_s, may now be found by comparing how much energy would enter the receiver pass-band in the absence of PSS modulation and when PSS modulation has taken place. For simplicity, it is assumed that for a pulse-width T_p, the receiver has a total bandwidth of $2/T_p$, since this is consistent with the generally accepted rule of thumb that the receiver bandwidth at the 3dB points is given approximately by $B = 1/T_p$.

Figure 5 shows the apparent reflectivity of a resonant (i.e. $d = \lambda/4$) PSS when the incident signal is a periodic train of 1μsec rectangular pulses of sinusoidal carrier with $f_c = 10$ GHz and a pulse duty cycle of 1%. When the PSS is in a non-resonant condition ($d \neq \lambda/4$), then it presents periodic reflection coefficients of -1 (active layer on, $G(t) = \infty$) and $-e^{j2\beta d}$ (active layer off, $G(t) = 0$) to the incident signal. The other two curves in Figure 5 show the PSS reflectivity performance for incident signals at frequencies of 11 and 15 GHz. These curves are identical to those for incident signals at 9 and 5 GHz, respectively, thus confirming the symmetrical nature of the PSS frequency response. It should also be noted that for high values of switching frequency, the predicted values of PSS reflectivity agree with those calculated from $20\log_{10}(|\cos \beta d|)$.

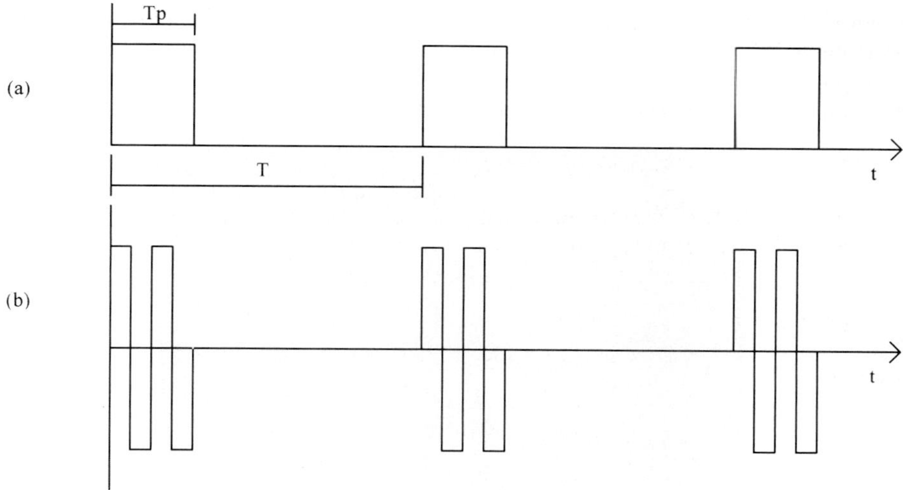

Figure 4   *Operation of the phase switched screen, (a) incident wave, (b) reflected wave (M = 2. For clarity, the sinusoidal carrier frequency signal itself is not shown)*

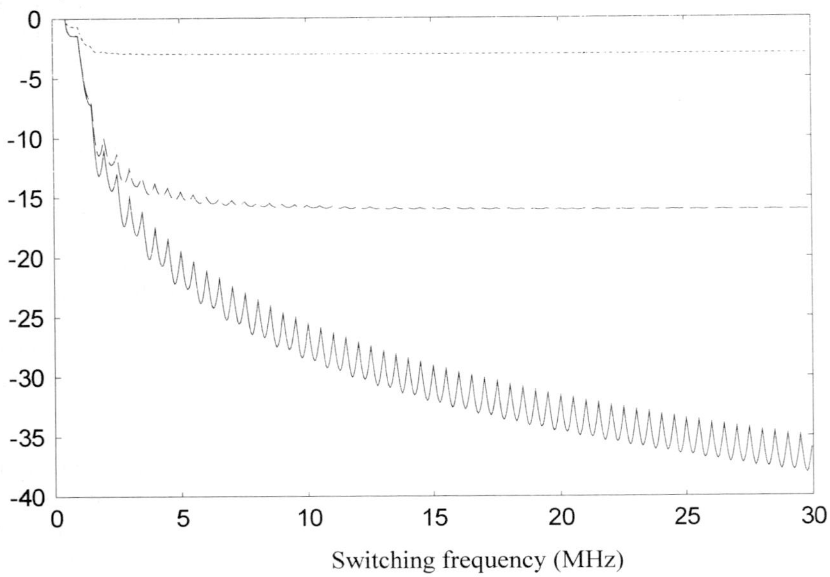

Figure 6   *Reflectivity of planar single layer PSS with 1:1 bipolar square-wave switching (1 μsec pulse length, 1 % duty cycle, 2 MHz receiver bandwidth)*
______ f_c = 10 GHz, __ __ f_c = 11 GHz, ... f_c = 15 GHz

4.   Practical implementation of a PSS

To realise a practical PSS we need a surface that can be rapidly switched between two impedance states. Potentially this could be achieved using some type of functional material such as an active conducting polymer; however, our present active layers consist of a two-dimensional grid of half wavelength dipoles that are loaded at their centres with *pin* diodes, as shown in Figure 7.

Figure 7   *Layout of PSS active layer showing pin diode-loaded bow-tie elements*

Under no-bias conditions the *pin* diodes present high microwave impedance and the dipoles may be considered open circuit. Hence the grid of dipoles is non-resonant and represents a high impedance equivalent surface. When the diodes are biased 'on', however, their high frequency impedance becomes low and the dipoles are resonant at the half-wavelength frequency. Hence the grid of shorted dipoles becomes a strong scatter and represents a low impedance surface.

Because of its topology, the structure shown in Figure 7 will only be effective for one plane of polarisation. By combining two such structures, rotated with respect to each other by 90° however, a dual polarised PSS may be realised. One element from the resulting active layer is shown in Figure 8 and typical measured reflectivity characteristics for two values of 'on' time are shown in Figure 9.

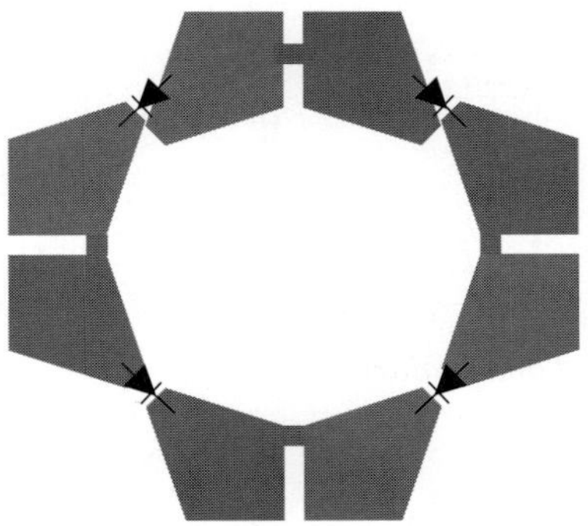

Figure 8   *Active surface element for dual-polarised PSS*

5.   **Multi-layer PSS**

In the same way that multiple layers may be used in passive RAM to increase bandwidth, the use of several active layers in a PSS increases its functionality, with the added advantage of dynamic reconfiguration. At normal incidence, the time-averaged reflection coefficient $\rho(f)$ of an N layer PSS is given by

$$\rho(f) = -(a_0 + a_1 e^{j2\beta_0 d} + a_2 e^{j4\beta_0 d} + \dots a_{N-1} e^{j2(N-1)\beta_0 d}) \tag{11}$$

This expression is *exact* for the ideal PSS (i.e. active layer conductances switched repetitively between the values of 0 and ∞) since no multiple reflection phenomena occur within the structure. The coefficients a_n are the normalised 'on' times τ_n/T associated with each active layer and these are related by $\sum_{n=0}^{N-1} a_n = 1$. Hence, by choosing appropriate values for a_n, a multi-layer PSS may be configured dynamically to have different reflectivity characteristics. As a simple example, consider a PSS having two active layers and a PEC back-plane; this corresponds to the case for N = 3. Assuming a symmetrical structure, then $a_0 = a_2$ and since $a_0 + a_1 + a_2 = 1$, this means that a_0 and a_1 are related by [4]

$$a_1 = 1 - 2a_0, \quad 0 \le a_0 \le 0.5 \tag{12}$$

Figure 10 shows the application of (12) to define the 'on' times for a PSS containing two active layers. When $a_0 = 0.5$, the reflectivity response has a single null at the frequency where $d = \lambda/4$. As the value of a_0 decreases, however, the null bifurcates to give two nulls that move apart as 'mirror-images'. Thus a reflectivity null may be positioned anywhere within the frequency range $c/8d \le f \le 3c/8d$, where $c = 3 \times 10^8$ m/s. If $a_0 = 0$, and $G_1G_2 \ne Y_0$, then the reflectivity characteristic only has a single null which fills in as $a_1 \to 0$ or 1. This case is analogous to that shown in Figure 2. When the PSS has more active layers, then the use of appropriate 'on' time relationships deduced by induction from (12) result in the formation of multiple 'mirror-image' reflectivity nulls which can also be tuned dynamically over a wide band of frequencies.

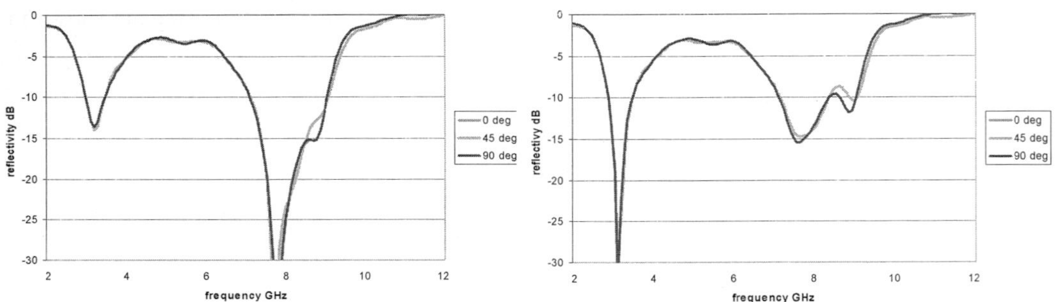

Figure 9   *Measured performance of dual-polarised PSS*

Reflectivity (dB)

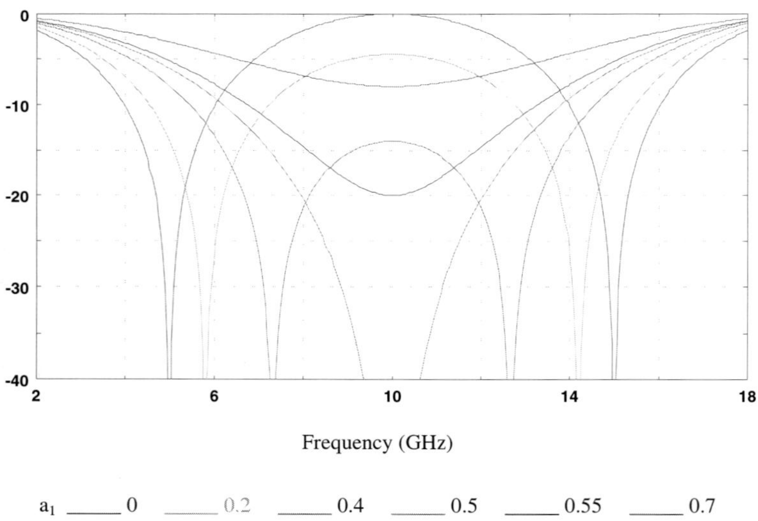

Frequency (GHz)

a_1 ______ 0 ______ 0.2 ______ 0.4 ______ 0.5 ______ 0.55 ______ 0.7

Figure 10   *Control of null position and depth in a two active layer PSS by varying the layer 'on' times*

References

[1]   Rozanov K N 2000, *IEEE Trans. Antennas Propagat,* **48,** 1230-1234.
[2]   Chambers B 1999, *Smart Materials and Structures,* **8,** 64-72 .
[3]   Chambers B and Tennant A 2002, *IEE Proc. Radar Sonar Navig,* **149,** 243-247.
[4]   Chambers B 1997, *Electron. Lett.,* **33,** 2073-2074.
[5]   Tennant A and Chambers B 2003, *Electron. Lett.,* **39,** 121-122.
[6]   Chambers B and Tennant A 2002, *IEEE Trans. Electromag Compat,* **44,** 434-441.

Session F2

Dielectric Composites

Co-chairs: A.N. Lagarkov
B. Chambers

ON ELECTRODYNAMICS OF ONE-DIMENSIONAL RANDOM SYSTEMS AND RELATED MAGNETO-OPTICAL PHENOMENA

A.P. Vinogradov[1], S.G. Erokhin[2], A.B. Granovsky[2], M. Inoue[3], A. Lisyansky[4], and
A. M. Merzlikin[1]

[1] Institute of Theoretical and Applied Electromagnetism (ITAE), Scientific Association JIHT, Russian Academy of Sciences

[2] Faculty of Physics, Lomonosov Moscow State University, Leninski Gory, Moscow 119992, Russia

[3] Department of Electrical and Electronic Engineering, Toyohashi University of Technology, 1-1, Hibari-Ga-Oka, Tempaku, Toyohashi 441-8580, Japan

[4] Department of Physics, Queens College of CUNY, Flushing, New York 11367, USA

In quantum mechanics it is well established that one-particle wave function is localized in any infinite disordered one-dimensional system [1-3]. The similarity of the Schrodinger and Maxwell equations suggests that light should also be localized in one-dimensional disordered system. This expectation was confirmed by computer simulations [4].

Besides the computer simulation there is the mathematically rigorous proof [3] of this fact based on Furstenberg's theorem [3]. The proof reduces to the statement that all solutions (with probability one) of the involved equations have "the exponential growth" related with localization [1-3]. Unfortunately, both the clear physical understanding of causes leading to this "exponential growth" and the physics of Ishii's speculations [3] have been still absent.

It is assumed the mesoscopic picture of the light localization is as follows [1]. Until the system size L is less than the free path of the light l then it is implied that the effective permittivity and permeability can be introduced. The non-coherent (diffuse) scattering of waves is described by the imaginary part ε''_{eff} of the effective permittivity [1]. At distances greater than $l \sim \lambda / \sqrt{\varepsilon''_{eff}}$ almost all of electromagnetic energy is transferred by diffusively scattered waves. The energy transfer on scale greater than $l \sim \lambda / \sqrt{\varepsilon''_{eff}}$ is governed by diffusion equation. Further, the effect of coherent backscattering [4] leads to decrease of diffusion coefficient as the systems size grows. Vanishing of the diffusion coefficient is treated as localization. The coherent backscattering should be destroyed in the presence of magnetic field because the system becomes nonreciprocal [2]. Thus, external magnetic field should delocalize the waves. This fact is accepted as the explanation of negative magnetoresistance effect in the case of electron transport. The presented picture is of the general character independent of system dimensionality d.

Unfortunately, consideration of one-dimensional system of ferrite layers placed in magnetic field that is perpendicular to them shows that the mathematical formulation of the problem coincides with one for dielectric layers but employing circular polarization instead of linear polarization. In other words magnetic field does not destroy localization of electromagnetic waves in $d=1$ case.

We present here a simple physical approach to the problem of light localization in one-dimensional system that is consistent with mathematical speculations. We expand a concept of band structure on random systems of finite thickness L identifying such an associated band structure with one of a periodic system where our finite system stands for an elementary cell. The band structure as well as attenuation of wave could be described in terms of the Lyapunov index $\gamma^{Sp} = \mathrm{Im}\, k_{eff} = \mathrm{Im}\left\{\arccos\left[\mathrm{Tr}\left(T_{cell}\right)/2\right]\right\}/L$, where T_{cell} is a T-matrix of the cell. We relate the Anderson localization of light with total band gap growth, which is observed in our computer simulation, as L increases. The measure $\tau = \lim_{\Omega \to \infty}\left\{(1/\Omega)\int_0^\Omega \left[1 - sign\left(\gamma^{sp}\right)\right]d\omega\right\}$ of bands of transparency vanishes exponentially with characteristic length $l_{loc}(\omega = \infty)$ Fig. 1. As a cause of the band gaps it is the Bragg reflection that plays the key role in localization. Thus, contrary to the common viewpoint, we show that localization due to disorder is caused by the Bragg reflection.

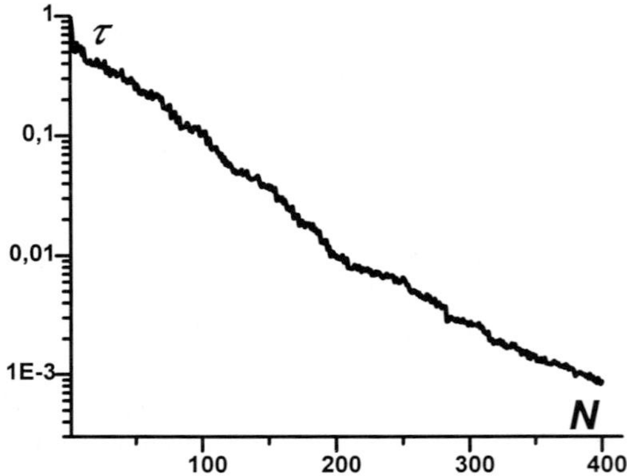

Fig. 1 *The dependence* $\tau(N)$, $N = L_{cell}/d$, *the permittivity* ε *equiprobably takes values from the interval* $(2, 37)$.

These band gaps are real band gaps with zero density of state. The localized states come from free bands as the frequency width of these bands as well as the group velocity of delocalized waves inside them comes to zero. Thus, the localized states are really a standing waves existing in accidentally appearing resonators. The mirrors in such resonators are induced by two half-spaces filled by random one-dimensional systems. Any finite part of the systems containing the localized state is an open resonator. As the amplitudes of fields in the resonator are inversely proportional to the transmission coefficient of mirrors we can obtain a system with very strong fields inside it by increasing the thickness of the system.

Such a treatment of localized states permits us to understand the nature of the Faraday effect enhancement in random systems. Such an enhancement was observed in the computer simulation [5] as well as in the experiment [6]. Indeed, if we fill the resonator supporting a localized state with a magneto-optical material we can anticipate a significant increase of the Faraday rotation. This increase was experimentally observed [6] and an optimal configuration, which turns out to be a Fabri-Perot resonator, was found with genetic algorithm [7].

As a result of our calculations it follows that there are a lot of optimal configurations leading to an increase of the Faraday and Kerr effects. An account of losses leads to weakening of the effects. The losses differently affect the properties of optimal configurations. Therefore one can increase the transmission coefficient with decrease of the Faraday rotation or vise versa.

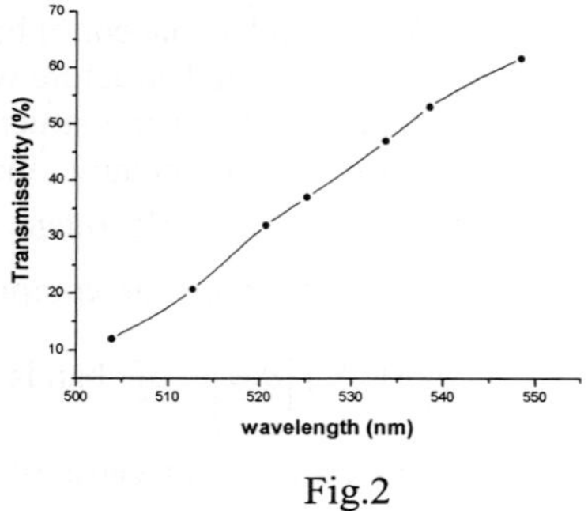

Fig.2

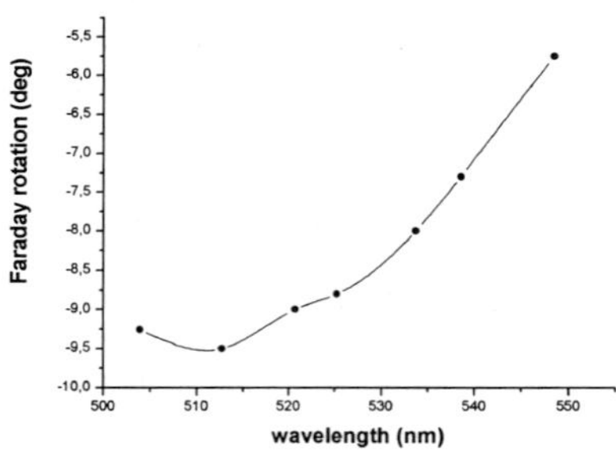

Fig. 3

For instance, in the case of Faraday effect geometry we consider magnetophotonic crystal made of the 295 nm thick magneto-optical layer of Bi:DyIG placed between six-layer sandwiches of SiO_2/Ta_2O_5. The thickness values of dielectric layers are varied to govern the resonance frequency of the system. We found two orders of magnitude enhancement of the Faraday rotation accompanied by the rather high transmittance (Fig. 2,3).

To observe the Kerr effect enhancement we have to seek an asymmetric configuration. In this case we consider the magneto-optical (Bi:DyIG) layer placed between six-layer and twelve-layer sandwiches of SiO_2/Ta_2O_5. The thickness ratio 1:2 of the dielectric mirrors is the optimal one for the configuration. We found that at $\lambda = 512,5 nm$ the reflectivity is about 18,8% and the Kerr rotation is $\vartheta_K = 18,6°$, that is also two orders of magnitude larger than the Kerr rotation for single Bi:DyIG layer without dielectric layers. The technology for the Kerr cell manufacturing can be simplified by substitution of the twelve-layer sandwich with a silver mirror.

This work was partly supported by Russian Foundation for Basic Research.

References

1. P. Sheng, *Introduction to wave scattering, localization, and mesoscopic phenomena*, Academic Press, London, 1995
2. N. F. Mott, Adv. Phys. V. 16, (1967) v. 50, No. 7, 865-945 (2001)
3. K. Ishii, Prog. Theor. Phys. Suppl. V. 53, p. 77-138 (1973)
4. V. I. Klyatskin, "Embedding Method in the Wave Propagation Theory", Moscow: Nauka, 1986
5. M. Inoue, T. Fujii, "A theoretical analysis of magneto-optical Faraday effect of YIG films with random multilayer structures", J. Appl. Phys. 81(8) p. 5659-5661, (1997)
6. M. Inoue, K. Arai, T. Fujii, M Abe, "One-dimensional magnetophotonic crystal", J. Appl. Phys. Vol. 85 p. 5768-5770 (1999)
7. M. Inoue, K. Arai, T. Fujii, M Abe, "Magneto-optical properties of one-dimensional photonic crystals composed of magnetic and dielectric layers", J. Appl. Phys. Vol. 83 p. 6768-6770 (1998)

A Study on the Dependence of Resonance Frequency on the Thickness of Anisotropic Composites with Long Conductive Fibers

L. Liu[1], S. M. Matitsine, K. M. Hock, and Y. B. Gan,
Temasek Laboratories, National University of Singapore, Singapore

K. N. Rozanov
Institute for Theoretical and Applied Electromagnetics, RAS, Moscow, Russia

Abstract. *The resonant frequency of long conductive fibers embedded in an anisotropic composite as a function of layer thickness is investigated both numerically and experimentally. An empirical law is suggested to fit the results obtained. The law involves a critical thickness value, below which the layer can no longer be treated as a bulk material for a fiber of given length and thickness. The values of the critical thickness were found from numerical results for two particular composites.*

Introduction

Composites with polymer matrix and long high-conductive fibers have been extensively used as electromagnetic materials. The EM materials have found applications in the areas such as EMC/EMI, FSS, etc [1,2]. The resonance frequency of effective permittivity as an important performance indicator of these materials has been investigated by experimental, analytical and numerical methods [3, 4].

When an electric dipole is embedded in a dielectric layer, its resonance frequency is known to depend not only on the permittivity, ε, but also on the thickness, t, of the layer. Munk [1] employed the dielectric image approach to study this dependence. He used the concept of effective dielectric constant, ε_e, which is attributed to a layer of finite thickness to account for the effect of t on the resonance frequency of dipoles, as shown in Eq. (1) below. An approximate guideline was obtained for dielectric layer with $\varepsilon \approx 3$ and $t \geq 0.05\lambda\varepsilon$: (i) $\varepsilon_e = \varepsilon$, when the dielectric substrates are on both sides of the dipoles; (ii) $\varepsilon_e = (1 + \varepsilon)/2$, when the dipoles are located on the surface of the layer. The resonant frequencies of long aluminum fibers embedded in a sheet composite of aluminum flakes have been studied analytically and experimentally in [5].

In the cited studies, the substrate for the fibers was assumed to have isotropic permittivity. However, dielectrics with multilayer structure usually have uniaxial anisotropy of permittivity. An example is composites with non-spherical conducting inclusions that are advantageous to obtain high permittivity values. Inclusions shaped as flakes or fibers are prone to rotation when confined in the thin coating layer during the spraying process. As a result, the composites have much higher permittivity along in-plane direction as compared to that along the out-of-plane direction. Therefore, the study of resonant frequency of long conductive fiber inside an anisotropic medium is of interest.

The objective of this study is to investigate the resonant frequency of long conductive fibers in an anisotropic composite layer as a function of the layer thickness. In practice, the anisotropic composite can be obtained by mixing carbon fibers or aluminum flakes with silicone matrix. Since the particles used have much smaller size as compared to both the wavelength and the conductive fibers, the two-phase composite can be taken as a homogenous material in the problem under study. For an isotropic bulk medium with permittivity ε, the resonance frequency of fiber embedded is $f_{res} = f_0/\sqrt{\varepsilon}$, where f_0 is the free space resonance frequency of the fiber. In a uniaxial anisotropic bulk medium, $f_{res} = f_0/\sqrt[4]{\varepsilon_{//}\varepsilon_\perp}$ [6], where $\varepsilon_{//}$ and $\varepsilon_\perp$ are the in-plane and out-of-plane permittivity of the medium, respectively. Smaller frequency shift will be expected when the composite is thin, since the effective permittivity around fibers is lower than the bulk value. In any case, no quantitative relation is available from literatures [1, 6].

---

[1] Corresponding author, email: tslll@nus.edu.sg

In this paper, the Finite Element Method (FEM) is used to investigate the dielectric resonance of anisotropic composite slabs with long conductive fibers as inclusions. The simulation results are verified by measurement. Curve fitting method is used to find an empirical relation for the thickness dependence of resonance frequency.

Experiment

Composite substrates were prepared by mixing inclusions with liquid silicone and solvent. Commercial aluminum flakes with diameter of about 15 μm and thickness of 3 μm or carbon fibers with length of 0.5 mm and thickness of 8 μm, also commercially available, were used as inclusions. The mixture was sprayed onto a substrate to form a first layer of about 0.1 mm to 0.2 mm in thickness and 20 cm by 20 cm in size. When the solvent has completely evaporated, another layer was sprayed onto it. The process was repeated till the desired thickness was obtained. Finally, the sample sheet was removed from the substrate. To prepare the composite samples with long resonant fibers embedded, small amount of copper fibers with diameter of 0.1 mm and length of 10 mm were dropped randomly on the surface of a thin, sprayed substrate, followed by spraying another layer over to double the substrate's thickness. The amount of the fibers was so small that they are considered to be isolated. A sketch of the sample under study is shown in Fig. 1.

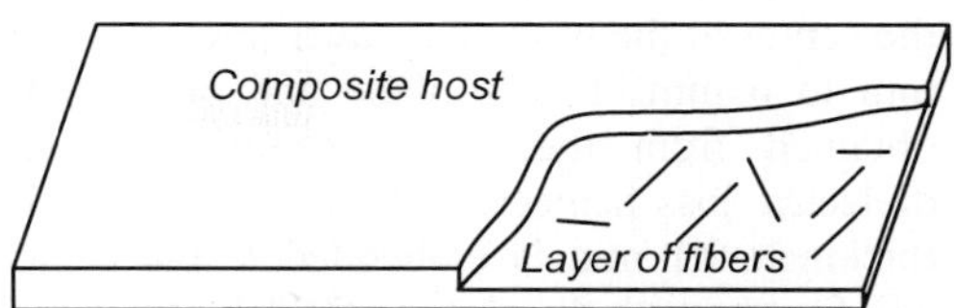

Figure 1. Schematic structure of composite

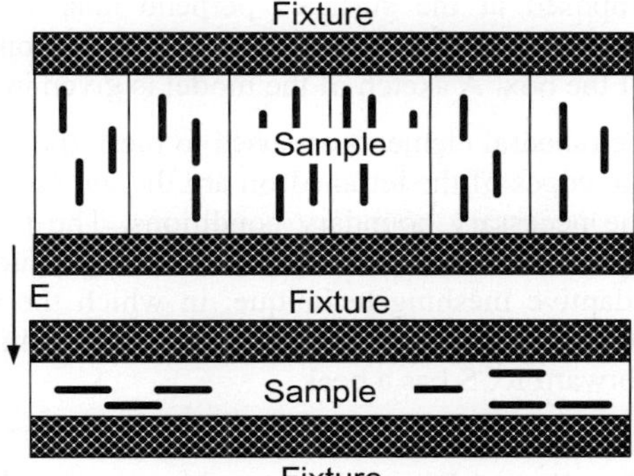

Figure 2. Impendence measurement in parallel and perpendicular directions

To measure the permittivity of composites without copper fibers, we used the Agilent 4291B RF Impedance/Material Analyzer operating from 1 MHz to 1.8 GHz, equipped with a capacitance measurement cell. To measure the out-of-plane permittivity, a sheet sample was cut into slices of 7 mm by 7 mm by 0.3 mm and placed into the cell. To measure the in-plane permittivity, the sample was first cut into slices of 7 mm by 3 mm by 0.3 mm. The slices were then rotated by 90° and glued together to form a sample of 7 mm by 7 mm by 3 mm in size (Fig. 2). The top and bottom surfaces of the sample were polished and coated with conductive glue to ensure that they have good electrical contact with the fixture. According to the specification of the analyzer, the measurement accuracy achievable for permittivity below 1GHz is about 15%.

The resonance frequency of copper fibers in a composite layer was determined by the near-field free space method employing the reflectivity measurement of the sample from 2 GHz to 20 GHz, with the use of a vector network analyzer (VNA), broadband horn antennas, lens, and radar absorbing material board with an aperture. The normal incident wave has only TEM component inside a cylinder of diameter 15 cm. The resonant frequency of the sample with copper fibers can be found from the frequency dependence of the transmission coefficient. The free space technique was also used to measure the in-plane permittivity of the samples without fibers. The results obtained are in good agreement with the capacitance method.

The details of the experimental techniques are given in [6].

Numerical

The commercial FEM software High Frequency Structure Simulator 8 (HFSS^{TM}) from Ansoft is used in the numerical study. The model includes a box of size 40mm×10mm×30mm filled with air. A layer of t×10mm×30mm in size is located at the center of the base. The layer thickness, t, is varied from 0.1 mm to 6 mm. The in-plane and out-of-plane permittivity is obtained from the measurement of actual samples, with dielectric loss neglected. A copper fiber of length 10 mm and thickness 0.1 mm is embedded at the center of the layer. A TEM wave with the electric field E parallel to the fiber and the wave vector k perpendicular to the layer surface is used to illuminate the model. The PML boundary conditions are imposed at the surfaces perpendicular to the wave vector. Periodic boundary conditions are applied on the other surfaces of the box. A sketch of the model is given in Fig. 3.

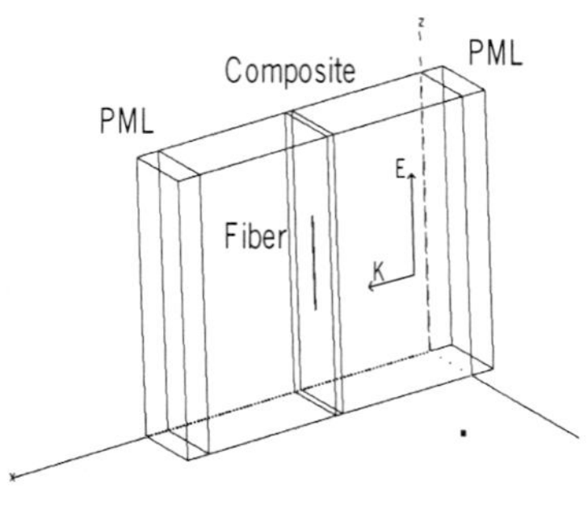

Fig. 3. FEM model of conductive fibers in an anisotropic layer

Tetrahedral elements are used to mesh the model. The electric fields at the vertices and midpoint of the edges of the tetrahedron are the unknowns to be calculated by solving the matrix equation with the necessary boundary conditions. The field within each tetrahedron is interpolated from these values and the basis functions. To produce the optimal element size, HFSS uses an iterative adaptive meshing technique, in which the mesh is automatically refined in critical regions. The resonance frequency of the fiber imbedded in the layer was obtained as the frequency at which the forward RCS has a peak.

Results and Discussion

The results obtained for the resonance frequency of copper fibees of length 10 mm embedded in an anisotropic layer as a function of the layer thickness are given in Fig. 4. The figure presents two sets of data, one related to an aluminum flake composite (triangles), and another to a carbon fiber composite (boxes). The measurements show that both composites have uniaxial anisotropy of permittivity, with in-plane permittivity $\varepsilon_{//}$=33 and out-plane permittivity $\varepsilon_{\perp}$=8 for the Al-flake composite and $\varepsilon_{//}$=8 and $\varepsilon_{\perp}$=3 for the C-fiber composite. Numerical results are represented by empty squares and triangles, while the experimental data are the filled symbols. The close agreement between the result of FEM and the measurement is evident.

In uniaxial materials, the resonance frequency is determined by $\sqrt[4]{\varepsilon_{//}\varepsilon_{\perp}}$ instead of $\sqrt{\varepsilon}$ [6]. This is in agreement with the asymptotical behavior of the data obtained for large t. As $t\rightarrow0$, the resonance frequency for both composites tends to the resonance frequency of the fiber located in free space, f_0. For fibers of length 10 mm and thickness 0.1 mm used in this study, f_0 =13.5 GHz [8].

After Munk [1], we describe the resonance frequency, f_{res}, of a fiber embedded in a layer of a finite thickness in terms of the effective permittivity, ε_e. This value differs from the average permittivity of an inhomogeneous layer that is conventionally obtained from reflectivity or transmittance measurements. The physical meaning of ε_e is that it determines the overall energy of the electric field scattered by the fiber to account for the finite thickness of the layer. Thus defined, the effective permittivity is closely related to the capacitance of the fiber. It can be determined by microwave measurement as:

$$\varepsilon_e = \left(f_{res}/f_0\right)^2, \tag{1}$$

and depends on t, $\varepsilon_{//}$, and $\varepsilon_{\perp}$, as well as on the fiber length and thickness. Figure 5 plots the thickness dependence of the effective permittivity calculated from the data in Fig. 4, with the same notation.

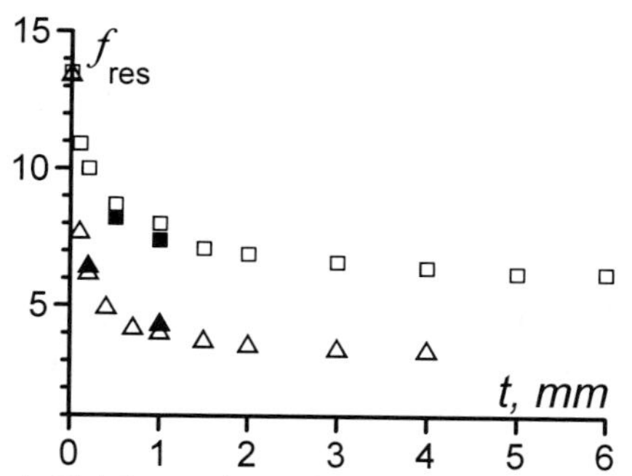
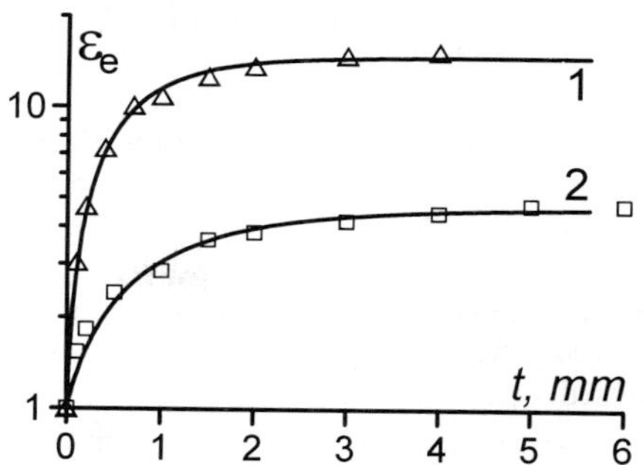

Fig. 4. Thickness dependence $f_{res}(t)$ for Al flakes (triangles) and carbon fibers (boxes) composites

Fig. 5. Thickness dependency of the effective permittivity retrieved from the data in Fig. 4.

For very thin layers, the resonance frequency is close to that in vacuum, hence, $\varepsilon_e = 1$. With increasing thickness, the effective permittivity increases and tends to the value typical for bulk samples, i.e. $\varepsilon_e = \sqrt{\varepsilon_{//}\varepsilon_{\perp}}$. Based on this, we fit the thickness dependence of effective permittivity by a simple exponential law:

$$\varepsilon_e = \sqrt{\varepsilon_{//}\varepsilon_{\perp}} - \left(\sqrt{\varepsilon_{//}\varepsilon_{\perp}} - 1\right)\exp\left(-t/t_0\right), \tag{2}$$

The fitting curves obtained from the numerical data with the only fitting parameter, t_0, are shown in the figure by solid lines. It is seen that law (2) provides precise fitting of the numerical data. The fitting of the numerical data with law (2) is much better than that achievable with an exponential law applied to the thickness dependence of resonant frequency as proposed in [5].

The value of t_0 can be referred to as the critical thickness, since a layer filled with fibers of a given length and thickness, with the thickness less than t_0, cannot be treated as a bulk medium, and therefore, cannot be described in terms of effective material parameters. In view of this, the values of t_0 for actual composites are of certain interest. The fitting procedure produces $t_0 = 0.68$ for the Al-flake composite and $t_0 = 1.21$ for the C-fiber composite. Therefore, t_0 decreases as the layer permittivity increases. In addition, the critical thickness must be a function of the fiber dimensions, as well as the $\varepsilon_{//}/\varepsilon_{\perp}$ ratio. More studies are necessary to reveal these dependences. Notice that the data obtained does not support the rule suggested in [1]: the thickness of $0.05\lambda_e$ on either side of the fibers appears to be insufficient for the layer to be considered as a bulk medium, at least for the fibers and permittivity values used in this study.

Acknowledgements

The authors are grateful to Dr. Kong Lingbing for the help in the permittivity measurement, to Mrs. Fang Xiao Jia and Mr. A. Lagoisky for the preparation of samples, to Mr. Tan Szu Hau and Dr. V. B. Moustafaev for the assistance in measurements, to Dr. Qing Anyong, Mr. Xu Xin for fruitful discussions. A. Lagarkov and K. Rozanov appreciate RFBR for partial support of the work according to Agreements no. 00-15-96570 and 03-02-16247.

References

[1] B. A. Munk, (2000), *Frequency Selective Surface*, Wiley-Interscience Publication, p. 393.
[2] J. W. Molyneux-Child, (1997), *EMC shielding materials*, Oxford, Boston, Reed Educational & Professional Publishing, p. 63
[3] A. N. Lagarkov and A. K. Sarychev (1996), *Phys. Rev. B* **53** (10), 6318.
[4] D. P. Makhnovskiy, L. V. Panina, D. J. Mapps, and A.K. Sarychev (2001) *Phys. Rev.B* 64(9), 134205.
[5] A. P. Vinogradov, D. P. Machnovskii, K. N. Rozanov (1999), *J. Communic. Technol. and Electr.* 44 (3), 317
[6] S. M. Matitsine, K. M. Hock, L. Liu, Y. B. Gan, A. N. Lagarkov, and K. N. Rozanov (2003), *Journal of Applied Physics*, submitted.
[7] A. M. Nicolson and G. F. Ross (1968) *IEEE Trans. Instrum. Meas.*, IM-17, 395.
[8] R. S. Elliott, *Antenna theory and design*, 1981, p. 304

Self-Adaptive Materials with External Support System

L. R. Arnaut
Center for Electromagnetic and Time Metrology
National Physical Laboratory
Teddington, Great Britain

Abstract: Self-adaptive material systems (SAMs) and their effective constitutive characterization are reviewed. The effect of a PEC shield for the SAM support system on the optimal control of the SAM is analysed.

Introduction

Synthetic electromagnetic (EM) materials can be engineered to yield improved reflection, absorption or polarization transformation properties for incident waves. However, due their typically passive and fixed nature, there remain fundamental limitations on their capabilities and tolerances. Even materials that are intended for operation in a 'static' environment suffer from inevitable limitations on manufacturing tolerances and from drift due to contamination and physical or chemical instability (ageing). Self-adaptive material (SAM) systems (Fig. 1) are a class of fully-electric 'smart' materials that offer the prospect of overcoming such limitations [1]. They permit the construction of a hybrid active/passive effective medium whose constitutive parameters are adjustable after implementation so as to yield a composite medium that can be used as an EM reference material.

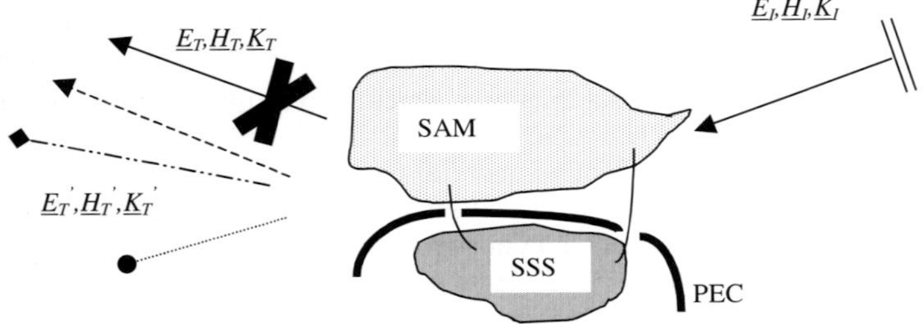

Figure 1: Concept of a self-adaptive material (SAM) with external SAM support system (SSS) shielded by a PEC. The naturally scattered wave ($\underline{E}_t,\underline{H}_t,\underline{k}_t$) (solid outward arrow) from the SAM in its passive state is suppressed by its activating its sources, yielding instead the desired artificial response ($\underline{E}_t',\underline{H}_t',\underline{k}_t'$).

In [1]-[2], the focus was on implementations whose SAM support system (SSS) is fully embedded (free-standing SAM). Here, we report on SAMs with an external support system, which is easier to realize in practice. Typically, the SSS is now shielded from the sensing and actuating dipole arrays by a PEC plane, which affects the radiation characteristics.

Controlled sensor-actuator dipole pairs as basic SAM elements

Fundamental to SAM operation is the response of a coupled pair of dipole sources above a PEC plane (Fig. 2), in which the radiation by a primary source s (e.g., scatterer) is controlled by a secondary source a. The transfer function $H_{as,opt}^{PEC}$ of the optimal controller depends on the particular choice of cost function, which is formulated in terms of the desired EM characteristics of the *overall* radiated field.

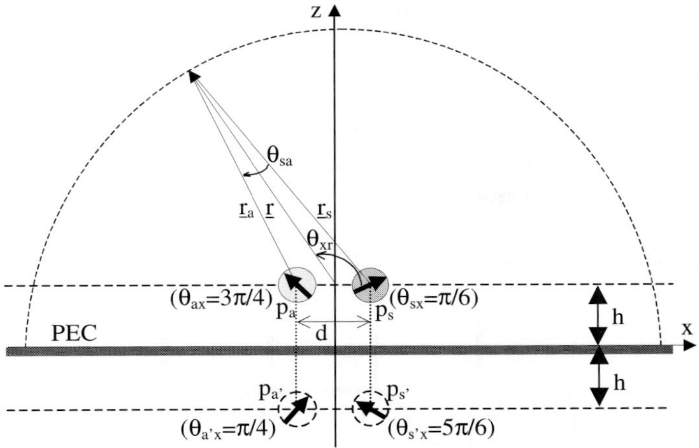

Figure 2: Adaptive control of radiation by primary (master) source s by secondary (slave) source s, with relative source spacing d and observation point at distance r. The source pair is parallel to, and at a height h above an optional PEC ground plane which can be replaced by image dipole sources a' and s'.

A typical implementation of a fully-electric SAM contains a planar PEC shield between the two interleaved dipole arrays and the support system consisting of the control hardware, power hybrids, beam formers, etc (Fig. 3). The support system presents itself as a load impedance to the array elements. The SAM exhibits two independent 'look' directions, one for the incident ($\underline{r}_s$) and one for the controlled ($\underline{r}_a$) radiation.

The basic function of a SAM is the compensation of the naturally (i.e., passively) scattered field. To this end, we require the minimization of the magnitude of the radiated power

$$W_{tot} = \frac{1}{2}\begin{bmatrix} \underline{p}_s^T & \underline{p}_a^T & \underline{p}_{s'}^T & \underline{p}_{a'}^T \end{bmatrix} \cdot \begin{bmatrix} \underline{\underline{A}}_{ss} & \underline{\underline{A}}_{sa} & \underline{\underline{A}}_{ss'} & \underline{\underline{A}}_{sa'} \\ \underline{\underline{A}}_{as} & \underline{\underline{A}}_{aa} & \underline{\underline{A}}_{as'} & \underline{\underline{A}}_{aa'} \\ \underline{\underline{A}}_{s's} & \underline{\underline{A}}_{s'a} & \underline{\underline{A}}_{s's'} & \underline{\underline{A}}_{s'a'} \\ \underline{\underline{A}}_{a's} & \underline{\underline{A}}_{a'a} & \underline{\underline{A}}_{a's'} & \underline{\underline{A}}_{a'a'} \end{bmatrix} \cdot \begin{bmatrix} \underline{p}_s^* \\ \underline{p}_a^* \\ \underline{p}_{s'}^* \\ \underline{p}_{a'}^* \end{bmatrix}$$

in a single direction of observation $\underline{1}_r$, in which

$$\underline{\underline{A}}_{\alpha\beta} = (j\omega\mu_0)^{-1}\underline{\underline{G}}_{ee_\alpha}^T \cdot \underline{\underline{\varepsilon}} \cdot \underline{\nabla} \cdot \underline{\underline{\varepsilon}} \cdot \underline{\underline{G}}_{ee_\beta}^*, \quad \underline{\underline{G}}_{ee_\alpha} = \frac{\exp(-jkr_\alpha)}{4\pi\varepsilon_o r_\alpha^3}\left[(jkr_\alpha)^2(\underline{1}_{r_\alpha}\underline{1}_{r_\alpha} - \underline{\underline{I}}) + (1 + jkr_\alpha)(3\underline{1}_{r_\alpha}\underline{1}_{r_\alpha} - \underline{\underline{I}})\right]$$

for α, $\beta = s$, a, s' or a', where $\underline{\underline{G}}_{ee}$ is the electric-electric dyadic Green function. Zero radiated power in a single direction in the far field can be achieved, provided the transfer function of the controller for one pair of dipoles, defined by $p_{a,opt} = H_{as,opt}^{PEC} p_s$, is chosen as

$$H_{as,opt}^{PEC.} = -\frac{\left(F_a A_{as} F_s^* + F_a A_{as'} F_{s'}^*\right) + \left(F_{a'} A_{a's} F_s^* + F_{a'} A_{a's'} F_{s'}^*\right)}{\left(F_a A_{aa} F_a^* + F_a A_{aa'} F_{a'}^*\right) + \left(F_{a'} A_{a'a} F_a^* + F_{a'} A_{a'a'} F_{a'}^*\right)}$$

where

$$A_{\alpha\beta} = \frac{k^4 \exp[-jk(r_\alpha - r_\beta)]}{16\pi^2\varepsilon_o\sqrt{\mu_o\varepsilon_o}\,r^2}\sin\theta_{\alpha r}\sin\theta_{\beta r}; \qquad F_\alpha = \exp(\pm jkh\sin\theta_{xr})$$

in which the upper or lower sign corresponding to a real or image source α, respectively. This defines the impressed source current and hence the radiated field. The result for a free-standing SAM [2] is retrieved when putting $\underline{p}_{\alpha'}=\underline{p}_{\beta'}=\underline{0}$. For the case of dipoles parallel or perpendicular to the PEC plane,

$$H_{as,opt}^{PEC} = -\frac{\sin\theta_{sr} + C_s \exp(j2kh\sin\theta_{xr}\cos\phi_r)\sin\theta_{s'r}}{\sin\theta_{ar} + C_a \exp(j2kh\sin\theta_{xr}\cos\phi_r)\sin\theta_{a'r}} \exp(jkd_x\sin\theta_{xzr}\cos\phi_r)$$

where the image coefficient C_α has the value $+1$ or -1 if the wire dipole α is oriented perpendicular ($\theta_{\alpha x}=\pi/2$ or $3\pi/2$) or parallel ($\theta_{\alpha x}=0$ or π) to the PEC plane, respectively.

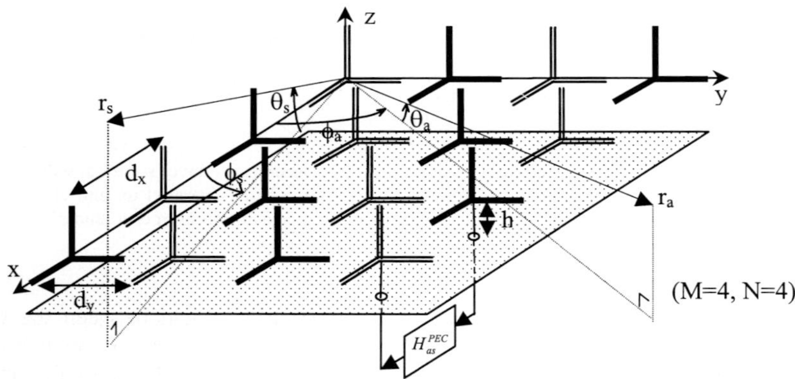

Figure 3: Planar chequered array of interleaved primary and secondary sources above PEC groundplane.

For the total radiated power integrated across a hemisphere above the PEC plane, the terms $F_\alpha A_{\alpha\beta} F_\beta^*$ are to be replaced by

$$A'_{\alpha\beta} = \frac{k^4}{16\pi\varepsilon_o\sqrt{\mu_o\varepsilon_o}} \int_0^\pi F_\alpha F_\beta^* \sin\theta_{\alpha r} \sin\theta_{\beta r} \exp\left[-jk(r_\alpha - r_\beta)\cos\theta_{xr}\right] \sin\theta_{xr} d\theta_{xr}$$

which yields $H_{as,opt}^{PEC'}$. The total integrated radiated power, referenced to that of the primary dipole source and its image, is then

$$\frac{W_{tot,min}^{PEC'}}{W_s^{PEC}} = 1 - \frac{(A'_{as}+A'_{as'})+(A'_{a's}+A'_{a's'})}{(A'_{aa}+A'_{aa'})+(A'_{a'a}+A'_{a'a'})} \cdot \frac{(A'_{sa}+A'_{sa'})+(A'_{s'a}+A'_{s'a'})}{(A'_{ss}+A'_{ss'})+(A'_{s's}+A'_{s's'})}$$

which is a measure of the residual radiation (inefficiency) for the SAM.

Fig. 4 shows the complex-valued optimal transfer function $H_{as,opt}^{PEC'}$ and the associated minimum total radiated power $W_{tot,min}^{PEC'}/W_s^{PEC}$ for mutually parallel dipoles oriented parallel or perpendicular to the PEC plane, at quarter-wave distance ($h=\lambda/4$). The total radiated power is significantly larger than for free-standing SAMs, because of the effect of the ground plane. This implies that a larger density of sensor/actuator pairs is needed for a specified level of performance. The effect of impedance loading of the dipoles has been analysed in [2].

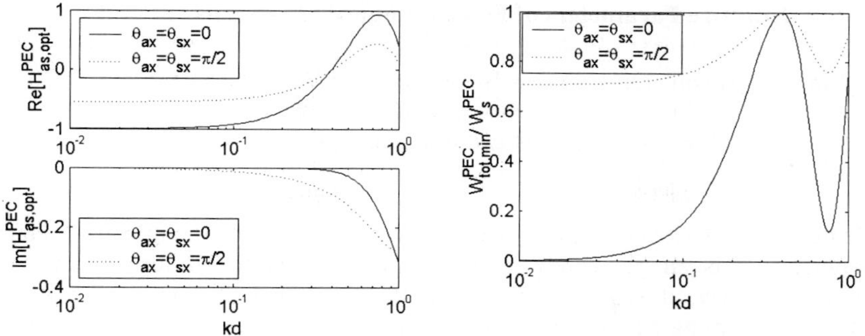

Fig. 4: Optimal transfer function and total radiated power.

Effective Permittivity

Based on the analysis in [2], the effective relative permittivity $\varepsilon_{eff,r}$ of an electric SAM is obtained mutatis mutandis as (cfr. [2] for definition of symbols)

$$\varepsilon_{eff,r} = \varepsilon'_{eff,r} - j\varepsilon''_{eff,r} = \frac{\left(2d_x d_y s_z\right)^2 \exp\left[j2\tan^{-1}\left(\frac{H_{as}^{0'}}{1+H_{as}^{0''}} \right) \right]}{\left(PP^t\right)^2 \left(1-s_x^2\right)^2 \left[\left(1+H_{as}^{0'}\right)^2 + \left(H_{as}^{0''}\right)^2 \right]}.$$

where $H_{as}^0 = H_{as}^{0'} - jH_{as}^{0''}$ relates to $H_{as,opt}^{PEC}$. As an example, Fig. 5 shows the effective permittivity for SAM particles in a dielectric matrix for which $\varepsilon_{host,r}=5-j0.05$.

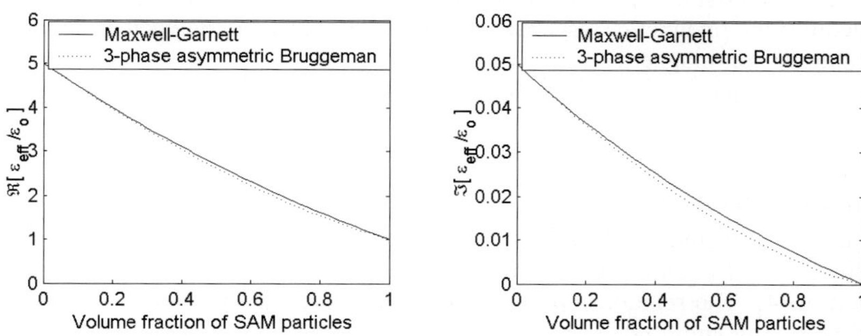

Figure 5: Effective permittivity of SAM.

References

[1] Arnaut, L.R., Self-adaptive material systems; in: S. Zouhdi, A. Sihvola, and M. Arsalane (eds.): *Advances in Electromagnetics of Complex Media and Metamaterials*, NATO Sci. Ser. II vol. 89. Kluwer: Dordrecht, NL (2003), pp. 421-438.

[2] Arnaut, L.R., Adaptive control and optimization of electromagnetic radiation, attenuation and scattering using self-adaptive material systems, *IEEE Trans. Antennas Propag.*, vol. 51 no. 7 (2003), to appear.

[3] Culshaw, B., *Smart Structures and Materials*. Artech House, 1996.

Effective Parameters of Fiber Composite Materials

Xin Xu[1], Anyong Qing[1], Yeow Beng Gan[1], and Yuan Ping Feng[2]
[1]Temasek Laboratories, NUS, Singapore
[2]Physics Department, NUS, Singapore

1. Introduction

The Effective Medium Theory (EMT) [1] has been widely used to obtain the effective parameters of composite materials. It is simple and can give reasonable prediction for certain types of composite materials. However, it is limited by the assumptions that the inclusion particles are sparsely distributed and coupling between particles are weak. For composite materials with dense and/or strong inclusion particles, more rigorous approaches that consider the multiple scattering effects are required. For example, the T-matrix method combined with the configurational averaging technique and the quasi-crystalline approximation (TCQ) [2],[3] has been used to obtain the effective propagation constant of composite materials with spherical or spheroidal inclusions. However, it is very difficult for the T-matrix method to handle scatterers with extremely small (<0.01) or large (>100) aspect ratio. As such, the TCQ model fails to predict the effective properties of composite materials with fiber inclusions, since the aspect ratio of fibers is usually larger than 100.

This study focuses on the effective properties of a fiber composite slab comprising a host material and fiber inclusions. The configuration is as shown in Fig. 1. The fibers are randomly but uniformly distributed in planes parallel to the slab surface. It is assumed that the host medium and fibers are non-magnetic. In addition, the fibers are very thin as compared to its length and wavelength of interest, so that circumferential currents are not likely to be excited to create the artificial magnetic effects. Consequently, the magnetic response of the fibers can be neglected, rendering the composite material to be non-magnetic.

The unit incident plane wave is polarized along the x-direction (time factor $e^{j\omega t}$ is assumed and suppressed):

$$\mathbf{E}^{inc}(\mathbf{r}) = \hat{x}e^{-j\mathbf{k}\cdot\mathbf{r}}, \tag{1}$$

where $\mathbf{k} = k_0\hat{\mathbf{z}}$ is the free space wave vector.

A semi-analytic-numerical method is developed to solve the problem. The effective permittivity of the fiber composite material is obtained from its effective transmission coefficient using the well-known formulation in [4]. The relationship between the effective transmission coefficient and the averaged forward scattering amplitude (AFSA) is derived following Bohren and Huffman [4]. The AFSA is numerically obtained using the method of moment and averaging technique. Numerical computation of the ASFA allows consideration of multiple scattering of the fibers.

2. Theory

Consider a collection of identical fibers. According to [4], the scattered field from the jth fiber centered at $\mathbf{r}_j$ is given by

$$\mathbf{E}_j^{sca}(\mathbf{r}) = \frac{e^{-jk_0|\mathbf{r}-\mathbf{r}_j|}}{k_0|\mathbf{r}-\mathbf{r}_j|}\mathbf{F}_j(\hat{\mathbf{e}}_j)e^{-j\mathbf{k}\cdot\mathbf{r}_j}, \tag{2}$$

where $\mathbf{e}_j = \dfrac{\mathbf{R}_j}{R_j} = \dfrac{\mathbf{r}-\mathbf{r}_j}{|\mathbf{r}-\mathbf{r}_j|}$, $\mathbf{F}_j(\hat{\mathbf{e}}_j)$ is the vector scattering amplitude of the jth fiber along the $\hat{\mathbf{e}}_j$ direction, in the presence of other fibers.

Accordingly, the total scattered field by all fibers is

$$\mathbf{E}^{sca}(\mathbf{r}) = \sum_j \mathbf{E}_j^{sca}(\mathbf{r}) = \sum_j \frac{e^{-jk_0 R_j}}{k_0 R_j} \mathbf{F}_j(\hat{\mathbf{e}}_j) e^{-j\mathbf{k}\cdot\mathbf{r}_j} \, , \qquad (3)$$

Now, consider a semi-infinite fiber composite slab $0 \le z \le h$, $-\infty < x < \infty$, $-\infty < y < \infty$. The fibers are randomly but uniformly distributed in the plane parallel to the slab surface. This assumption is consistent with the sample preparation process, in which the samples are prepared through spraying the fiber-liquid mixture layer by layer till the desired sample thickness is obtained.

In this paper, we are concerned with the effective properties of the composite slab. Therefore, the composite slab is regarded as a homogeneous medium. In this case, the summation in (3) becomes an integral, given by

$$\mathbf{E}^{sca}(\mathbf{r}) = \int_0^h dz' \int_{-\infty}^{\infty} dx' \int_{-\infty}^{\infty} dy' \frac{e^{-jk_0|\mathbf{r}-\mathbf{r}'(x',y',z')|}}{k_0|\mathbf{r}-\mathbf{r}'(x',y',z')|} \mathbf{F}(\hat{\mathbf{e}}) e^{-jk_0 z'} \, , \qquad (4)$$

where $\hat{\mathbf{e}} = \dfrac{\mathbf{r}-\mathbf{r}'(x',y',z')}{|\mathbf{r}-\mathbf{r}'(x',y',z')|}$ is the unit vector, $\mathbf{F}(\hat{\mathbf{e}})$ is the scattering amplitude contributed by a unit volume, which can be written as

$$\mathbf{F}(\hat{\mathbf{e}}) = \lim_{V \to 0} \frac{\mathbf{F}_V(\hat{\mathbf{e}})}{V} \, , \qquad (5)$$

where $\mathbf{F}_V(\hat{\mathbf{e}})$ is the scattering amplitude contributed by the volume V.

According to Bohren and Huffman [4], the integral in (4) can be approximately evaluated in a straightforward manner by the method of stationary phase. The result is

$$\mathbf{E}^{sca}(\mathbf{r} = r\hat{\mathbf{z}}) = -j \frac{2\pi h}{k_0^2} \mathbf{F}(\hat{\mathbf{e}} = \hat{\mathbf{z}}) e^{-jk_0 z} \, . \qquad (6)$$

This scattered field is a plane wave, with wave number k_0, and propagating in the $\hat{\mathbf{z}}$ direction.

For the problem concerned, the scattered field is solely due to the fibers in the composite slab. Therefore, $\mathbf{F}(\hat{\mathbf{e}} = \hat{\mathbf{z}})$ can be approximated as

$$\mathbf{F}(\hat{\mathbf{e}} = \hat{\mathbf{z}}) = \frac{\mathbf{F}_V(\hat{\mathbf{e}} = \hat{\mathbf{z}})}{V} = n\overline{\mathbf{F}}(\hat{\mathbf{e}} = \hat{\mathbf{z}}), \qquad (7)$$

where $\mathbf{F}_V(\hat{\mathbf{e}} = \hat{\mathbf{z}})$ is the total forward scattering amplitude due to all fibers in the volume V, and $\overline{\mathbf{F}}(\hat{\mathbf{e}} = \hat{\mathbf{z}})$ is the AFSA, n is the number of fibers per unit volume.

Substituting (7) into (6) leads to

$$\mathbf{E}^{sca}(\mathbf{r} = r\hat{\mathbf{z}}) = -j \frac{2\pi h n}{k_0^2} \overline{\mathbf{F}}(\hat{\mathbf{e}} = \hat{\mathbf{z}}) e^{-jk_0 z} \, , \qquad (8)$$

Next, consider the fiber composite slab as an effective homogeneous dielectric medium with complex relative permittivity $\varepsilon_r = \varepsilon_r' - j\varepsilon_r''$. The transmission coefficient is readily obtained as

$$\tilde{\tau} = \frac{\mathbf{E}^{inc}(\mathbf{r}) + \mathbf{E}^{sca}(\mathbf{r})}{\mathbf{E}^{inc}(\mathbf{r})} \, , \qquad (9)$$

On the other hand, according to [4], we have the relationship between the transmission coefficient and the relative permittivity of the slab:

$$\tilde{\tau} = \frac{4\sqrt{\varepsilon_r}}{\left(\sqrt{\varepsilon_r}+1\right)^2} \frac{e^{jk_0h}}{e^{jk_0\sqrt{\varepsilon_r}h} - \left(\dfrac{1-\sqrt{\varepsilon_r}}{1+\sqrt{\varepsilon_r}}\right)^2 e^{-jk_0\sqrt{\varepsilon_r}h}} . \tag{10}$$

Finally, it remains to compute the AFSA $\overline{\mathbf{F}}(\hat{\mathbf{e}} = \hat{\mathbf{z}})$ as defined in (7). On the first look, it appears to be impossible to solve for AFSA accurately, due to the presence of infinite number of fibers. However, we can approximate this value as the average of the total forward scattering amplitude of a finite number of fibers, for a sufficiently large number of fibers. The total forward scattering amplitude of a finite number of fibers can be easily calculated using numerical methods such as the method of moment [5]. The reaction integral formulation with piecewise sinusoidal basis functions [6] is chosen for this study.

The error in this approximation is due mainly to the edges of the fibers. To minimize this error, the AFSA is calculated by excluding a reasonable amount of edge fibers.

3. Numerical Results and Discussions

The fibers are distributed in a circular slab with radius r and thickness h. The host medium is assumed to be air. The fibers are identical of radius a and length l.

In order to obtain reliable results, a sufficient amount of fibers should be included in the numerical calculation. During the averaging process, we also need to exclude the fibers near the edges, in order to minimize the edge effects. We conducted a numerical experiment to assess the number of fibers near the edges to be excluded in the averaging process. We considered 1000 PEC fibers with $l = 0.5cm$ and $a = 5\mu m$. These fibers are distributed randomly in a circular slab with $r = 10cm$ and $h = 0.05cm$. Figure 2 shows the AFSA at 14 GHz. It is observed that when the average is taken on less than 150 fibers, the results are not stable. By taking more fibers into the averaging process, the results tend towards a stable value that can be regarded as the correct ASFA for an infinite composite slab. Due to the edge effect, we should exclude as many fibers near the edges as possible, while keeping a sufficiently large number of fibers to obtain a stable average, e.g. 300 wires in figure 2.

This numerical experiment is based on a particular random distribution of the fibers. The effect of random distribution will give rise some deviations on the field results. The amount of deviation can be estimated by calculating the fields over many different random distributions. As a result, the AFSA should be represented by its mean value and deviation.

In the numerical experiment, we use 300 fibers near to the center of the slab to compute the AFSA. The effective permittivity of the composite sheet material is obtained from the calculations, with the result as shown in figure 3.

4. Conclusions

A simple model is proposed to calculate the effective permittivity of the fiber composite materials. It combines both analytical and numerical methods, thus it is able to include the interactions between fibers.

Acknowledgement

The authors would like to thank Dr. Chao-Fu Wang of Temasek Laboratories for his patient discussion on the Moment Method.

Reference

[1] T. C. Choy, *Effective Medium Theory: Principles and Applications*, Oxford: Clarendon Press, 1999

[2] V. K. Varadan, B. N. Bringi, and V. V. Varadan, Coherent electromagnetic wave propagation through randomly distributed dielectric scatterers, *Phys. Rev. D*, Vol.19. No. 8, pp. 2480-2489, 1979

[3] A. Qing, X. Xu, and Y. B. Gan, Effective wave number of composite materials with oriented randomly distributed inclusions, 2003 IEEE AP-S Int. Symp. (submitted)

[4] C. F. Bohren, and D. R. Huffman, *Absorption and Scattering of Light by Small Particles*, New York: Wiley. 1998

[5] J. H. Wang, *Generalized Moment Methods in Electromagnetics*, New York: Wiley, 1991

[6] J. H. Richmond, Radiation and scattering by thin-wire structures in the complex frequency domain, in E. K. Miller, L. Medgyesi-Mitschang, and E. H. Newman, Ed., *Computational Electromagnetics*, New York: IEEE Press, 1992, pp. 156-169

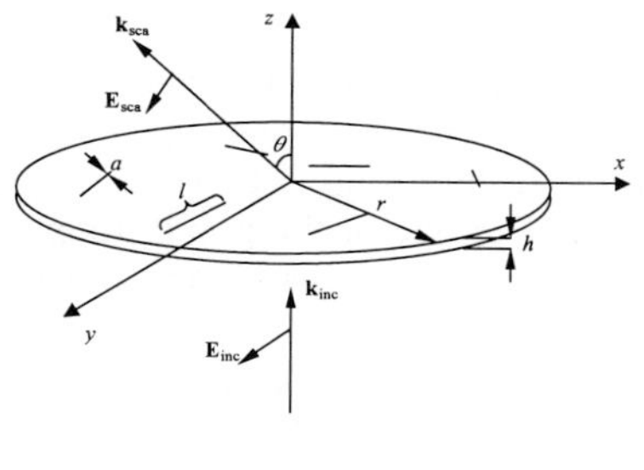

Figure 1: Configuration

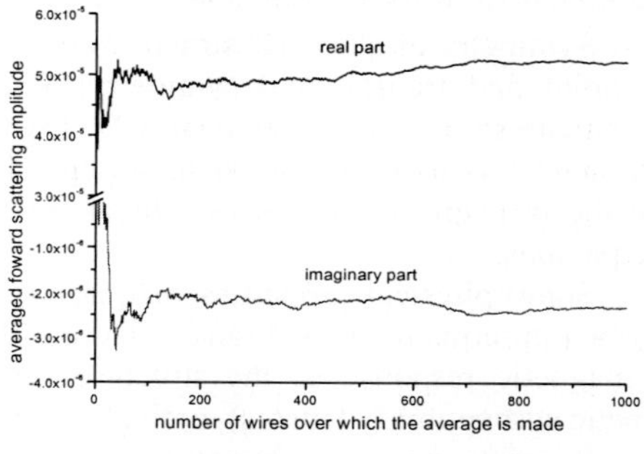

Figure 2: Averaged Forward Scattering Amplitude

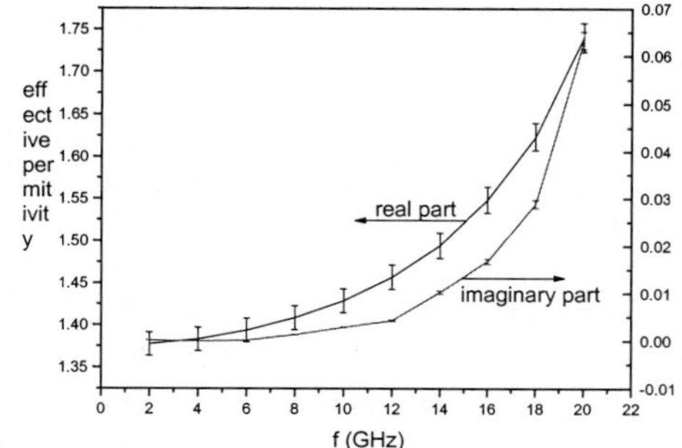

Figure 3: Effective Permittivity of a Collection of PEC Fibers

Second-rank tensors for quasi-1D periodical electromagnetic structures described by magnetic line groups

V. Dmitriev

University Federal of Para, Belem, Para, Brazil

I. Introduction

Quasi-1D periodical structures have periodicity along their principal axis and some rotation-reflection point symmetry. Examples of such structures are polymers and nanotubes [1] where the lateral extension of the structures is small, and also multilayer films [2] and 1D photonic crystals where the lateral extension can be irrelevant to the physical processes, for example to electromagnetic wave propagation in the direction normal to the layers.

Symmetry of quasi-1D structures is described by the line groups [3] which take into account both point and translational symmetry. When the wave vector is small one can consider an approximate symmetry of the quasi-1D structures, namely, the factor group L/T where L is a line group and T is its translational subgroup. The factor group L/T is one of the point groups which leave the principal rotational axis invariant. These groups are axial groups of any order of the principal axis.

Some physical properties of quasi-1D structures defined by the factor group L/T can be described in terms of second rank polar and axial tensors. These properties for example are electric and magnetic response of the structures on electromagnetic waves (dielectric permittivity and magnetic susceptibility tensors), optical activity, electrical conductivity.

Recently, periodical structures with the broken Time reversal symmetry have attracted much attention of researchers. Due to asymmetry in electron interaction with different excitations (photons, phonons, etc) having opposite signs of wave vectors, many new interesting physical effects appear in these structures.

General matrix forms of the polar and axial second-rank tensors for 13 *nonmagnetic* line group families have been calculated in [4]. But this information *is not complete* because it does not include magnetic structures, i.e. the structures with the broken Time reversal symmetry.

The point symmetry of periodical quasi-1D *magnetic* structures is described by an infinite number of magnetic axial point groups [5]. They are comprised by 7 families of axial point groups of the first category, 7 families of the second category and 17 groups of the third category.

Using invariance of quasi-1D structures with respect to the corresponding Space-Time reversal symmetry operations, we calculate the matrix forms of the polar and axial second-rank tensors for some cases. The method allows one to make a catalogue of the tensors for all the possible Space-Time reversal symmetries of these structures. Notice that the line groups can also be used for investigation of anisotropic 2D and 3D structures for specified directions with high symmetry.

II. Nonmagnetic line groups

A line group L is a symmetry group of a physical structure periodic along a line (we choose this line along the z-axis). Any element of L can be written as $\{R|t+v\}$ defined by $\{R|t+v\}\mathbf{r}=R\mathbf{r}+(t+v)a\mathbf{e}_z$, where $t=0,\pm 1, \ldots$, $0\leq v<1$, $\mathbf{e}_z$ is the unit vector in the z-direction, a is the translation period, R is a rotation, a reflection or a rotoreflection which leaves the z-axis invariant. The operator R can be a rotation through $2\pi/n$ around the z-axis (denoted by C_n), a rotation through π around an axis orthogonal to the z-axis (U_x for rotation around the x-axis and U_y for rotation around the y-axis), the reflection σ_v in a plane containing the z-axis (σ_x for the reflection in the plane x=0 and σ_y for the reflection in the plane y=0), the reflection in the plane perpendicular to the z-axis (σ_h), or any combination of these operations. The pure translations $\{E|t\}$ where E is the unit element of the group, form a discrete 1D translational group T which is a subgroup of L. The pure rotation-reflections $\{R|0\}$ are the point groups P. The possible point groups are C_n, S_{2n}, D_n, C_{nv}, D_{nd} and D_{nh} (in the Schöenflies notations) where n=1, 2, 3,…, ∞. P is a subgroup of L only if the latter is symmorphic, i.e. if it contains neither glide planes nor screw axes, so that v=0 in every element of L. In a nonsymmorphic line group, v≠0 in some elements.

For simplicity discussing the tensors of the second-rank for the line groups, we shall use the corresponding isogonal point groups. It is because the isogonal point groups define the structure of the "macroscopic" tensors and the translational parts of the line groups are irrelevant in our long-wavelength approach. The isogonal point group of a symmorphic group is the factor group L/T where L is a line group and T is its translational subgroup. Defining the isogonal point group of a nonsymmorphic group, one should replace screw axes and glide planes by the usual axes and planes without translations. Isogonal point groups for linear groups are in fact *axial* point groups [5]. Therefore calculating the second-rank tensors we can restrict ourselves by consideration of axial point groups.

III. Magnetic point groups

There exist three categories of magnetic point groups. In the nonmagnetic state, physical objects are described by magnetic groups of the first category G. Sometimes, these groups are called nonmagnetic ones. The group G consists of a unitary subgroup H (it contains the usual rotation-reflection operations) and products of the Time reversal operator Θ with all the elements of H (the Time reversal operator Θ changes the sign of time t→-t). The full group is then H+ΘH including Θ. In the case of magnetic groups of the second category **G** (denoted by bold-face type), there is no elements with time reversal operator Θ, and Θ is not an element of the group. The magnetic groups of the third category G(H) contain in addition to the unitary operators (rotation-reflections) of the unitary subgroup H, an equal number of antiunitary operators which are the product of the antiunitary operator Θ and the usual geometrical symmetry operators. These combined operators form a conjugate class of the subgroup H and lead to the existence of antiaxes, antiplanes and anticenter of symmetry. The unitary operators of a magnetic group of the third category form a unitary subgroup of index 2. It means that in every group of the third category there are equal number of elements with and without Θ. The operator Θ itself is not an element of magnetic groups of the third category.

IV. Method of calculation of second-rank tensors

The method of calculation of the allowed forms of the second-rank tensors for nonmagnetic line groups used in [4] is based on the polar- and axial-vector representations of the line groups and group projector technique. In contrast to this, we shall make the direct calculations proceeding from the invariance of the polar and axial tensors with respect to the rotation-reflections, Time-reversal and combined rotation-reflections and Time reversal symmetries of the quasi-1D magnetic structures. Notice that there are some discrepancies in the calculated forms of the tensors presented in [4] for nonmagnetic groups and of those calculated by our method.

The symmetry transformation properties of the tensors of the second-rank are well known [6]. Under rotation-reflections, the polar tensors [a] the axial ones [A] are transformed as follows:

$$[a]'=[R][a][R]^{-1}, \qquad [A]'=(det[R])[R][A][R]^{-1}, \tag{1}$$

where [R] is the 3D matrix representation of the rotation-reflection operations and det means determinant.

The combined rotation-reflection and Time reversal operations give the following laws of the tensor transformations:

$$[a]'=\pm[R][a]^{t}[R]^{-1}, \quad [A]'=\pm(det[R])[R][A]^{t}[R]^{-1}, \tag{2}$$

where t stands for transposition. For the "pure" Time reversal operator, the matrix [R] in relations (2) is the unit one.

Table I. Polar [a] and axial [A] tensors of the second rank for magnetic line groups whose isogonal groups are D_{nh} of the first category, $\mathbf{D}_{nh}$ of the second category, $D_{nh}(C_{nh})$, $D_{nh}(D_n)$ $D_{nh}(C_{nv})$, $D_{nh}(D_{kd})$ and $D_{nh}(D_{kh})$ of the third category.

N°	Magnetic group and its generators	Polar tensor [a]	Axial tensor [A]	N°	Magnetic group and its generators	Polar tensor [a]	Axial tensor [A]
1a	D_{1h} U_x, σ_h, θ	$\begin{vmatrix} a_{11} & 0 & 0 \\ 0 & a_{22} & 0 \\ 0 & 0 & a_{33} \end{vmatrix}$	$\begin{vmatrix} 0 & 0 & 0 \\ 0 & 0 & A_{23} \\ 0 & -A_{23} & 0 \end{vmatrix}$	4a	$D_{1h}(D_1)=$ $C_{2v}(C_2),$ $U_x, \theta\sigma_h$	$\begin{vmatrix} a_{11} & 0 & 0 \\ 0 & a_{22} & a_{23} \\ 0 & -a_{23} & a_{33} \end{vmatrix}$	$\begin{vmatrix} A_{11} & 0 & 0 \\ 0 & A_{22} & A_{23} \\ 0 & -A_{23} & A_{33} \end{vmatrix}$
1b	D_{2h} $C_2, U_x, \sigma_h, \theta$	$\begin{vmatrix} a_{11} & 0 & 0 \\ 0 & a_{22} & 0 \\ 0 & 0 & a_{33} \end{vmatrix}$	0	4b	$D_{2h}(D_2)$ $C_2, U_x, \theta\sigma_h$	$\begin{vmatrix} a_{11} & 0 & 0 \\ 0 & a_{22} & 0 \\ 0 & 0 & a_{33} \end{vmatrix}$	$\begin{vmatrix} A_{11} & 0 & 0 \\ 0 & A_{22} & 0 \\ 0 & 0 & A_{33} \end{vmatrix}$
1c	D_{nh} (n=3,4,...∞) $C_4, U_x, \sigma_h, \theta$	$\begin{vmatrix} a_{11} & 0 & 0 \\ 0 & a_{11} & 0 \\ 0 & 0 & a_{33} \end{vmatrix}$	0	4c	$D_{nh}(D_n)$ (n=3,4,.. ∞) $C_4, U_x, \theta\sigma_h$	$\begin{vmatrix} a_{11} & 0 & 0 \\ 0 & a_{11} & 0 \\ 0 & 0 & a_{33} \end{vmatrix}$	$\begin{vmatrix} A_{11} & 0 & 0 \\ 0 & A_{11} & 0 \\ 0 & 0 & A_{33} \end{vmatrix}$
2a	$\mathbf{D}_{1h}$ U_x, σ_h	$\begin{vmatrix} a_{11} & 0 & 0 \\ 0 & a_{22} & 0 \\ 0 & 0 & a_{33} \end{vmatrix}$	$\begin{vmatrix} 0 & 0 & 0 \\ 0 & 0 & A_{23} \\ 0 & A_{32} & 0 \end{vmatrix}$	5a	$D_{1h}(C_{1v})=$ $C_{2v}(C_s)$ $\sigma_y, \theta\sigma_h$	$\begin{vmatrix} a_{11} & 0 & a_{13} \\ 0 & a_{22} & 0 \\ -a_{13} & 0 & a_{33} \end{vmatrix}$	$\begin{vmatrix} 0 & A_{12} & 0 \\ A_{12} & 0 & A_{23} \\ 0 & -A_{23} & 0 \end{vmatrix}$
2b	$\mathbf{D}_{2h}$ C_2, U_x, σ_h	$\begin{vmatrix} a_{11} & 0 & 0 \\ 0 & a_{22} & 0 \\ 0 & 0 & a_{33} \end{vmatrix}$	0	5b	$D_{2h}(C_{2v})$ $C_2, \sigma_y, \theta\sigma_h$	$\begin{vmatrix} a_{11} & 0 & 0 \\ 0 & a_{22} & 0 \\ 0 & 0 & a_{33} \end{vmatrix}$	$\begin{vmatrix} 0 & A_{12} & 0 \\ A_{12} & 0 & 0 \\ 0 & 0 & 0 \end{vmatrix}$
2c	$\mathbf{D}_{nh}$ (n=3,4,...∞) C_4, U_x, σ_h	$\begin{vmatrix} a_{11} & 0 & 0 \\ 0 & a_{11} & 0 \\ 0 & 0 & a_{33} \end{vmatrix}$	0	5c	$D_{nh}(C_{nv})$ (n=3,4, ..∞) $C_4, \sigma_y, \theta\sigma_h$	$\begin{vmatrix} a_{11} & 0 & 0 \\ 0 & a_{11} & 0 \\ 0 & 0 & a_{33} \end{vmatrix}$	0
3a	$D_{1h}(C_{1h})=$ $C_{2v}(C_{1h}),$ $\theta U_x, \sigma_h$	$\begin{vmatrix} a_{11} & a_{12} & 0 \\ -a_{12} & a_{22} & 0 \\ 0 & 0 & a_{33} \end{vmatrix}$	$\begin{vmatrix} 0 & 0 & A_{13} \\ 0 & 0 & A_{23} \\ A_{13} & -A_{23} & 0 \end{vmatrix}$	6a	$D_{2h}(D_{1d})=$ $D_{2h}(C_{2h})$ $U_y, \theta\sigma_h$	$\begin{vmatrix} a_{11} & 0 & a_{13} \\ 0 & a_{22} & 0 \\ -a_{13} & 0 & a_{33} \end{vmatrix}$	0
3b	$D_{2h}(C_{2h})$ $C_2, \theta U_x, \sigma_h$	$\begin{vmatrix} a_{11} & a_{12} & 0 \\ -a_{12} & a_{22} & 0 \\ 0 & 0 & a_{33} \end{vmatrix}$	0	6b	$D_{nh}(D_{kd})$ (k=2,3, ∞, n=2k) $\theta C_4, U_y, \theta\sigma_h$	$\begin{vmatrix} a_{11} & 0 & 0 \\ 0 & a_{11} & 0 \\ 0 & 0 & a_{33} \end{vmatrix}$	0
3c	$D_{nh}(C_{nh}),$ (n=3,4,.. ∞) $C_4, U_x, \theta\sigma_h$	$\begin{vmatrix} a_{11} & a_{12} & 0 \\ -a_{12} & a_{11} & 0 \\ 0 & 0 & a_{33} \end{vmatrix}$	0	7a	$D_{2h}(D_{1h})=$ $D_{2h}(C_{2v})$ $\theta C_2, \sigma_y, \sigma_h$	$\begin{vmatrix} a_{11} & 0 & 0 \\ 0 & a_{22} & 0 \\ 0 & 0 & a_{33} \end{vmatrix}$	$\begin{vmatrix} 0 & 0 & 0 \\ 0 & 0 & A_{23} \\ 0 & A_{23} & 0 \end{vmatrix}$
				7b	$D_{nh}(D_{kh})$ (k=2,3, ∞, n=2k) $\theta C_4, \sigma_y, \sigma_h$	$\begin{vmatrix} a_{11} & 0 & 0 \\ 0 & a_{11} & 0 \\ 0 & 0 & a_{33} \end{vmatrix}$	0

If the structure under consideration has a Space-Time reversal symmetry, the second-rank tensors [a] and [A] describing some physical properties of the structure must be invariant under the corresponding transformations. It means that the relations (1) and (2) become identities. Therefore, they can be used for calculations of the allowed forms of the tensors. For a given group, to reduce the volume of calculations, one can use in (1) and (2) not the all elements of the group but only the so-called generators of this group [6].

The complete Tables of the tensors for all 81 families of linear magnetic groups require much space. Here, we shall give only several examples of our calculations. Let us apply to the high symmetry groups D_{nh}. There exist 5 line groups whose isogonal groups are D_{nh}. The number of corresponding *magnetic* line groups is 28. Dealing with tensors of the second rank, we can consider only the magnetic axial point groups isogonal to these 28 line groups, namely, D_{nh} of the first category, $\mathbf{D_{nh}}$ of the second category, $D_{nh}(C_{nh})$, $D_{nh}(D_n)$, $D_{nh}(C_{nv})$, $D_{nh}(D_{kd})$ and $D_{nh}(D_{kh})$ of the third category. The results of calculations by formulae (1) and (2) for the line magnetic groups associated with D_{nh} are presented in Table I. Generators of the groups are also given in this Table. For some line groups, their equivalent notations adopted in the theory of point groups are also written, for example, $D_{1h}(D_1)=C_{2v}(C_2)$,

V. Discussion and conclusions

A general form of the second-rank tensors contains 9 independent parameters. As can be seen from Table I, the calculated tensors due to symmetry have a small number of independent parameters. For example, the tensor [a] for the symmetry $D_{1h}(C_{1h})$ has 4 parameters and the tensor [A] has only 2 independent parameter. The tensors with axes of the order n=3 and higher orders have the same form. It is in accordance with a known theorem: any axis of symmetry of the third order and of the higher orders for a tensor of the second rank is transformed into the axis of infinite order. Therefore, the tensors for n≥3 coincide with those for continuous groups with n= ∞.

A general information of the physical properties of the structures with different symmetries can be obtained by inspection of the tensors. For example, optical activity is allowed in the structures where the diagonal elements of the axial tensor [A] are no null. Scrutiny of these tensors can give also information of the orientation of the optical axes. Such analysis for nanotubes which are described by the nonmagnetic line groups is given in [1]. On the other hand, optical activity is prohibited in the structures with symmetries where the tensor [A] is zero.

Another information accessible from our calculations is nonreciprocal properties of the structures. Nonreciprocal effects (such as for example, Faraday effect) are possible in those structures, which are described by a nonsymmetrical about the main diagonal polar tensor [a] which is associated in this case with the tensor of dielectric permittivity [ε] or magnetic permeability [μ].

Summarising we have described a method for calculation of the tensors of the second rank for magnetic quasi-1D structures using their symmetry properties. Some examples of the calculations are given in Table I. The obtained tensors with reduced number of independent parameters simplify the theoretical analysis of the corresponding physical problems. These tensors can provide a useful guideline for synthesis of new structures with prescribed electromagnetic properties, in particular of nonreciprocal and tunable electromagnetic components based on magnetic photonic crystals. The information obtained by symmetry methods is exact. Therefore, another possible application of these results is checking the tensors calculated by other methods.

References

1. M. Damnjanovi, I. Miloševi, T. Vukovi and R. Sredanovi, "Full symmetry, optical activity, and potentials of single-wall and multiwall nanotubes", Phys. Rev. B, **60**, 2728-2739, 1999

2. P. Weinberger, "Multilayer systems and symmetry", Philosophical Magazine B, **75**, 509-533, 1997

3. M. Vujii, I. B. Boovi, and F. Herbut, "Construction of the symmetry groups of polymer molecules", J. Phys. A: Math. Gen., **10**, 1271-1279, 1977

4. I. Miloševi, "Second-rank tensors for quasi-one-dimensional systems", Physics Letters A, **204**, 63-66, 1995

5. M. Damnjanovi, M. Vujii, "Subgroups of weak-direct products and magnetic axial point groups", J. Phys. A: Math. Gen., **14**, 1055-1063, 1981

6. A. A. Barybin, V. A. Dmitriev, Modern Electrodynamics and Coupled-Mode theory: Application to Guided-Wave Optics, Rinton Press, Princeton, 2002

Impedance Spectroscopy of Composite Materials

Rosario A. Gerhardt
School of Materials Science and Engineering
Georgia Institute of Technology
Atlanta, GA 30332-0245
USA

Impedance Spectroscopy is a well known technique in the solid state ionic community, where it has been used for over thirty years[1]. The main premise of this method is that any system, be it a device, object or material may be represented via equivalent circuits. In the case of solid state ionic materials, this method was instrumental in demonstrating the deleterious effect of grain boundaries on the effective dc conductivity of large grain sized oxygen ion conductors[2-4], where the material was modeled as a series of parallel R-C networks. More recently, it has been shown that when the grains are sub-micron in size, the properties of the grain boundary region are lower than that in the bulk[5]. In figure 1, simulated complex impedance spectra for these two extremes are shown together with two other cases, illustrating the variety of spectra one may get from a simple material system. This method continues to be used as one of the most important tools for monitoring the properties of newly prepared materials for fuel cells and batteries.

In contrast, the emphasis on the characterization of novel antennas and RF shielding materials centers around the response of these complex materials at the frequencies at which they will be used. This is essential for determining the expected performance at the frequencies of interest, but it neglects the contributions of the underlying microstructure. Many of these novel electromagnetic applications require addition of conducting inclusions into a dielectric matrix or the stacking of many layers of very different materials. In order to detect chemical inhomogeneities or other defects, it is necessary to use as broad a range of frequencies as possible so as to detect all the conducting paths and polarization mechanisms present.

In composite materials, there can be as many as four major causes of frequency dispersion: (1) space charge layer at the interface between the conducting and the insulating phase, (2) space charge layer at the grain boundaries within the matrix, (3) space charge at the boundaries between the fillers and (4) dielectric or magnetic relaxations and/or resonances. The actual response will be dependent on the composite material's individual resistance, capacitance and/or inductance values. When the matrix and the fillers are both dielectric, one expects little or no frequency dispersion[6] unless the dielectric relaxations or resonances give rise to detectable frequency dispersions[7]. When there is a large disparity in the properties of the individual phases that make up the composite, there are many more opportunities for frequency dispersion[8-14].

In this presentation, we will discuss the usage of impedance and/or dielectric spectroscopy (IS/DS) as a tool for characterizing microstructural features at all length scales in a wide variety of composite materials. Our recent work on SiC-AlN, SiC-BN, BN-B$_4$C, SiCw-mullite, SiCw-Al$_2$O$_3$, TiB$_2$-Al$_2$O$_3$, and Al$_2$O$_3$-CeO$_2$ composites will be reported. It will be demonstrated that the ability to detect certain microstructural features depends on the ratio of the properties of the main component phases and the ac equipment specifications. Examples will include detection of compositional dependence, whisker orientation dependence, interconnectivity degree in percolated systems and key signatures associated with each type of composite. While data acquisition is a non-issue these days, data interpretation is more complex. Proper data collection and data interpretation of a specific set of materials can lead to in-line process monitoring, quality control monitoring, mechanical damage monitoring and environmental degradation monitoring. ID/DS has also been used to characterize conducting polymers[15-18], semiconductors[19] and metallic alloys[20-21]. If time permits, examples of these materials will also be given.

40

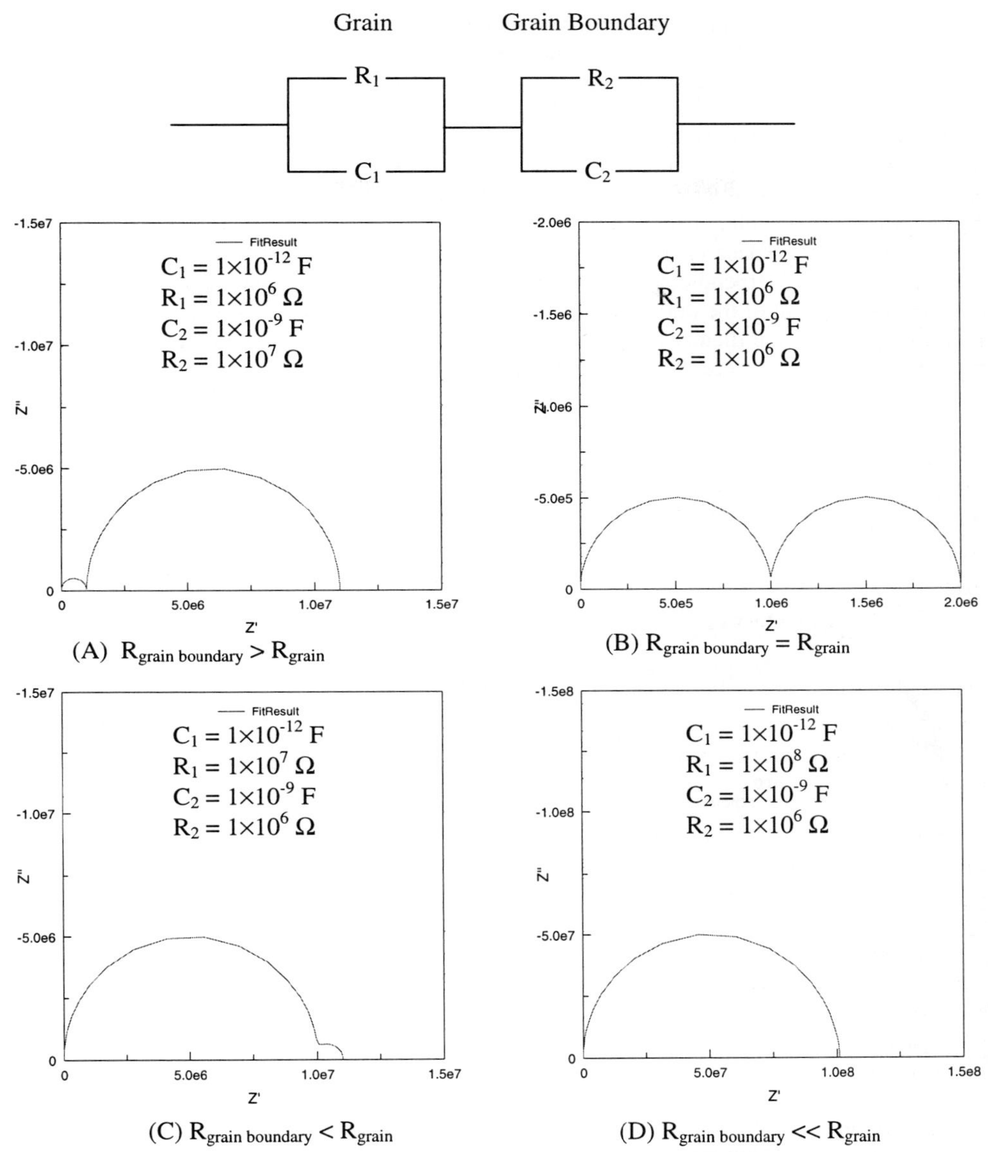

Figure 1. Simulated impedance spectra for a material sample containing grain boundaries that have (a) a higher resistance than the bulk grains; (b) the same resistance as the bulk grains; (c) a smaller resistance than the bulk; and (d) a lot smaller resistance than the bulk grains.

To illustrate the advantages of using multiplane analysis[22], which permits the transformations of the impedance data into other dielectric functions, and to help separate the contributions of the more conducting phase from the insulating phase, Fig. 2 depicts the measurements of a composite containing 10% SiC whiskers (designated as SiCw)dispersed in a mullite matrix. Mullite is a highly insulating material while SiC whiskers are expected to be semiconducting. Since the specimens were prepared by hot-pressing, the whiskers tend to align with their long axes perpendicular to the hot pressing direction. It is clear from the figures that the SAME composite displays different electrical response depending on whether the field is parallel or perpendicular to the whisker long axes. The complex impedance plots indicate the presence of two processes. For the parallel to the whiskers measurement, the impedance of the bulk phase is so small that it cannot be seen at the same magnification as the semicircle which results from the space charge behavior. This is supported by the frequency explicit imaginary impedance plot, where the higher conductivity oriented sample shows a larger peak. In contrast, for the perpendicular to the whisker measurement, both semicircles are visible. The dielectric constant response also shows a large enhancement due to the presence of the space charge layers in the composite. Only the sample that was measured with the electric field perpendicular to the long axes of the whiskers even approaches the expected bulk dielectric constant of the pure mullite (~6.6). It is also interesting to point out that the tan delta response emphasizes the dielectric relaxation, which results in a larger tan delta response for the samples whose whiskers were aligned perpendicular to the field.

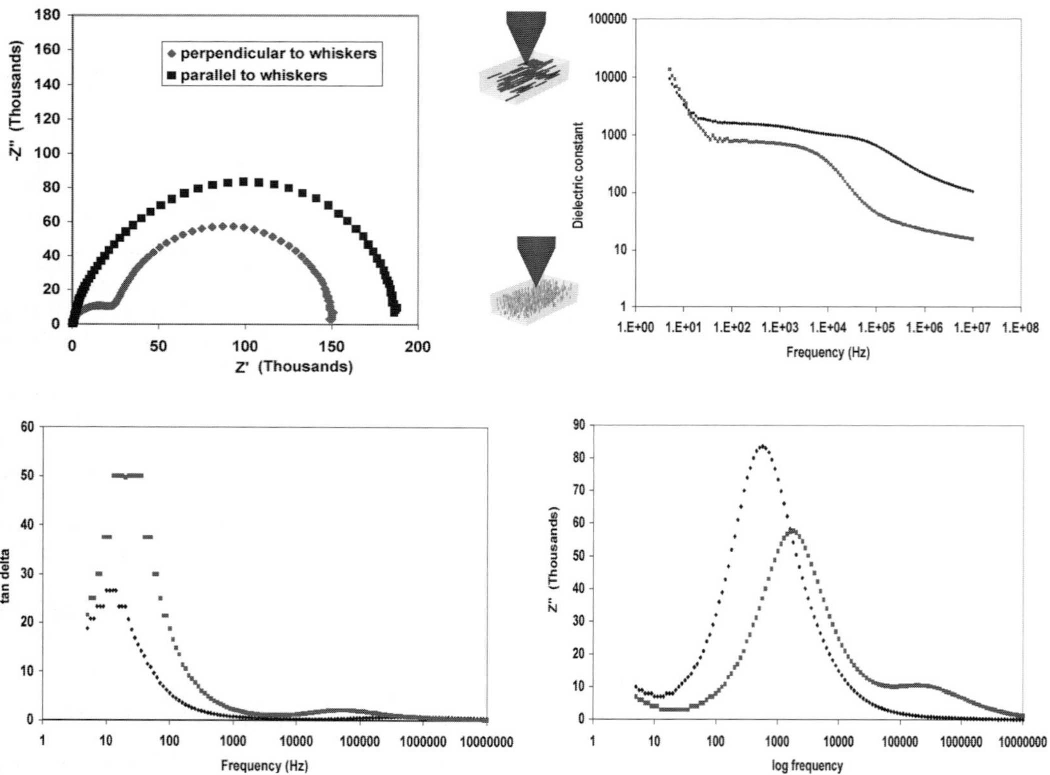

Figure 2. Complex impedance plot (-Z" vs. Z'), Dielectric constant, tan delta and imaginary impedance curves plotted as a function of frequency.

<u>Acknowledgements</u>

This work was funded by the US National Science Foundation under DMR-0076153. The author acknowledges the contributions of various students and collaborators: J. Runyan, R. Ou, D. Maybury, D. Mebane, L. Ferranti, R. Ruh, J. Rhodes, R. McCauley, K.V. Logan and N.N. Thadhani.

References

[1] J.E. Bauerle, "Study of Solid Electrolyte Polarization by a Complex Admittance Methods," *J.Phys.Chem.Solids 30,* 2657-69(1969).

[2] R.A. Gerhardt and A.S. Nowick, "The Grain Boundary Conductivity Effect in Ceria doped with Various Trivalent Cations. Part I - Electrical Behavior." J. *Amer.Ceram. Soc. 69 [9],* 641-646 (1986).

[3] R. Gerhardt, A.S. Nowick, M.E. Mochel and I. Dumler: "The Grain Boundary Conductivity Effect in Ceria Doped with Trivalent Cations. Part II - Microstructure and Microanalysis." *J. Amer. Ceram. Soc. 69 [9],* 647-651 (1986).

[4] N. Bonanos, B.C.H. Steele, E.P. Butler, W.B. Johnson, W. Worrell, D.D. Macdonald and M.C.H. McKubre, "Applications of Impedance Spectroscopy," in Impedance Spectroscopy:Emphasizing Solid Materials and Systems, ed. J.R. Macdonald, John Wiley, 1987.

[5] J-H Wang, D.S. McLachlan and T.O. Mason, "Brick-Layer Model Analysis of Nanoscale-to-Microscale Cerium Dioxide," *J.Electroceram. 3,* 7-16(1999).

[6] R. Anderson, R. Gerhardt, J.B. Wachtman, JR., S. Becher and D.G. Onn, "Thermal, Mechanical and Dielectric Properties of Mullite-Cordierite Composites," *Advances in Ceramics, vol. 26:* 265-278 (1989).

[7] R. Gerhardt, W.K. Lee and A.S. Nowick: "Dielectric and Anelastic Relaxation in SC-Doped Ceria," *J. Phys. Chem. Solids 48 [6],* 563-570 (1987).

[8] J.R. Kokan, R.A. Gerhardt , R. Ruh and D.S. McLachlan, "Dielectric Spectroscopy of Insulator-Conductor Mixtures," M*atRes.Symp.Proc. vol. 500,* 341-346 (1998).

[9] J.Runyan, R.A. Gerhardt and R. Ruh, "Electrical Properties of BN Matrix Composites – Part I: Analysis of the McLachlan Equation in Modeling the Conductivity of $BN-B_4C$ and BN-SiC Composites", *J.Am.Ceram.Soc. 84[7],* 1490-1496(2001).

[10] J.Runyan, R.A. Gerhardt and R. Ruh, "Electrical Properties of BN Matrix Composites – Part II: Dielectric Relaxations in Boron Nitride- Silicon Carbide Composites," *J.Am.Ceram.Soc.. 84[7],* 1497-1503(2001).

[11] R.A. Gerhardt, J. Runyan, C. Sana, D.S. McLachlan and R. Ruh, "Electrical Properties of BN Matrix Composites – Part III: Observations Near Percolation Threshold in $BN-B_4C$ Composites", *J.Am.Ceram Soc. 84[10],*2335-2342(2001).

[12] R.A. Gerhardt and R. Ruh, "Volume Fraction and Whisker Orientation Dependence of the Electrical Properties of SiC Whisker Reinforced Mullite Composites," *J.Am.Ceram Soc. 84[10],*2328-2334(2001).

[13] R. Ou, R.A. Gerhardt and R. Ruh, "Electrical Properties of SiC-AlN Composites," P*roc. 27th Annual Cocoa Beach Conf.,* in press(2003).

[14] R.A. Gerhardt, R. Ou, J. Colton, C. Marrett and A. Moulart, "Assessment of Percolation and Homogeneity in ABS/Carbon Black Composites," N*inth International Conference on Composites Engineering, Proc. ICCE/9,* 231 (2002).

[15] R. Ou, R. Samuels and R. A. Gerhardt, "In-plane Impedance Spectroscopy of Doped Polyaniline Films," *J.PlasticFilm and Sheeting 17[4],*184-196(2001).

[16] R. Ou, R. Samuels and R.A. Gerhardt, "The Structure and Electrical Properties of Polyaniline," *Mat.Res.Symp.Proc. 699,* 333-338(2002).

[17] R. Ou, R. Samuels and R.A. Gerhardt, "A Structure-Electrical Property Study of Anisotropic Polyaniline Films," *J.Poly.Sc.: Pol.Phys.,* in press (2003).

[18] R. Ou, R. Samuels and R.A. Gerhardt, "Structure and Electrical Properties of Undoped and Oriented Poly(Phenylene Vinylene) Films," submitted to *J.Poly. Sc. & Eng.*

[19] J.R. Kokan, R.A. Gerhardt and C-H Su, "Dielectric Spectroscopy of ZnSe Grown by Physical Transport," M*at.Res.Symp.Proc. vol. 487,* 517-522 (1998).

[20] K. Pinkos, C. Laboy and R.A. Gerhardt, "Effect of Grain Boundaries and Indentation Load on the Electrical Properties of Nickel Base Superalloys," *Mat.Res.Symp.Proc. 699,* 289-294(2002).

[21] X. Zou, T. Makram and R.A. Gerhardt, "Detection of Compositional Fluctuations in Waspaloy Exposed to High Temperatures," *Mat.Res.Symp.Proc. 699,* 301-306(2002).

[22] R. Gerhardt, "Dielectric and Impedance Spectroscopy Revisited: Distinguishing Localized Relaxation From Long Range Conductivity," *J. Phys. Chem. Solids. 55[12]* 1491-1506 (1994).

Effective Wave Number of Composite Materials with Randomly Distributed Inclusions

Anyong Qing, Xin Xu, and Yeow Beng Gan
Temasek Laboratories, National University of Singapore, Singapore

1 Introduction

We follow the notation system in [1] unless specified.
In [1], we have assumed that the inclusion particles are aligned. However, this assumption is not true. In practice, the inclusion particles are oriented randomly. We therefore have to take into account the random orientation of inclusion particles to model the effective wave number of composite materials more accurately.

2 Rotational Addition Theorems

The rotation is represented by the triplet Euler angles $(\alpha\beta\gamma)$ as shown in Fig.1, where $oxyz$ is the laboratory (global) coordinates and $o'x'y'z'$ is the particle (local) coordinates. The particle coordinates are obtained by successive rotation of the laboratory coordinates $oxyz$ about oz to Ox_1y_1z with angle α, rotating ox_1y_1z about oy_1 to ox_2y_1z' with angle β, and rotating ox_2y_1z' about oz' to $o'x'y'z'$ with angle γ.

The addition theorem for the spherical harmonic function $Y_{mn}(\theta,\phi)$ is [2, 3]

$$Y_{m'n}(\theta',\phi') = \sum_{m'=-n}^{n} D_{mm'}^{(n)}(\alpha\beta\gamma)Y_{mn}(\theta,\phi), \tag{1}$$

where

$$Y_{mn}(\theta,\phi) = \lambda_{mn}P_n^m(\cos\theta)e^{im\phi}, \tag{2}$$

$\lambda_{mn} = \sqrt{\frac{2n+1}{4\pi}\frac{(n-m)!}{(n+m)!}}$, $D_{mm'}^{(n)}(\alpha\beta\gamma)$ is the Wigner D-functions [2,4]

$$D_{mm'}^{(n)}(\alpha\beta\gamma) = e^{im\alpha}d_{mm'}^{(n)}(\beta)e^{im'\gamma}, \tag{3}$$

From (1), we have

$$Y_{mn}(\theta,\phi) = \sum_{m'=-n}^{n} D_{m'm}^{(n)}(-\gamma-\beta-\alpha)Y_{m'n}(\theta',\phi'), \tag{4}$$

From (3), we have

$$D_{m'm}^{(n)}(-\gamma-\beta-\alpha) = e^{-im'\gamma}d_{m'm}^{(n)}(-\beta)e^{-im\alpha}, \tag{5}$$

It is shown in [2] that

$$d_{m'm}^{(n)}(-\beta) = d_{mm'}^{(n)}(\beta), \tag{6}$$

Therefore,

$$D_{m'm}^{(n)}(-\gamma-\beta-\alpha) = e^{-im'\gamma}d_{m'm}^{(n)}(-\beta)e^{-im\alpha} = e^{-im'\gamma}d_{mm'}^{(n)}(\beta)e^{-im\alpha}$$
$$= \left[e^{im\alpha}d_{mm'}^{(n)}(\beta)e^{im'\gamma}\right]^* = \left[D_{mm'}^{(n)}(\alpha\beta\gamma)\right]^* \tag{7}$$

Accordingly,

$$Y_{mn}(\theta,\phi) = \sum_{m'=-n}^{n} \left[D_{mm'}^{(n)}(\alpha\beta\gamma)\right]^* Y_{m'n}(\theta',\phi'), \tag{8}$$

Now, define the modified scalar and vector spherical wave functions as

$$\overline{\psi}_{mn}(\mathbf{r}) = z_n(kr)Y_{mn}(\theta,\phi) = \lambda_{mn}\psi_{mn}(\mathbf{r}), \tag{9}$$
$$\overline{\mathbf{M}}_{mn}(\mathbf{r}) = \nabla\times[\mathbf{r}\overline{\psi}_{mn}(\mathbf{r})] = \lambda_{mn}\mathbf{M}_{mn}(\mathbf{r}), \tag{10}$$
$$\overline{\mathbf{N}}_{mn}(\mathbf{r}) = \tfrac{1}{k}\nabla\times\overline{\mathbf{M}}_{mn}(\mathbf{r}) = \lambda_{mn}\mathbf{N}_{mn}(\mathbf{r}), \tag{11}$$

Consequently, we have the rotational addition theorems for the modified scalar and vector spherical wave functions addition theorems

$$\overline{\psi}_{mn}(\mathbf{r}) = \sum_{m'=-n}^{n} \left[D_{mm'}^{(n)}(\alpha\beta\gamma) \right]^{*} \overline{\psi}_{m'n}(\mathbf{r}'), \tag{12}$$

$$\overline{\mathbf{M}}_{mn}(\mathbf{r}) = \sum_{m'=-n}^{n} \left[D_{mm'}^{(n)}(\alpha\beta\gamma) \right]^{*} \overline{\mathbf{M}}_{m'n}(\mathbf{r}'), \tag{13}$$

$$\overline{\mathbf{N}}_{mn}(\mathbf{r}) = \sum_{m'=-n}^{n} \left[D_{mm'}^{(n)}(\alpha\beta\gamma) \right]^{*} \overline{\mathbf{N}}_{m'n}(\mathbf{r}'), \tag{14}$$

3 Averaged T-matrix over Orientation

Now, define

$$\mathbf{D}(\alpha\beta\gamma) = \begin{bmatrix} \mathbf{D}^{1}(\alpha\beta\gamma) & & & \\ & \ddots & & \\ & & \mathbf{D}^{n}(\alpha\beta\gamma) & \\ & & & \ddots \end{bmatrix}, \tag{20}$$

where

$$\mathbf{D}^{n}(\alpha\beta\gamma) = \left[D_{mm'}^{(n)}(\alpha\beta\gamma) \right], \tag{21}$$

From (5) and (14), it is easy to observe that

$$\mathbf{D}^{-1}(\alpha\beta\gamma) = \mathbf{D}^{+}(\alpha\beta\gamma), \tag{22}$$

The T-matrix [5-7] is written as

$$\overline{\mathbf{T}} = \overline{\mathbf{B}} \cdot \overline{\mathbf{A}}^{-1}, \tag{23}$$

$$\overline{\mathbf{T}}' = \overline{\mathbf{B}}' \cdot \overline{\mathbf{A}}'^{-1}, \tag{24}$$

It is easy to show that

$$\overline{\mathbf{B}} = \begin{bmatrix} \mathbf{D}(\alpha\beta\gamma) & \\ & \mathbf{D}(\alpha\beta\gamma) \end{bmatrix}^{*} \cdot \overline{\mathbf{B}}' \cdot \begin{bmatrix} \mathbf{D}(\alpha\beta\gamma) & \\ & \mathbf{D}(\alpha\beta\gamma) \end{bmatrix}^{+}, \tag{25}$$

$$\overline{\mathbf{A}} = \begin{bmatrix} \mathbf{D}(\alpha\beta\gamma) & \\ & \mathbf{D}(\alpha\beta\gamma) \end{bmatrix}^{*} \cdot \overline{\mathbf{A}}' \cdot \begin{bmatrix} \mathbf{D}(\alpha\beta\gamma) & \\ & \mathbf{D}(\alpha\beta\gamma) \end{bmatrix}^{+}, \tag{26}$$

where the superscript + stands for conjugate transpose.

Therefore,

$$\begin{aligned}
\left[\overline{\mathbf{T}}_{mn,\mu v} \right] &= \begin{bmatrix} \mathbf{D}(\alpha\beta\gamma) & \\ & \mathbf{D}(\alpha\beta\gamma) \end{bmatrix}^{*} \cdot \left[\overline{\mathbf{T}}'_{m'n,\mu'v} \right] \cdot \begin{bmatrix} \mathbf{D}^{*}(\alpha\beta\gamma) & \\ & \mathbf{D}^{*}(\alpha\beta\gamma) \end{bmatrix}^{-1}, \\
&= \begin{bmatrix} \mathbf{D}(\alpha\beta\gamma) & \\ & \mathbf{D}(\alpha\beta\gamma) \end{bmatrix}^{*} \cdot \left[\overline{\mathbf{T}}'_{m'n,\mu'v} \right] \cdot \begin{bmatrix} \mathbf{D}(\alpha\beta\gamma) & \\ & \mathbf{D}(\alpha\beta\gamma) \end{bmatrix}^{T}
\end{aligned} \tag{27}$$

The averaged T-matrix over all possible orientations of the scatter is written as [4, 7]

$$\left\langle \left[\overline{\mathbf{T}}_{mn,\mu v} \right] \right\rangle = \frac{1}{8\pi^{2}} \int_{0}^{2\pi} d\alpha \int_{0}^{2\pi} d\gamma \int_{0}^{\pi} \left[\overline{\mathbf{T}}_{mn,\mu v} \right] \sin\beta d\beta , \tag{28}$$

From [2], we have

$$\frac{1}{8\pi^{2}} \int_{0}^{2\pi} d\alpha \int_{0}^{2\pi} d\gamma \int_{0}^{\pi} D_{mm'}^{*(n)}(\alpha\beta\gamma) D_{\mu\mu'}^{(v)}(\alpha\beta\gamma) \sin\beta d\beta = \delta_{m\mu}\delta_{m'\mu'}\delta_{nv} \frac{1}{2n+1} , \tag{29}$$

Therefore,

$$\left\langle \left[\overline{\mathbf{T}}_{mn,\mu v} \right] \right\rangle = \delta_{m\mu}\delta_{nv} \frac{1}{2n+1} \sum_{m'} \delta_{m'\mu'} \left[\overline{\mathbf{T}}'_{m'n,\mu'v} \right], \tag{30}$$

In the laboratory (global) coordinates, we have

$$\left[\begin{array}{c}\bar{\mathbf{f}}_{mn} \\ \bar{\mathbf{g}}_{mn}\end{array}\right] = \left[\bar{\mathbf{T}}_{mn,\mu\nu}\right] \cdot \left[\begin{array}{c}\bar{\mathbf{a}}_{\mu\nu} \\ \bar{\mathbf{b}}_{\mu\nu}\end{array}\right], \tag{31}$$

while in the particle (local) coordinates, we have

$$\left[\begin{array}{c}\bar{\mathbf{f}}'_{m'n} \\ \bar{\mathbf{g}}'_{m'n}\end{array}\right] = \left[\bar{\mathbf{T}}'_{m'n,\mu'\nu}\right] \cdot \left[\begin{array}{c}\bar{\mathbf{a}}'_{\mu'\nu} \\ \bar{\mathbf{b}}'_{\mu'\nu}\end{array}\right], \tag{32}$$

It is easy to show that

$$\left[\begin{array}{c}\mathbf{a}_{mn} \\ \mathbf{b}_{mn}\end{array}\right] = \left[\begin{array}{cc}\mathbf{\Lambda}_{mn} & \\ & \mathbf{\Lambda}_{mn}\end{array}\right] \cdot \left[\begin{array}{c}\bar{\mathbf{a}}_{mn} \\ \bar{\mathbf{b}}_{mn}\end{array}\right] = \mathbf{\Lambda} \cdot \left[\begin{array}{c}\bar{\mathbf{a}}_{mn} \\ \bar{\mathbf{b}}_{mn}\end{array}\right], \tag{33}$$

$$\left[\begin{array}{c}\mathbf{f}_{mn} \\ \mathbf{g}_{mn}\end{array}\right] = \left[\begin{array}{cc}\mathbf{\Lambda}_{mn} & \\ & \mathbf{\Lambda}_{mn}\end{array}\right] \cdot \left[\begin{array}{c}\bar{\mathbf{f}}_{mn} \\ \bar{\mathbf{g}}_{mn}\end{array}\right] = \mathbf{\Lambda} \cdot \left[\begin{array}{c}\bar{\mathbf{f}}_{mn} \\ \bar{\mathbf{g}}_{mn}\end{array}\right], \tag{34}$$

$$\left[\mathbf{T}_{mn,\mu\nu}\right] = \mathbf{\Lambda} \cdot \left[\bar{\mathbf{T}}_{mn,\mu\nu}\right] \cdot \mathbf{\Lambda}^{-1}, \tag{35}$$

$$\left[\mathbf{T}'_{mn,\mu\nu}\right] = \mathbf{\Lambda} \cdot \left[\bar{\mathbf{T}}'_{mn,\mu\nu}\right] \cdot \mathbf{\Lambda}^{-1}, \tag{36}$$

Accordingly,

$$\left\langle\left[\mathbf{T}_{mn,\mu\nu}\right]\right\rangle = \delta_{m\mu}\delta_{n\nu} \frac{1}{2n+1} \sum_{m'} \delta_{m'\mu'} \left[\mathbf{T}'_{m'n,\mu'\nu}\right], \tag{37}$$

4 T-Matrix Configurational-Average Quasi-crystalline Approximation (TCQ) model for Composite Materials with Random Orientation of Inclusion Particles

For composite materials with inclusion particles aligned, we have [1]

$$\left[\begin{array}{c}\mathbf{Y}_{mn} \\ \mathbf{Z}_{mn}\end{array}\right] = \left[\begin{array}{cc}\left(\mathbf{T}_{11}^{i}\right)_{mn,m'n'} & \left(\mathbf{T}_{12}^{i}\right)_{mn,m'n'} \\ \left(\mathbf{T}_{21}^{i}\right)_{mn,m'n'} & \left(\mathbf{T}_{22}^{i}\right)_{mn,m'n'}\end{array}\right] \cdot \left[\begin{array}{cc}\left(\mathbf{I}'_{A}\right)_{m'n'}^{\mu\nu} & \left(\mathbf{I}'_{B}\right)_{m'n'}^{\mu\nu} \\ \left(\mathbf{I}'_{B}\right)_{m'n'}^{\mu\nu} & \left(\mathbf{I}'_{A}\right)_{m'n'}^{\mu\nu}\end{array}\right] \left[\begin{array}{c}\mathbf{Y}_{\mu\nu} \\ \mathbf{Z}_{\mu\nu}\end{array}\right], \tag{38}$$

Averaging over orientation leads to

$$\left[\begin{array}{c}\mathbf{Y}_{mn} \\ \mathbf{Z}_{mn}\end{array}\right] = \left\langle\left[\mathbf{T}_{mn,\mu\nu}\right]\right\rangle \cdot \left[\begin{array}{cc}\left(\mathbf{I}'_{A}\right)_{m'n'}^{\mu\nu} & \left(\mathbf{I}'_{B}\right)_{m'n'}^{\mu\nu} \\ \left(\mathbf{I}'_{B}\right)_{m'n'}^{\mu\nu} & \left(\mathbf{I}'_{A}\right)_{m'n'}^{\mu\nu}\end{array}\right] \left[\begin{array}{c}\mathbf{Y}_{\mu\nu} \\ \mathbf{Z}_{\mu\nu}\end{array}\right], \tag{39}$$

Similarly, we make use of the property that the determinant of the truncated coefficient matrix in (39) should vanish for a nontrivial solution. Similarly, the differential evolution strategy is applied to search for the effective wave number k_e.

5 Preliminary Numerical Results

We have applied the above approach to determine the effective wave number of a composite material with spheroidal inclusion. Fig. 3 shows the relationship between the attenuation coefficient defined in [8] and frequency. More results will be shown later.

6 Conclusions

In this paper, the TCQ approach is applied to predict the effective wave number of composite materials with randomly distributed inclusions. The rotational addition theorems for scalar and vector spherical wave functions, and T-matrix are derived. Averaging over orientation on the governing equation for aligned TCQ model leads to the governing equation for the current problem that requires the averaged T-matrix of a single scatterer over orientations.

References

[1] A. Qing, X. Xu, and Y. G. Gan, Effective Wave Number of Composite Materials with Oriented Randomly Distributed Inclusions, *2003 IEEE AP-S Int. Symp.*, June 2003, Ohio

[2] A. R. Edmonds, *Angular Momentum in Quantum Mechanics, Princeton*, New Jersey: Princeton University Press, 1960

[3] S. Stein, Addition theorems for spherical wave functions, *Q. Appli. Math.*, vol. 19, no. 1, pp. 15-24, 1961

[4] M. I. Mishchenko, J. W. Hovenier, and L. D. Travis, *Light Scattering by Nonspherical Particles: Theory, Measurements, and Applications,* San Diego: Academic, 2000

[5] P. C. Waterman, Matrix formulation of electromagnetic scattering, *Proc. IEEE*, vol. 53. no. 8, pp. 805-812, 1965

[6] P. C. Waterman, Symmetry, unitarity, and geometry in electromagnetic scattering, *Phys. Rev. D*, vol. 3, no. 4, pp. 825-839, 1971

[7] A. Qing, X. Xu, and Y. B. Gan, *Modeling of Composite Materials with Simple Inclusions*, Technical Report TL-EM-02-005, Temasek Laboratories, National University of Singapore, August 1, 2002

[8] V. V. Varadan, and V. K. Varadan, Multiple scattering of electromagnetic waves by randomly distributed and oriented dielectric scatterers, *Phys. Rev. D*, vol. 21, no. 2, pp. 388-394, 1980

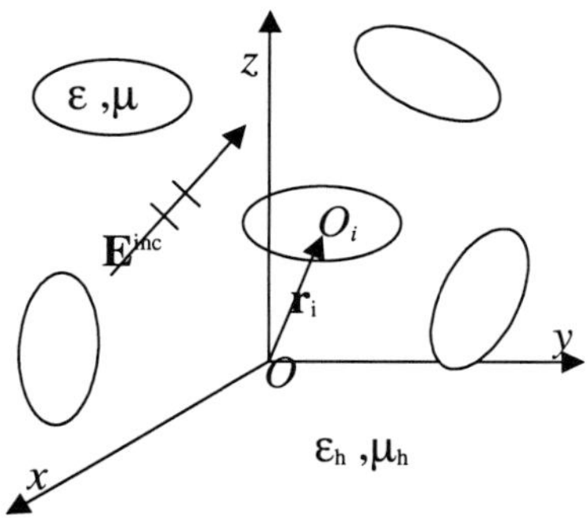

Figure 1: N randomly distributed scatterers

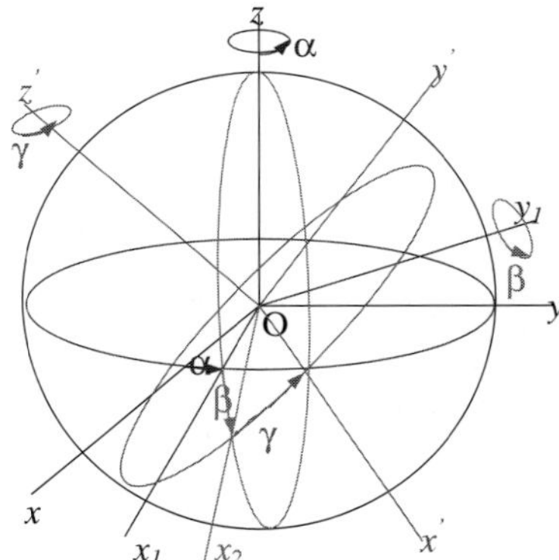

Figure 2: Coordinate Rotation

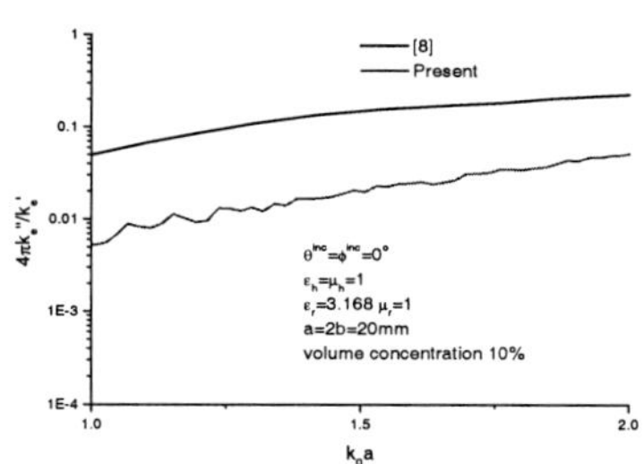

Figure 3: Attenuation Coefficient of Composite Materials with Lossless Spheroidal Inclusions

Session F3

Nano Composites

Chair: K. Lewis

Optical Properties of Engineered Metal-Dielectric Nano-Composite Materials

Keith Lewis and Richard Bennett
QinetiQ, Malvern Technology Centre
Great Malvern, Worcestershire, WR14 3PS, UK

1. General review

In recent years, it has become possible to realise nano-composite materials that have optical and dielectric properties significantly different from those of their constituent components. Optical effects can be produced across wavelength regimes ranging from the visible to far-infrared, that are also scalable to other frequency regimes, notably in the microwave region. In some cases, these materials have been used to explore quantum confinement effects as in the work of Charvet et al [1]. In that study, the spectral positioning of photoluminescence detected in annealed samples of Si/SiO_2 was found to be closely connected with the amount of excess silicon in the composite, and in consequence, with the mean size of the thermally grown nanocrystallites. Other workers have been interested in exploiting the properties of nano-composites containing metals as phase separated components in dielectric hosts such as silver in silica [2]. Here the hosts can encompass a number of materials, ranging from polymers to inorganic dielectrics such as elemental semiconductors, oxides, nitrides, carbides etc. Provided that the metals are incorporated as dispersed phases, where there is no significant electrical connectivity between the included particles, the materials are insulating and the dielectric properties of the ensemble can be described on the basis of effective medium models.

Such materials can be produced as thin films by a variety of different techniques, ranging from sol-gel deposition, solution casting, melt extrusion, and various reactive co-deposition techniques, through to ion implantation. The particle size of the included phase can be controlled by exploiting processes which control the degree to which Ostwald ripening occurs whilst the nano-composite is being formed. For films produced by co-deposition, this is usually effected by choice of growth temperature as shown for the case of $Au:VO_2$ nanocomposites in figure 1. In contrast, the dimensions of the metallic phase inclusions produced by ion implantation are controlled by subsequent annealing processes. This has the benefit of allowing the production of discrete layers of the dispersed phase, buried at a depth determined by the energy of ion implantation.

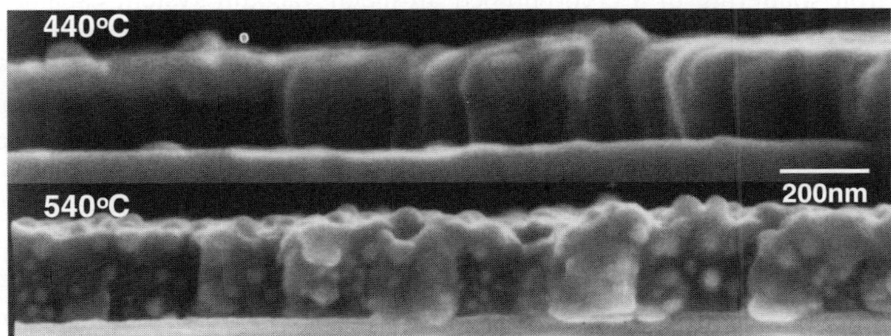

Figure 1 Cross-sectional scanning electron micrograph comparing morphology of $Au:VO_2$ nanocomposite films deposited at 440°C and 540°C. The resolution of the SEM is insufficient to resolve the phase separated gold in the 440°C case

For polymer hosts, the processes by which metallic precipitates are mobile can be important in practical applications. Cole et al [3] studied the effect of annealing of gold islands in a gold:poly t-butyl acrylate (PTBA) host which has a glass transition at 45°C. The gold diffuses through the polymer by Brownian motion, although particle mobility was decreased by 2-3 orders of magnitude in comparison with predictions of Stokes-Einstein theory. Whilst the degree of interaction between the polymer and the gold is relatively weak, it is still sufficient to affect the mobility of polymer chains that are close to the gold particles.

Local field effects can give rise to significant enhancements in the non-linear optical properties of the nanocomposite. Under the proper conditions, it is possible to achieve results in which the non-linear susceptibility of the composite exceeds those of the separate components. For example, a 3-fold enhancement was obtained for a layered composite of a dye-doped polymer (AF 30) and barium titanate [4]. However in a case where a nanocomposite was produced containing gold and the laser dye HITCI, which are both reverse saturable absorbers, there was a destructive cancellation of the two contributions to the third order susceptibility $\chi^{(3)}$ in the spectral region associated with the gold plasmon resonance [5, 6]. This occurred despite the fact that the imaginary terms of $\chi^{(3)}$ for the separate components are both positive.

In the linear regime, the nature of the host and specifically the interface between the metal and the matrix can also control the effects produced. This has been explored in this work using examples based on the inclusion of gold in two significantly different hosts. The first, vanadium dioxide, exhibits a semiconductor to metallic phase transition (SMPT) at 68°C due to a solid state transformation between monoclinic and tetragonal phases. The temperature at which the phase transition occurs is a sensitive function of vanadium-oxygen ratio, and also varies when electrically active impurities are present. Since gold is immiscible with VO_2, the SMPT also occurs in $Au:VO_2$ nanocomposites at 68°C, provided the VO_2 matrix is stoichiometric. Thus by using a single materials system, it is possible to explore the consequences of incorporating gold into both semiconductor and metallic phase hosts, merely by changing the temperature of measurement. The second system explored, Au:Si, is important because of its characteristics as a degenerate eutectic materials system.

2. Co-deposition and assessment of gold-dielectric composites

The material used for deriving the experimental data discussed in the present work was produced by reactive magnetron sputtering in a cryo-pumped all-metal vacuum system capable of reaching base pressures of about 10^8 mbar. The growth chamber was fitted with separate magnetron sources for the gold and for the host material. Process conditions were established for the reactive sputtering of pure vanadium dioxide by determining the characteristics of the hysteretic reaction space, using in-situ mass spectrometry and X-ray photoelectron spectroscopy (XPS) to locate the ideal conditions for deposition of films with the correct binding energy characteristics of VO_2. The XPS technique also allowed the gold composition to be established. Deposition temperatures were varied between 25 and 450°C to explore the role of size effects for the included phase.

Films were assessed by visible and infra-red spectroscopy and by cross-sectional transmission electron microscopy. Figure 2 shows an example of a cross-section of a gold-vanadium dioxide nanocomposite film, highlighting the degree of phase separation achieved. Such micrographs tend to over-estimate the amount of gold present, since the thickness of the foil is greater than the dimensions of the individual gold inclusions and some of the VO_2 is being obscured. Optical constants (n, k) were derived from simultaneous fits to the measured transmission and reflection spectra, using film thickness data measured using stylus profilometry.

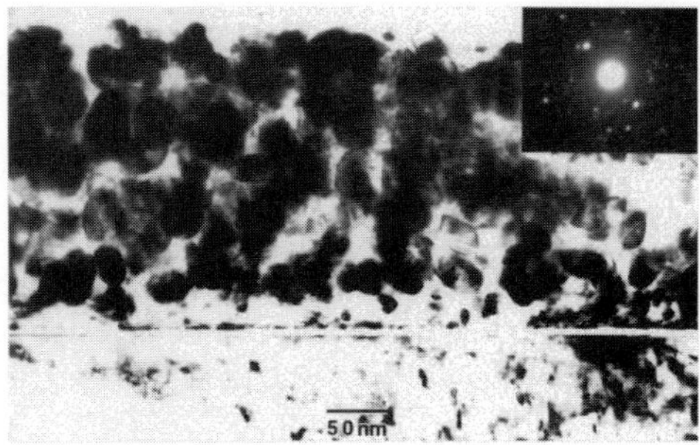

*Figure 2 Cross-sectional transmission electron micrograph of a
gold/vanadium dioxide nano-composite material deposited on a
semiconductor substrate. The film/substrate interface is clearly visible
whilst the dark regions are the gold component of the composite.*

3. Effective medium theories

Various effective medium models have been described in the literature, including Maxwell-Garnett [7] (MG), Bruggeman [8] and Looyenga [9] descriptions. The former approach was originally developed for dilute collections of small dipoles in a dielectric medium. At such concentrations a single resonance is obtained characteristic of that of isolated dipoles. As the concentration of the dispersed phase is increased, the resonance shifts to longer wavelengths due to electrical interactions between the inclusions. The Bruggeman model is more effective than Maxwell-Garnett theory in taking account of percolation effects as the concentration of particles is increased, leading to a network of connected inclusions. It defines a critical concentration for the onset of carrier percolation. In the Looyenga model, percolation effects occur as soon as the metallic inclusions are introduced into the dielectric.

It is instructive to compare the above effective medium models with the effectiveness of a layered model, in which the metallic component is considered as an ultra-thin one-dimensional film separated by layers of dielectric material a few atomic planes in thickness. Such films can be realised by layered deposition. In practice, the metallic component will usually be incorporated in discrete islands, since the thicknesses of the individual layers is much less than that required for the formation of continuous films.

4. Comparison of Effective Medium Theory and Experiment

Values of refractive indices have been obtained by fitting the measured optical characteristics of a number of Au:VO$_2$ nano-composites as a function of wavelength in the infra-red and the relevant dielectric constants calculated. Some typical examples for semiconductor phase material are highlighted in table (i) to illustrate the variation as a function of gold volume fraction.

Gold volume fraction	ε(exp)
1	-490.6
0.22	15.5
0.11	12.5
0.06	8.3
0	7.9

For semiconductor phase material, the most relevant model is the MG description, which relates the dielectric constant of the composite ε_{eff} to that of the host ε_h and inclusion phase ε_p according to:-

$$(\varepsilon_{eff} - \varepsilon_m) = f^*(\varepsilon_p - \varepsilon_h)(\varepsilon_{eff} + 2\varepsilon_h)/(\varepsilon_p + 2\varepsilon_h)$$

where

$\varepsilon = n^2 - k^2$ and f = volume fraction of metallic material

Figure 3 compares the predictions of this model with the 4μm data listed in table (i). Within the limits of experimental error associated with the extraction of refractive indices, the agreement is excellent demonstrating the degree to which the model is applicable to the case of semiconductor phase Au:VO₂.

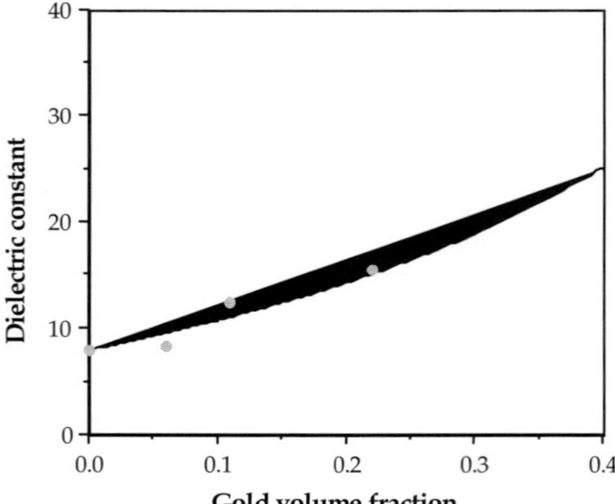

*Figure 3 Comparison of dielectric constants for Au:VO₂ semiconductor phase compositions
derived from Maxwell-Garnett theory (solid line) with experimental values (circles)*

For metallic phase material at temperatures above the phase transition, the most relevant approach is the Looyenga model, which allows for percolation effects at all compositions. Here the dielectric constant is given by:-

$$\varepsilon_{eff} = [f^*(\varepsilon_p^{1/3} - \varepsilon_h^{1/3}) + \varepsilon_h^{1/3}]^3$$

Once again, agreement with experiment is generally good as shown in figure 4, although it is difficult to extract accurate refractive index data from spectral measurements on metallic phase films, and only limited experimental data is available for comparison.

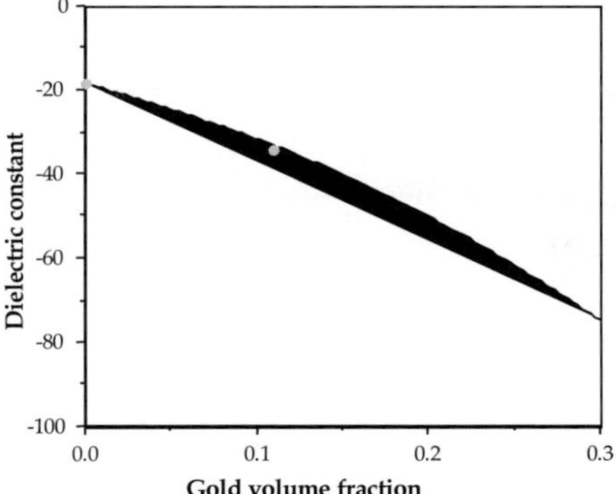

Figure 4 Comparison of dielectric constants for Au:VO₂ metallic phase compositions derived from the Looyenga model (solid line) with experimental results (circles)

For metallic phase material at temperatures above the phase transition, the measured reflectances of the Au:VO$_2$ nanocomposites are very similar to those of undoped tetragonal phase vanadium dioxide. This is a surprising result and indicates that the host oxide phase is unable to support the level of carrier percolation required to allow the nanocomposite to achieve the characteristics of a dilute noble metal, despite the level of electron delocalisation in the VO$_2$ and the short diffusion distances involved. The implication is that significant energy barriers to percolation exist at the oxide/gold interface.

5. Comparison of Layered Model and Experiment

An alternative approach to producing a dispersed-phase nanocomposites is by layering. For the case of Au:VO$_2$, separate discrete layers of VO$_2$ and Au can be deposited, relying on the degree of coalescence of the Au when it is deposited in very thin layers to achieve dispersion. Such behaviour is well-known for metals such as gold and silver and is the reason why it is impossible to realise very thin films of such materials with optical constants equivalent to those of bulk material. Indeed, Mandal et al [2] used such layering techniques for the preparation of their Ag:SiO$_2$ nanocomposite films and highlighted the use of temperature to control the degree of agglomeration of the dispersed phase.

Figure 5 compares measured transmission data for a nanocomposite film containing a volume fraction 0.11 of gold with values calculated for a film ensemble containing 200 repeat layer pairs of VO$_2$ and Au, with each VO$_2$ component being 10Å in thickness. When the calculations for the model are run using characteristics for the gold layers based on pure gold, it is clear that the layered model is not particularly successful in describing the optical characteristics of the dispersed nanocomposite. Here the layered model over-emphasises the role of the gold, since it is using optical constants for the gold that are typical of continuous films, rather than discrete metallic islands.

By modifying the optical constants of the gold layer so that it more closely approximates that of a dispersed layer containing discrete islands of gold, an improvement in the theoretical description can be obtained, but at the expense of introducing yet further degrees of freedom in the description of the nanocomposite.

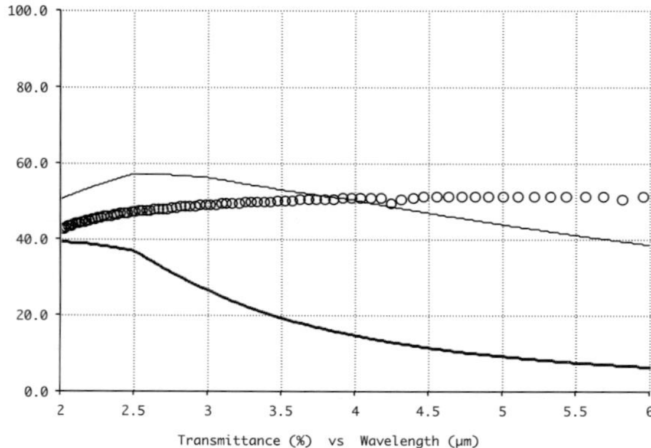

Figure 5 Comparison of calculated transmission spectra using the layered model with experimental data (circles) for a nanocomposite film containing a 0.11 volume fraction of gold. Au compositions for the theoretical plots are from top to bottom 0.02 and 0.04.

6. Properties of Au:Silicon Nanocomposites

The Au:Si system is of interest not only because of its technological importance, but also because it is a prototypical degenerate eutectic system, where there is negligible solubility of gold in silicon and of silicon in gold at ambient temperatures. Yet the system forms a eutectic alloy at temperatures in excess of 363°C. For this work, films were deposited by co-sputtering onto silicon substrates at a substrate temperature of 25°C. Compositions were determined by XPS and the optical properties of the films measured by Fourier transform infra-red spectroscopy. The XPS results also indicated that there was very little interaction between the gold and silicon on the basis of the positioning of the Au $4f_{7/2}$ peak at 84eV and Si 2p peak at 99eV, which are precisely the binding energies characteristic of the free elements. There was also no splitting of the Si 2p signal, a further indication that no change had occurred in the environment of the Si valence electrons.

The electrical properties of the Au:Si films were explored using simple 2-point probe measurements of film resistance. Some of the results of the experiments are summarised in figure 6 highlighting the complementary relation ship between the resistance of the films and the measured optical reflectance at 5μm.

These measurements indicate that the transition to a state where the nanocomposite is essentially metallic occurs at compositions of gold as low as 50%. This is in direct contradiction with the predictions of MG theory which indicate the transition to the metallic state should not be reached until a gold volume fraction in excess of 0.9 as shown in figure 7. The reasons for the difference are as yet unclear, although it is believed that it is related to the nanoscopic structure of the composite. Since the films were deposited at 25°C, the opportunity for enhanced adatom mobility during growth is very small and the individual dimensions of the dispersed particulates are likely to be very small. This would also imply that dipole-dipole interactions are very much enhanced and that the mean free path necessary for carrier hopping is very small.

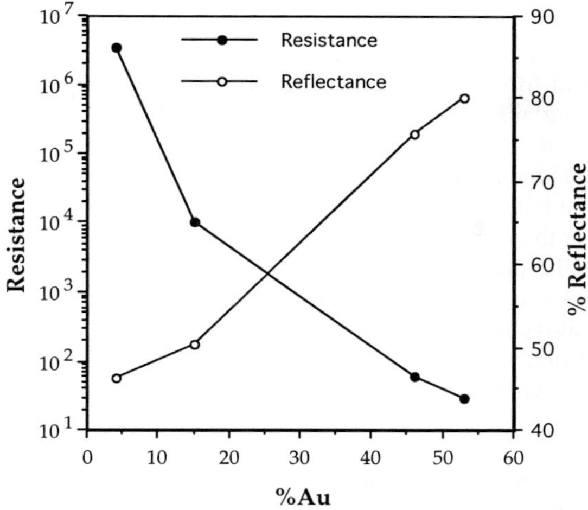

Figure 6 relationship between the optical reflectivity of Au:Si films deposited at 25°C and their electrical resistance measured using a simple 2-point probe technique.

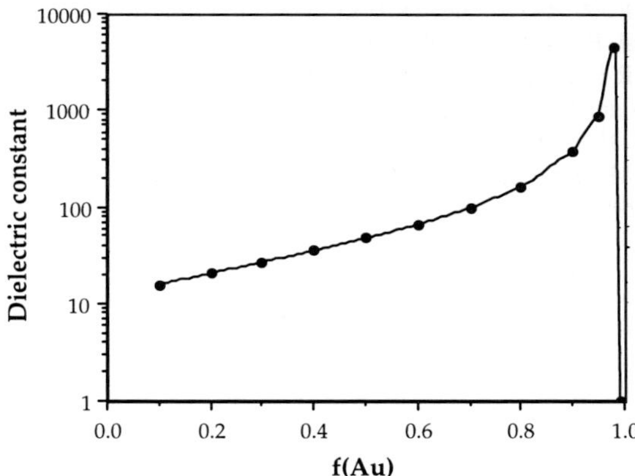

Figure 7 Variation of dielectric constant of a Au:Si nanocomposite in the infra-red at 10μm as a function of gold volume fraction, from MG theory.

Mandal et al [2] were able to describe the optical properties of their nanocomposite $Ag:SiO_2$ films deposited at 253K using MG theory, yet had to adopt a Bruggeman approach for films deposited at 300K. At lower frequencies, Merrill et al [10] have examined the application of effective medium theories in describing composite mixtures with spherical inclusions and found that most common theories collapse into Bruggeman's asymmetric formula when they are implemented using an iterative scheme to extend their validity to higher volume fractions. In their work, the MG model was found to under-estimate the rise in dielectric constant as the volume fraction of metallic phase inclusion (carbonyl iron) is increased in the rubber host matrix.

7. Conclusions

The optical characteristics of films of nanocomposites containing gold particulates in hosts of VO_2 and Si have been studied using material deposited by co-sputtering. The $Au:VO_2$ system provides opportunity to study different effective medium regimes at temperatures where the VO_2 can be either in its low temperature semiconductor phase, or its high temperature metallic phase. Here MG theory was found to be effective at temperatures below the phase transition, whilst the Looyenga mixing rule appeared to be effective for the metallic phase. Evidence was found for a significant energy barrier to carrier percolation in metallic phase material, which resulted in reduced levels of optical reflectance. Nanocomposites of gold in silicon could be deposited at room temperature despite the ability of the system to form eutectic alloys at temperatures in excess of $360°C$. The optical properties of such nanocomposites exhibited much higher optical reflectances than predicted by MG theory, indicating enhanced dipole-dipole interaction or increased opportunity for carrier hopping.

References
1. S. Charvet, R. Madelon, F. Gourbilleau, and R. Rizk: J Appl Phys <u>85</u> 4032 (1999)
2. S K Mandal, R K Roy and A K Pal: J Phys D: Appl Phys <u>36</u> 261 (2003)
3. D H Cole, K R Shull, L E Rehn and P Baldo: Phys Rev Letters <u>78</u> 5006 (1997)
4. R. W. Boyd, published as a book chapter in Laser Sources and Applications, edited by A. Miller and D.M. Finlayson, co-published by the Scottish Universities Summer School in Physics and the Institute of Physics Publishing, Bristol, 1996
5. D. B. Smith, G. Fischer, R. W. Boyd, and D.A. Gregory, J. Opt. Soc. Am. B, 14, 1625-1631, 1997
6. J. E. Sipe and R. W. Boyd, Phys. Rev. A 46, 1614, 1992
7. C M Garnett: Phil Trans R Soc London <u>203</u> 385 (1904)
8. D A G Bruggeman: Ann der Physik <u>24</u> 636 (1935)
9. H Looyenga: Physica <u>31</u> 401 (1965)
10. W M Merrill, R E Diaz, M M LoRe, M C Squires and N G Alexopoulos: IEEE Trans Antennas and Propagation <u>47</u> 142 (1999)

Nano-scale technologies as tools for transformation of traditional materials in functional and smart materials

I. T. Iakubov, A. N. Lagarkov, I. A. Ryzhikov, M. V. Sedova
Institute for Theoretical and Applied Electrodynamics RAS, Izhorskaya 13/19, 125412
Moscow, Russia; Phone: 7(095) 4859344; Fax: 7(095)4842633;
e-mail: iryzhikov2001@mail.ru

Recent progress in technology permits us to invade into nano-scales world. It means that now it is possible to modify the electromagnetic properties of materials controlling the microstructure of media. Indeed, changing the structure at nano-scale level we can not only strengthen or weaken some properties but also give new ones.

Below we provide examples where the possibility is realized.

1.High-frequency magnetic materials [1, 2, 3, 4]

Obtaining materials, which have high value of permeability (from several units up to several dozen) at GHz, is a task that challenges technologists since the forties of the last century. It is well known that the MW permeable properties are determined at high frequencies by ferromagnetic resonance.

Variations of the nitrogen concentration inside vacuum chamber, the applied magnetic field, the film thickness as well as of parameters of magnetron sputtering lead to modifications of Fe-based films at nano-scale. The AFM and MFM images show the variations in the granule structure and structure of magnetic domains and ripples. We manage to relate these modifications to MW properties of the films and obtain samples with a desired value of ferromagnetic resonance frequency lying in the range 1-10 GHz (Fig.1).

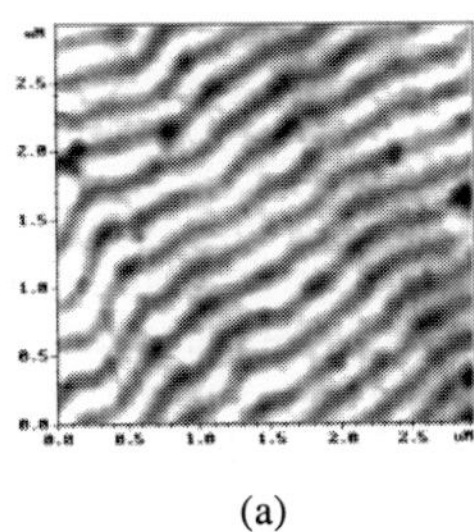

(a)

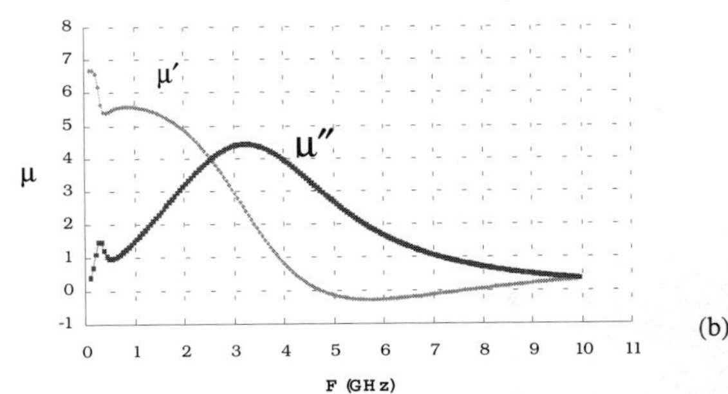

(b)

Fig.1. MFM image (a) and spectral characteristic (b) of thin iron film deposited on lavsan by ion-beam deposition. The period of the stripe domain structure is less than 250nm

Moreover, we can control magnetoresistance of the films e.g. made of permalloy and finemet. Changing the annealing temperature from 20°C to 600°C we can transfer a film from amorphous to nano-crystalline state.

This transfer is accompanied by the gradual increase of granular size. The process is controlled by TEM and by X-ray diffraction (Fig. 2).

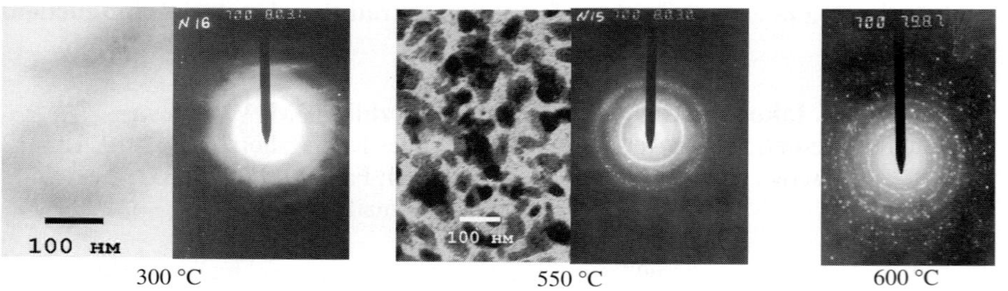

300 °C   550 °C   600 °C

Fig.2. TEM image and microdiffraction pattern of samples of finemet films by different annealing temperature

We find that the real part of impedance increases from its minimum value to maximum and decreases to minimum values as the annealing temperature increases from 100°C to 450°C (Fig.3). It is worth emphasizing that for primary nanocrystalline films such processes proceed at temperatures higher by 100 -150°C.

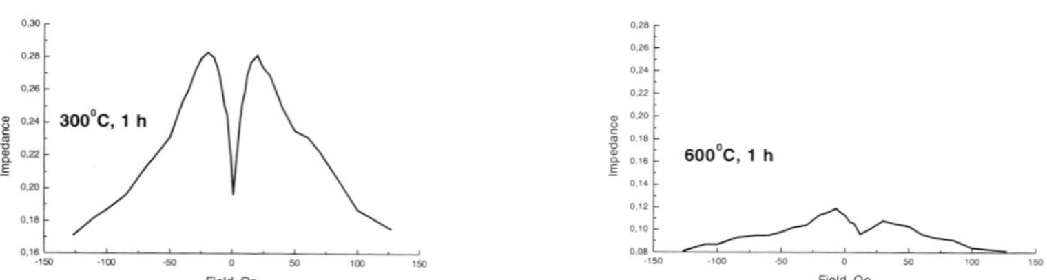

Fig.3. Modification of real part impedance R vs annealing temperature T by f=500 MHz for multilayer samples: finemet/SiO_2/Ti/Cu/Ti/SiO_2/finemet.

2. Photoconductor CdS-CdSe [5]

The main characteristic of photoconductor is the ratio of bright to dark currents. The higher is the ratio the better is the photoconductor. However, we cannot increase the ratio to infinity because a problem of stability that appears at high bright current. The losses can lead to a breakdown of the sample that is quite possible with a positive temperature dependence of photoconductivity. Selecting the condition of annealing we obtain samples with the bright/dark current ratio about 10^4 and with a negative temperature dependence of photoconductivity. The SEM shows that such samples represent systems of micrometer granules with pronounced nanoporous structure (Fig. 4). Consideration of such a pore as a nano-volume separating two quantum walls permits to explain the observed macroscopic properties.

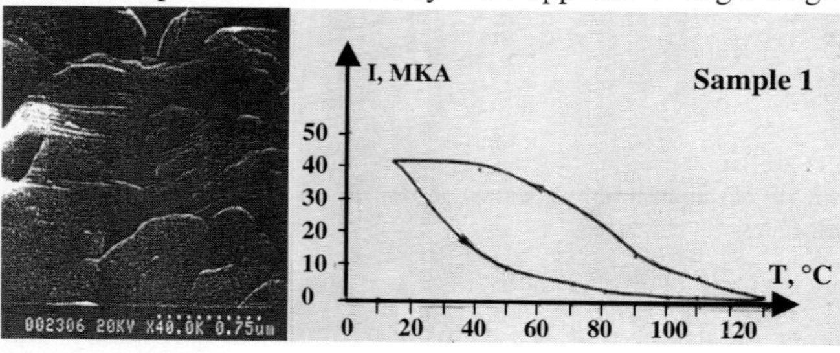

Fig.4. SEM image of the CdS-CdSe film structure and the dependence of the photocurrent *I* vs temperature *T* for CdS-CdSe sample with opposite direction of hysteresis

<u>3. Energy-saving ITO films [6]</u>

ITO films have wide applications, e.g. in manufacturing of energy - saving coatings, special displays etc. All these applications demand ITO films with different spectral properties. We find that the properties can be changed and moreover the desired properties can be obtained by means of doping the composition by some additives or annealing the films at different conditions. We have observed a new phenomenon that is a sharp variation of optical thickness of the film, which is caused by a collapse of nanoporosity that happens due to smoothing of granules shape. In the other words, this jump of optical thickness is a result of the change in the internal surface of the granules.

We established that the spectral characteristic (Fig. 5), conductivity and density of the ITO films unequivocal depend on nanogranular structure formed by annealing (Fig. 6).

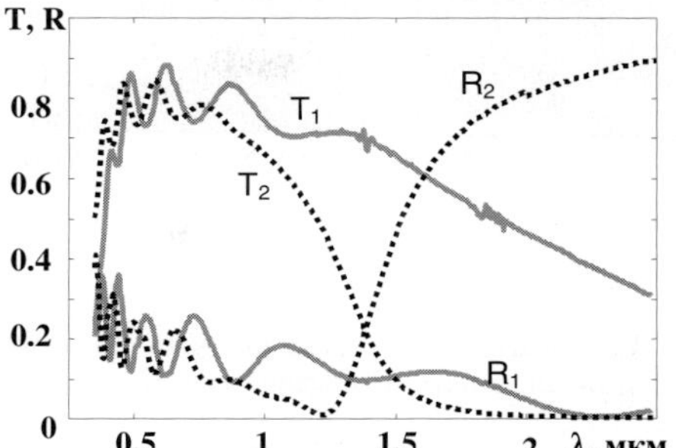

Fig.5. Transmission T and reflectance R of ITO films versus wavelength λ prepared at the glass substrate temperature t_s=100 °C: T_1, R_1 – before annealing (σ=1·10^5 Ω^{-1}·m^{-1}); T_2, R_2 – after annealing at 300 °C (σ=5·10^5 Ω^{-1}·m^{-1})

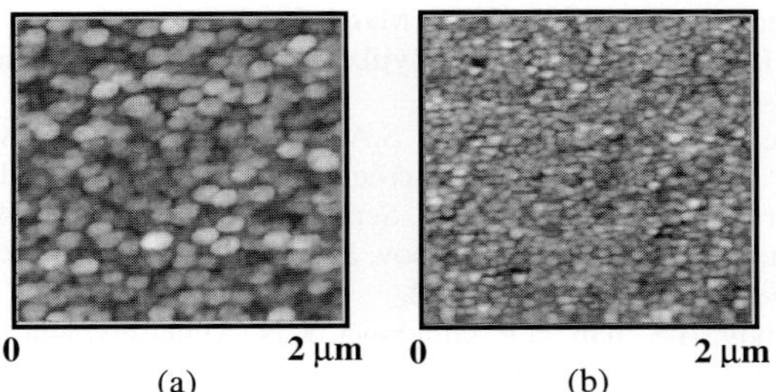

Fig.6. The AFM images of ITO films prepared at substrate temperature 100 °C (a) before annealing and (b) after annealing by 300 °C

<u>4. Frequency-selective Au-films [7]</u>

We have studied the optical properties of conducting coatings deposited on polymer glasses. As we cannot use an annealing as a tool we resort to ion beam polymer etching through nano-scale Au-mask. Varying the time of etching we can modify a surface structure that is controlled by AFM (Fig.7). The subsequent secondary deposition of functional Au-film on surfaces of such structures results in wanted electrophysical properties of the films (Fig. 8). In particular, we can obtain conducting but optically transparent films.

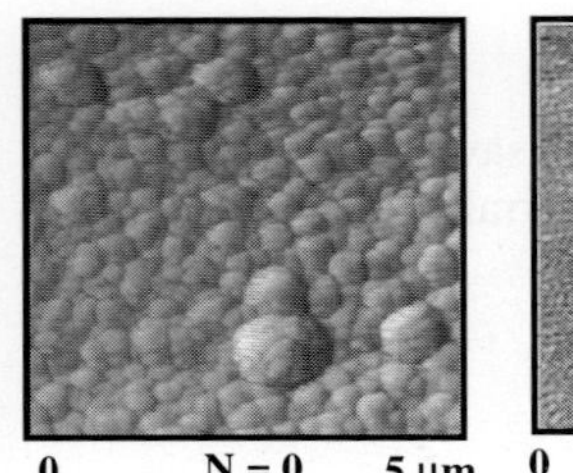
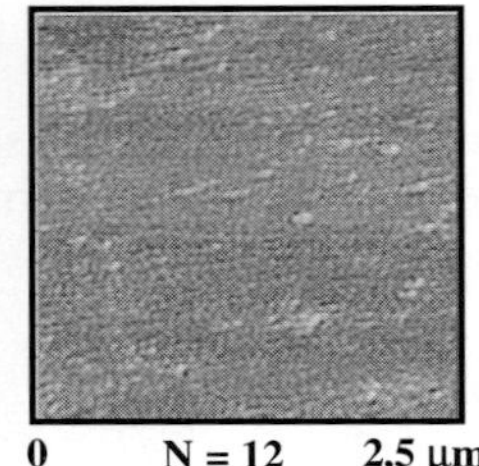

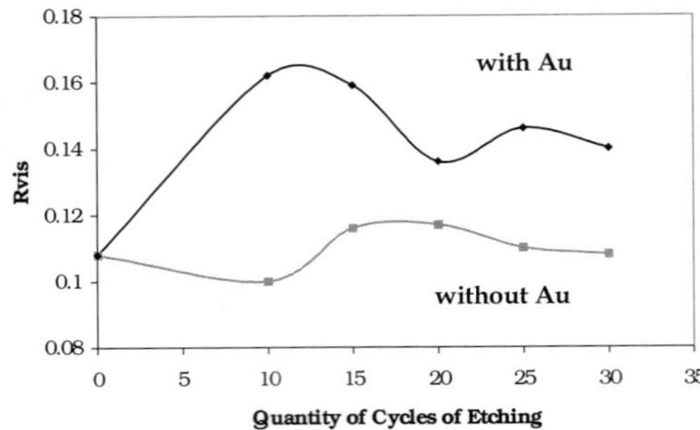

Fig.7. AFM images of samples series O_2/Au/CH_4 with different number of etching cycles (N)

Fig.8. Dependence of visual transmission coefficient vs depth of etching

References

1. T. Iakubov, A. N. Lagarkov, S. A. Maklakov, A. V. Osipov, K. N. Rozanov, I. A. Ryjikov, N. A. Simonov, and S. N. Starostenko. Experimental Study of Microwave Permeability of Thin Fe Films, MISM'2002, Moscow, June 20–24, 2002, Book of abstracts, p. 310, 2002.
2. T. Iakubov, A. N. Lagarkov, S. A. Maklakov, A. V. Osipov, K. N. Rozanov, I. A. Ryjikov, N. A. Simonov, and S. N. Starostenko. Experimental Study of Microwave Permeability of Thin Fe Films, J. Magn. Magn. Mater, Spec. Issue on MISM'2002, March 2003, in print.
3. Antonov A.S., Iakubov I.T., Rakhmanov A.L., Ryjikov I.A. Advanced magnetic materials, Ed. Y. Liu, Tsinghua University Press, 2003, vol. 2(in press).
4. I.A. Ryzhikov, L.A. Alekseeva, A.L. Djachkov, S.A. Maklakov, N.S. Perov, M.V. Sedova, and T.A. Furmanova. Nano and Giga Challenges in Microelectronics Research and Opportunities in Russia. Symposium and Summer Shool. Moscow, Russia, September 10-13. 2002, 224.
5. A. Ryzhikov, A.L. Rakhmanov, and Y.V. Trofimov. Proc. 10th Symp. Advanced Dysplay Technologies. Minsk,Belarus, September 18-21. 2001, 160 - 162.
6. I.A.Ryzhikov, A.A. Pukhov, A.S. Il'in, N.P. Glukhova, K.N. Afanasiev, and A.S. Ryzhikov. . Nano and Giga Challenges in Microelectronics Research and Opportunities in Russia. Symposium and Summer Shool. Moscow, Russia, Sept. 10-13. 2002, 222.
7. I.A. Ryzhikov, K.N. Afanasiev, E.A. Bondar, A.L. Djachkov, A.S. Il'in, M.V. Sedova, and L.P. Shadrina. . Nano and Giga Challenges in Microelectronics Research and Opportunities in Russia. ibid 225.

Microwave properties of nano-structured powders prepared by mechanical alloying

L.Z. Wu[a], H.B. Jiang[a], J. Ding[a], L.F. Chen[b], C. K. Ong[b], S. Y. Lim[c], C.R. Deng[c]

a. Department of Materials Science, National University of Singapore, Singapore
b. Physics Department, National University of Singapore, Singapore
c. DSO National Laboratories, 20 Science Park Drive, Singapore 118230, Singapore

Abstract

Mechanical alloying is a powerful tool for the fabrication of amorphous and nano-structured powder materials. Many magnetic materials fabricated by mechanical alloying have shown many unique properties. Our recent work has shown that mechanically alloyed fine nano-structured powders have the potential as microwave absorbers. In this work, we have fabricated micron and sub-micron $Fe_{100-x}M_x$ with M = 3d element or 4f element. SEM, TEM, VSM and Mössbauer spectroscopy were used for powder characterization. Microwave properties were investigated using HP Vector Network Analyzer. The EM wave absorbing properties of the mechanically alloyed powders were compared with commercial Carbonyl Iron powders and pure Fe powder. Results show that mechanical alloyed Fe-based powders may be promising for microwave absorbing applications.
Keywords: *Mechanical alloying; Microwave properties; Carbonyl Iron powder*

1. Introduction

Mechanical milling (or high energy ball milling) is a powerful method for the synthesis of nanocrystalline powders, which possess many unique physical and chemical properties [1, 2]. Mechanical milling has been widely used for the production of fine particles with size down to 0.1 μm [3]. Nanocrystalline and nanocomposite materials with particle size of 0.1-1 μm have been synthesized by mechanical alloying [4, 5]. Many nanostructured magnetic materials have exhibited excellent soft magnetic properties, which are suitable for many applications [6-8].

With the development of radar and microwave communication technology, and especially the need for antielectromagnetic interference coatings, self-concealing technology, the study of electromagnetic wave absorbing materials has increased in recent years [9]. Metallic magnetic materials can be used to make thinner electromagnetic wave absorbers because of their high saturation magnetization and high relative complex permeability. As it is well known, Carbonyl iron powders have been widely used as electromagnetic wave absorption materials. Recently, sub-micron nano-grained Fe-Si powders prepared by mechanical milling showed interesting magnetic and microwave properties for microwave absorbing purpose [8].

As reported previously [10, 11], particle size can be reduced significantly during mechanical milling, when a dispersion medium is present. In this work, several methods with mechanical milling were used to reduce the particle size. Microwave properties of the samples were studied and compared to those of commercially available fine carbonyl Fe powders.

2. Experimental

The starting powders were mixtures of Fe powder (purity>99%) and a alloying M powder with different M compositions in the form of $Fe_{90}M_{10}$ with M = Al, Si, Ni, Co, Dy, Nd, Gd. 20g of staring mixture powder were loaded together with thirteen steel balls of a diameter 15mm in a steel container. The powder and ball charge ratio was 1:9. The mechanical milling was performed using a Fritch 5 planetary mill for 36h at 300 rmp. The microstructure was studied by a Field Emission Scanning Electron Microscope (SEM Philips XL 30 FEG) and TEM.

In comparison, fine carbonyl Fe powder, as denoted as HQ, was studied. For the fabrication of finer powder, NaCl was used as the dispersion medium for $Fe_{90}Si_{10}$. A mixture of $Fe_{90}Si_{10}$ and NaCl with a weight ratio of 1:4 was mechanically milled for 36h. After milling, powder was obtained after removal NaCl by a simple washing process using deionized water. For the fabrication of Co nanoparticles, cobalt hydroxide powder was milled together with NaCl. Cobalt nanoparticles were obtained after hydrogen reduction and removal of NaCl.

30 vol% of powder samples were mixed with 70 vol% GY298 epoxy resin, and then the mixture were cured in a mold at 110°C for 1h. The relative complex permittivity ε_r and permeability μ_r at microwave frequencies were measured for toroidal samples of Φ7mm $\times$ Φ 3mm $\times$ Thickness (4-8mm) by the coaxial method using a Vector Network Analyzer (HP / Agilent 8722D). The frequency (f) dependence of reflection loss (RL) at a thickness (d) with the following formulae which characterize electromagnetic wave absorption was also calculated.

$$RL = 20\log\left|(Z_{in} - Z_0)/(Z_{in} + Z_0)\right| \qquad Z_{in} = Z_0\sqrt{u_r / \varepsilon_r}\,\tanh\{j(2\pi fd / c)\sqrt{u_r\varepsilon_r}\}$$

Where Z_{in} is the input impedance of absorber, Z_0 is the impedance of air, and c is the light speed.

3. Results and discussion

Fig. 1 shows the SEM micrographs of the commercial carbonyl Fe powder (HQ). The HQ powder consisted of small Fe particles with a mean particle size of ~1 μm and a narrow particle size distribution.

Fig. 1 SEM micrographs of HQ.

Fig. 2 shows SEM micrographs of mechanically milled Fe (a), Fe$_{90}$Si$_{10}$ (b), Fe$_{90}$Dy$_{10}$ (c), and fine Fe$_{90}$Si$_{10}$ (d) powders. The starting Fe powder consisted of coarse Fe particle with a mean particle size of ~150 μm. After mechanically milling for 36h, the mean particle size was reduced to 10-20 μm with irregular and isotropic shapes. For Fe$_{90}$Si$_{10}$, the morphology and mean particle size were similar to those of mechanically milled Fe powder with a slightly broader particle size distribution. Results showed that structure of Fe$_{90}$M$_{10}$ powders with M=Al, Ni and Co was similar to that of Fe$_{90}$Si$_{10}$. If M is a rare-earth element (Dy), the structure was much finer. Most particles were found in the range of 1-5 μm. The reason for the formation of much finer particles was the presence of an intermediate phase as found in Mössbauer investigation. We can demonstrate that particle size can be reduced by another way – milling with a dispersion medium. The milling with NaCl resulted in formation of fine Fe$_{90}$Si$_{10}$ powder with a particle size of 1-5 μm as shown in Fig. 2d.

Fig. 2 SEM micrographs of mechanically milled Fe (a), $Fe_{90}Si_{10}$ (b), $Fe_{90}Dy_{10}$ (c), fine $Fe_{90}Si_{10}$ (d)

In this work, Co nanoparticles were fabricated. Cobalt hydroxide powder was milled together with NaCl and followed with Hydrogen reduction process. Co nanoparticles were obtained after removal of NaCl. Fig. 3 shows TEM micrographs of Co nanoparticles reduced at different temperatures. It can be seen that Co particles with a particle size of 10-20 nm could be produced.

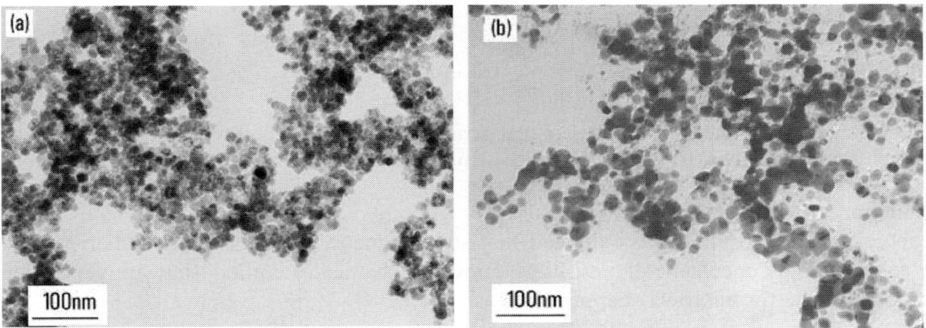

Fig. 3 TEM images of NaCl/Co powders reduced at different temperatures:
(a) reduced at $300^{\circ}C$; (b) reduced at $500^{\circ}C$.

As discussed in the introduction, Fe-based powders or Carbonyl iron powders may have a significant potential for applications in the area of magnetic shielding and microwave absorbing. In this work, complex relative permittivity and permeability were measured with HP vector network analyzer. 4 samples with the same host medium material (epoxy resin) and same host medium volume percentage (70%) were tested. Except the commercial carbonyl iron (HQ), all the other samples were fabricated by mechanical milling.

Fig. 4 shows real permittivity, real permeability and the calculated frequency dependence of the microwave reflection loss of four samples (Carbonyl Fe (HQ), mechanically milled Fe, fine Fe powder and fine $Fe_{90}Si_{10}$). A thickness of d=2.00mm was used for the calculation. The fine Fe and $Fe_{90}Si_{10}$ powders were obtained after milling with NaCl followed by removal with a washing process. The reflection loss figure shows an absorption loss peak near 10GHz and the absorption peak values range from 6 dB to 36 dB. The microwave-absorbing frequency bandwidths defined as the frequency width in which the absorption is larger than 8 dB are about 8 GHz for HQ and 4 GHz for fine $Fe_{90}Si_{10}$, respectively. Compared with bandwidths 1 GHz for the milled Fe, the fine Fe milled with dispersion medium NaCl had much larger bandwidth with 4 GHz. Such wide absorption bandwidths and high absorption loss peaks indicate the attractive potential microwave applications.

The microwave properties of the samples are closely related to the fabrication procedure. Results showed that fine $Fe_{90}Si_{10}$ and Fe milled with dispersion medium NaCl have larger real permeability than that of the milled Fe without NaCl, and have broad bandwidth and sharp peek.

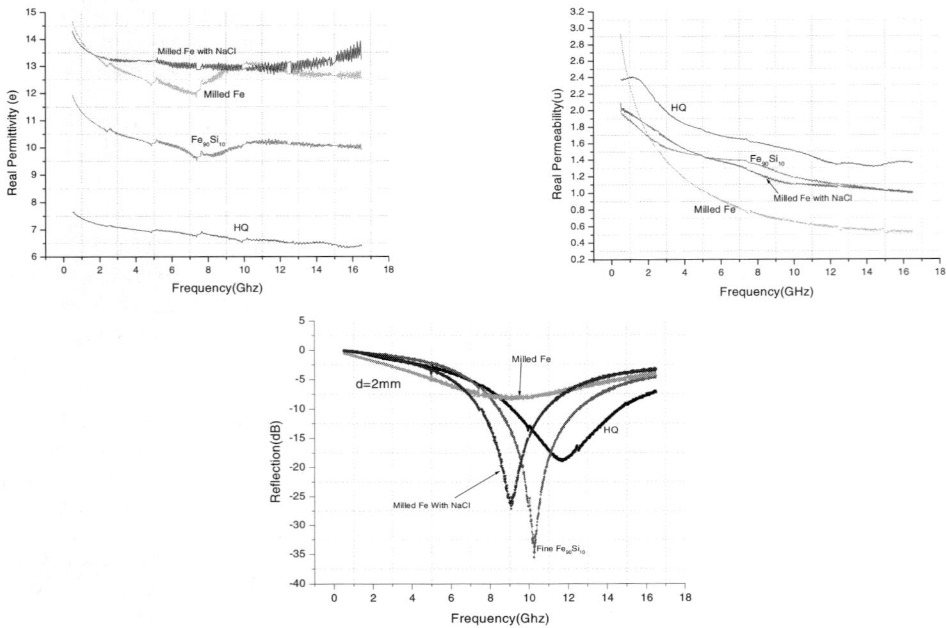

Fig. 4 Frequency dependencies of the complex real permittivity and permeability and reflection loss of resin composite of (i) carbonyl iron (HQ) (ii) Fine $Fe_{90}Si_{10}$ (iii) Milled Fe (IV) Fine Fe Milled with NaCl

4. Conclusion

In conclusion, our results have indicated that soft magnetic powders with smaller particle size can be fabricated by mechanically milling. These mechanically milled fine alloy powders are promising candidate for microwave applications.

References
[1]   C.C. Koch, Ann. Mater. Sci. 19 (1989) 121
[2]   J. Ding, P.G. McCormick, R. Street, J. Magn. Magn. Mater. 124 (1993) 1
[3]   N. Ichinose, *Superfine Particle Technology* (Spring-Verlag, Berlin, 1992)
[4]   Ch. Kuhrt and L. Schultz, J. Appl. Phys. 71, 1896 (1992)
[5]   L. Takas, Nanostructured Mater. 2, 241 (1993)
[6]   K. Suzuki, J.M. Cadogan, Phys. Rev. B 58 (1998) 2730
[7]   G. Buttino, M. Poppj, J. Magn. Magn. Mater. 170 (1997) 211
[8]   J. Ding, Y. Li, L.F. Chen, C.R. Deng, Y. Shi, Y.S. Chow, T.B. Gang, J. Alloys Compound 314 (2001) 262
[9]   H.S. Cho and S. S. Kim, IEEE Trans. Magn. 15, 3151 (1999)
[10]   J. Ding, W.F. Miao, P.G. McCormick, R. Street, Appl. Phys. Lett. 67 (1995) 3804
[11]   J. Ding, T. Tsuzuki, P.G. McCormick, J. Mater. Sci. 34 (1999) 5293
[12]   G. Li, G.-G. Hu, et al, J. App. Phy. 90 (2001) 5512
[13]   S. Yoshida, M. Sato, E. Sugawara, Y. Shimada, J. App. Phy. 85 (1999) 4636

Microwave characterization and modeling of single-walled carbon nanotube composites

Wu Junhua, Kong Lingbing and Hock Kai Meng
Temasek Laboratories, National University of Singapore
10 Kent Ridge Crescent, Singapore 119260
Tel.: (65) 6874-1006, Fax: (65) 6872-6840, email: tslwujh@nus.edu.sg

Abstract

We conducted dielectric measurements on composites comprising bundles of single-walled carbon nanotubes embedded in an epoxy matrix. The complex permittivity spectra exhibit dielectric resonance and electronic conduction behavior. Subsequently, we model the system based on two mechanisms: dielectric response and conduction. The former is described by the theory of ferroelectric resonance and the latter by the motion of the conducting electrons. Comparison between theoretical analysis and experimental permittivity data shows good agreement.

Introduction

Since the discovery of carbon nanotubes [1], single-walled carbon nanotubes (SWNT) have been investigated intensively for the physics underlying low-dimensional systems and technological applications such as hydrogen storage, visual display, battery and anti-static additive, thanks to their unique physico-chemical properties featuring small diameter, high aspect ratio, mechanical strength and flexibility [1-8]. We are interested in the electromagnetic response of the carbon nanotubes [9-12], particularly in the microwave frequency range, that may arise from the unique electronic structure of the SWNT, changeable from metallic to semiconducting, controlled by the diameter, length, chirality, even by doping and inclusion. Effective electromagnetic shielding and attenuating composite materials are achievable by dispersing such nanorods of small, excellent mechanical strength, high aspect ratio conducting cylinders into dielectric hosts, to reveal potentials as left-handed materials, microwave lenses, antennas, waveguides, light-weight high-strength electromagnetic interference shielding and attenuating materials, among others. We report here the preparation and characterization of such advanced electromagnetic composites comprising SWNT.

Experimental

The SWNT is a commercial product and can be used without further purification. The SWNT bundles have an averaged diameter of ~20 nm and are several microns in length. The preparation of the test samples was carried out by mixing weighed SWNT with epoxy resin matched with a hardener and moulding into disks of 13 mm in diameter and 1 mm thick, curing overnight at room temperature. SEM/EDX examined the morphology and composition, while x-ray diffraction (XRD) analyzed the structure of SWNT. The dielectric properties were measured with the electrical impedance method and free-space technique.

Results and discussion

Figure 1 is the XRD pattern of the SWNT nanotubes, scanned from 20° to 70°. The peaks reflect the basic structure of the graphite-like sheet of SWNT under strain, deviating from equilibrium.

Fig. 2 shows the hysteresis loop of the carbon nanotubes, indicating that the SWNT has a weak magnetism that may be attributed to the residual catalyst and/or the nanotube itself. The electromagnetic response of the material is dielectric and becomes semiconducting at a high concentration as it becomes more conductive. The existence of the electrical conduction is well-established in the range of low frequency, where the effect is more distinct, as shown in Fig. 3.

The dielectric properties of the nanotube-epoxy composites of different concentrations were measured by the electric impedance technique over 10MHz to 1.8GHz. Fig. 3 plots the real and imaginary permittivities of composites comprising bundles of SWNT embedded in an epoxy matrix, at concentrations of 4.8% to 33.3% by weight as a function of frequency and concentration, respectively. The complex permittivity spectra display dielectric resonance and electric conducting behavior. Both the real and imaginary parts decrease rapidly with frequency in the lower MHz range, then level off in mid-frequency range, and finally end with resonance. The resonance frequency decreases from 1.8GHz to 1.5GHz as the SWNT concentration increases from 4.8% to 33.3%. In general, permittivity increases with SWNT concentration, except for the resonance region. In the mid-frequency range, both the real and imaginary permittivity increase exponentially with concentration as discussed below, although they do not vary much with frequency.

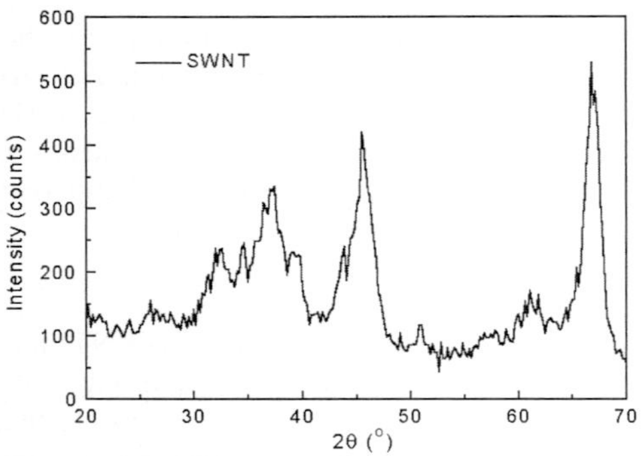

Fig. 1. XRD pattern of SWNT, with peaks reflecting the graphite-like sheet of SWNT under strain.

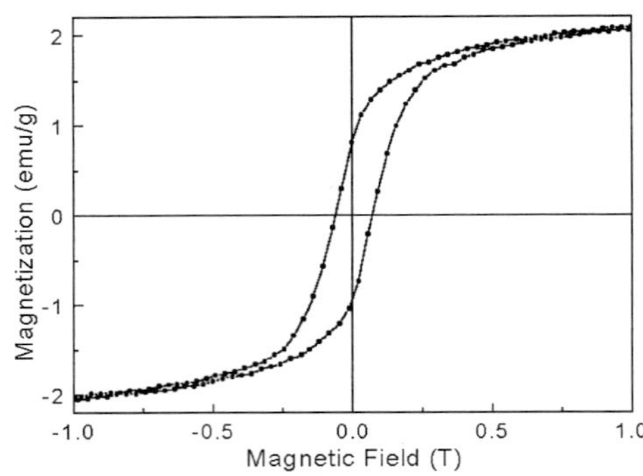

Fig. 2. Hysteresis loop of SWNT, showing a weak magnetism.

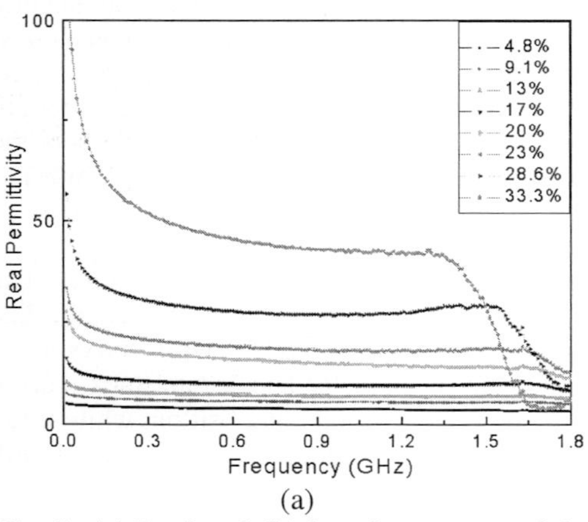

(a)

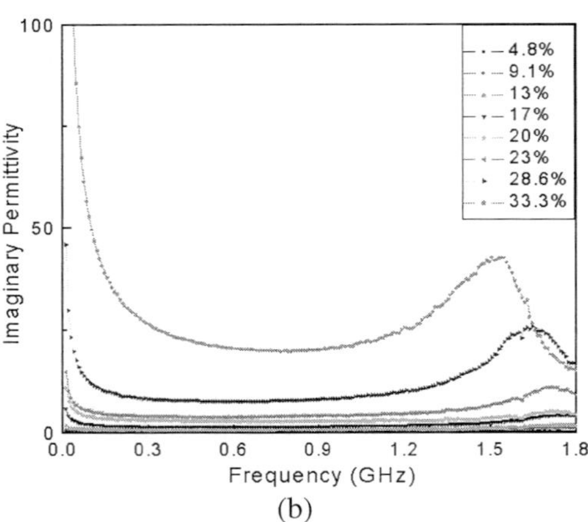

(b)

Fig. 3. (a) Real and (b) imaginary parts of the permittivities of the nanotube-epoxy composites of different SWNT concentrations over the frequency range of 10Mz to 1.8GHz.

To see the strong dependence of the permittivity on the loading percentage of the SWNT nanotubes, we select one specific frequency, 890 MHz, in the plateau range of the dielectric response. At this frequency, both the real and imaginary parameters of the complex dielectric permittivity demonstrate an exponential increase with the SWNT concentration, as given in Fig. 4. The real part increases by a factor of 11 and the imaginary part by a factor of 30, as the SWNT concentration increases to 33.3%. The apparent fitting of the permittivities as a function of the SWNT concentration with an exponential form is described by $\varepsilon' = 0.482 + 2.24e^{0.0882x}$ for real and $\varepsilon'' = 0.547 + 0.0482e^{0.180x}$ for imaginary permittivities (x=weight percentage of loaded SWNT), respectively. The relations manifest that the permittivities have a strong dependence on the concentration of SWNT in the epoxy matrix. We note that a good sample can be made with a loading of SWNT up to ~35% by weight, before it becomes brittle and easy to fracture when subject to an impedance measurement.

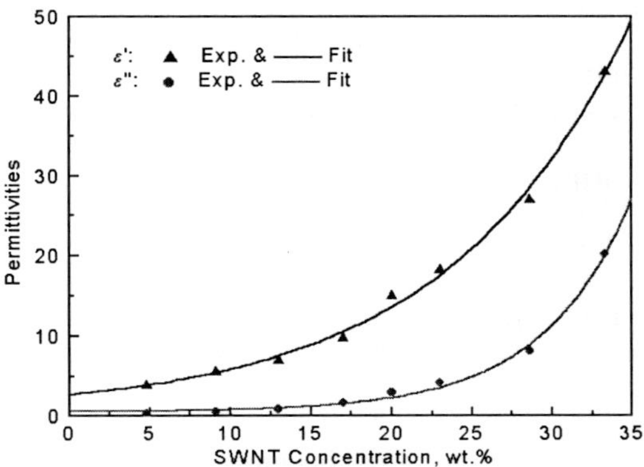

Fig. 4. (a) Real (in blue) and (b) imaginary parts (in red) of the permittivities of the nanocomposites of different SWNT concentrations at the frequency 890Mz

The dynamics of ferroelectric materials is the focus of considerable research and technological vigor for high-frequency microwave applications and for ferroelectric resonance attenuation of high-permeable dielectrics. We model the system of SWNT nanocomposites here on the basis of two mechanisms: dielectric response and conduction behavior. The former is described by the theory of ferroelectric resonance[13] and the latter by the motion of the conducting electrons. The dynamics of the dielectric relaxation reads [13]

$$\frac{d\mathbf{P}}{dt} = -\kappa\left(1+\lambda^2\right)\left(\mathbf{P}\times\mathbf{E}_e\right)+\frac{\lambda}{P}\mathbf{P}\times\frac{d\mathbf{P}}{dt}$$

for the macroscopic electric polarization $\mathbf{P}$ of magnitude P, with λ the damping parameter and $\mathbf{E}_e$ the effective electrical field. The gyroelectric constant κ is defined by $\kappa=\eta\rho$ with the splitting factor $\eta=a\alpha^{-1}$ (fine structure constant $\alpha=e^2/4\pi\varepsilon_0\hbar c$ and spectrum parameter $a=n^2/J(J+1)$ is on the order of one) and the gyroelectric ratio for the orbital motion of the electron, $\rho=e/2mc$ (e and m are the charge and mass of the electron, and c is the speed of light). λ is related to the diffusing electric dipole moment orientation associated with the time evolution equation of the non-equilibrium probability distribution governed by the Fokker-Planck equation.

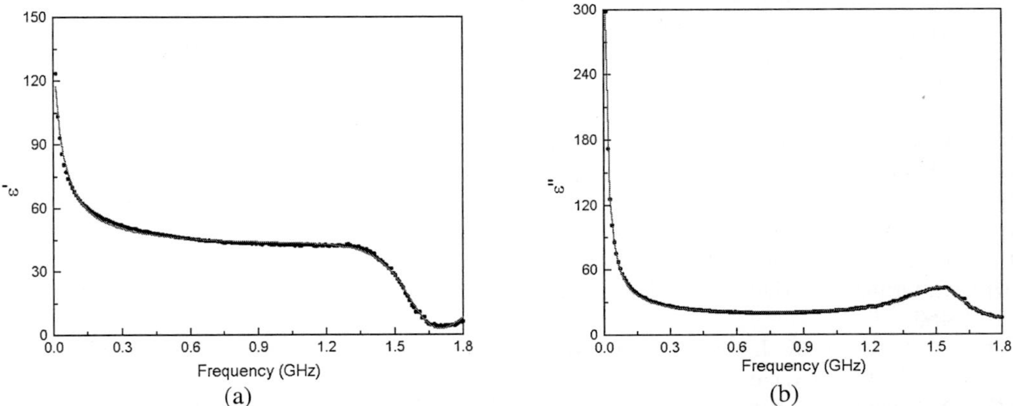

Fig. 5. Comparison of simulation (red lines) with experimental data (blue dots) of the permittivities of the SWNT composite at a concentration of 33.3%: (a) Real and (b) imaginary parts.

The dependence of the permittivity over the frequency is then given by the equations below:

$$\varepsilon_{xx}' = 1 + \kappa P_0 \frac{1 - \omega^2 \tau_r}{\left(1 - \omega^2 \tau_0^2\right)^2 + (\omega \tau_m)^2} + \frac{\sigma''}{\omega} \qquad \text{and} \qquad \varepsilon_{xx}'' = \kappa P_\alpha \omega \frac{1 + \omega^2 \tau_z^2}{\left(1 - \omega^2 \tau_0^2\right)^2 + (\omega \tau_m)^2} + \frac{\sigma'}{\omega}$$

by solving the exact analytical solutions for a general ellipsoid under a weak field approximation, where κ is a constant while P_0, P_α, τ_0, τ_r and τ_m are frequency-independent materials parameters, usually to be determined from experiments. σ' and σ'' are reduced real and imaginary electrical conductivities of linear frequency dependence [14]. Fig. 5 compares the simulation and experimental results for the case of the SWNT composite with a concentration of 33.3%, showing good agreement. We note that the real and imaginary conductivities derived from the fitting are on the order of 0.1 S m at 1 GHz for the nanocomposite of 33.3% SWNT, indicating that the system is semiconducting.

Conclusions

We have measured the dielectric properties, in the frequency range of 10 MHz – 1.8 GHz, of composites comprising bundles of SWNT embedded in an epoxy matrix, at concentrations of 4.8% to 33.3 % by weight. The complex permittivity spectra exhibit dielectric resonance and electric conduction behavior. Both the real and imaginary parts decrease rapidly with frequency in the lower MHz range, then level off in mid-frequency range, and finally end with resonance. The resonance frequency decreases from 1.8GHz to 1.5GHz as the SWNT concentration increases from 4.8% to 33.3%. In the mid-frequency range, both the real and imaginary parts of permittivity increase exponentially with concentration. We model the system based on two mechanisms: dielectric response and conduction behavior. The former is described by the theory of ferroelectric resonance and the latter by the motion of the conducting electrons. Theoretical analysis based on this model is in good agreement with the experimental permittivity data. The SWNT nanocomposite at a high concentration investigated is semiconducting.

Acknowledgements

The authors acknowledge the support of the Defence Science and Technology Agency (DSTA) of Singapore.

References

[1]   S. Ijima, Nature 354 (1991) 56; Physica B: Condensed Matter, 323 (2002) 1.
[2]   O. A. Shenderova, V. V. Zhirnov and D. W. Brenner, Critical Reviews in Solid State and Materials Sciences, 27 (2002) 227.
[3]   R. Saito, G. Dresselhaus and M.S. Dresselhaus (eds), Physical Properties of Carbon Nanotubes, Imperial College Press, London, 1998.
[4]   M. L. Cohen, Materials Science and Engineering: C, 15 (2001) 1.
[5]   P. Avouris, Chemical Physics, 281 (2002) 429.
[6]   M. Knupfer, Surface Science Reports, 42 (2001) 1
[7]   F. L. Darkrim, P. Malbrunot and G. P. Tartaglia, Int. J. of Hydrogen Energy, 27 (2002) 193
[8]   Y. P. Wu, E. Rahm and R. Holze, J. Power Sources, in press
[9]   L. Duclaux, Carbon, 40 (2002) 1751
[10]   C. Grimes, C. Mungle, D. Kouzoudis, S. Fang, P.C. Eklund, Chem. Phys. Lett. 319 (2000) 460
[11]   A. B. Kaiser, K. J. Challis, G. C. McIntosh, G. T. Kim, H. Y. Yu, J. G. Park, S. H. Jhang and Y. W. Park, Current Applied Physics, 2 (2002) 163
[12]   T. I. Jeon, K. J. Kim, C. Kang, S. J. Oh, J. H. Son, K. H. An, D. J. Bae, and Y. H. Lee, Appl. Phys. Lett., 80 (2002) 3403
[13]   Wu Junhua, Progress in Electromagnetics Res. Sym., Singapore, Jan. 2003, Proceedings, p.206
[14]   C. Leon, A. Rivera, A. Varez, J. Sanz and J. Santamaria, Phys. Rev. Lett., 86 (2001) 1279

Highly coercive $Sm(CoFe_{0.1}[Cu\text{-}Ni]_{0,0.09,0.12}Zr_{0.04}B_{0.04})_{7.5}$ nanocomposite melt-spun magnetic alloys for high temperature applications

S. S. MAKRIDIS, G. LITSARDAKIS,
Dept. of Electrical & Computer Engineering, Aristotle University, Thessaloniki, Greece

K. G. EFTHIMIADIS
Dept. of Physics, Aristotle University, Thessaloniki, Greece
I. PANAGIOTOPOULOS
Dept. of Materials Science and Engineering, University of Ioannina, 45110, Greece

D. NIARCHOS
Institute of Materials Science, NCSR Demokritos, Athens, Greece

G. C. HADJIPANAYIS
Dept. of Physics and Astronomy, University of Delaware, Newark, DE 19716, U.S.A

Abstract

We have been studying melt-spun Sm-Co 1:7.5 type alloys with boron substitution that present a nanocomposite microstructure and high coercivities in both as-spun and short time annealed ribbons. In this work, we examine four different compositions, namely $Sm(Co_{0.82}Fe_{0.1}Zr_{0.04}B_{0.04})_{7.5}$, $Sm(Co_{0.73}Fe_{0.1}Cu_{0.09}Zr_{0.04}B_{0.04})_{7.5}$, $Sm(Co_{0.70}Fe_{0.1}Cu_{0.12}Zr_{0.04}B_{0.04})_{7.5}$ and $Sm(Co_{0.70}Fe_{0.1}Ni_{0.12}Zr_{0.04}B_{0.04})_{7.5}$ in order to understand the role of Cu or Ni in the development of microstructure and high coercivity. Melt-spun ribbons have been obtained from arc-melted bulk samples and were subsequently annealed in argon atmosphere for 10-60 min at 800-870°C. In as-spun ribbons, the hexagonal $TbCu_7$ crystal structure type has been determined from X-ray diffraction patterns, while fcc-Co has been identified as a secondary phase. In annealed ribbons, the 1:7 phase transforms into 2:17 and 1:5 phases. TEM pictures show a homogeneous nano-crystalline microstructure with average grain size of 30-80 nm. Coercivity values of 15-27 kOe are obtained from hysteresis loops traced at non-saturating fields of 50 kOe. The coercivity decreases with temperature, but it is high enough to maintain values higher than 5 kOe at 380°C. After long annealing, the Co phase increases. Although the grains grow larger (50-120 nm), with favorable annealing conditions, the microstructure remains uniform, while loop squareness is improved and the coercivity is not degraded. Samples with Cu have high coercivity values, and replacement of Cu by Ni has increased the magnetization and improved the loop squareness without affecting adversely the coercivity.

Introduction

$Sm(Co,Fe,Cu,Zr)_z$ alloys are established as permanent magnet materials with high energy product and high Curie temperature. Optimization studies for use at high temperatures have been performed recently and resulted in compositions with Hc=10 kOe at 450°C [1]. In these materials, coercivity is obtained by a precipitation hardening process, which comprises solid solution treatment at 1100-1200°C for 4-24 h, isothermal aging at 850°C for 10-24 h and slow cooling to 400°C at cooling rate <1°C/min. This prolonged and complicated heat treatment is necessary for the development of the typical cellular - lamellar microstructure, with 2:17 rhombohedral type structure cells, Cu-rich cell boundaries and Zr-rich lamellae perpendicular to the c-axis. The presence of Cu and Zr is considered as a requirement for the formation of the complex microstructure. Substitution of Ni for Cu has not prevented the formation of the cellular - lamellar microstructure, but the resulting coercivity is very low at room temperature, although it increases significantly at higher temperatures [2]. By melt spinning only and slow cooling, without the conventional solid solution and aging treatments, the development of the cellular-lamellar microstructure with high coercivity (28 kOe) has been achieved [3]. Precipitation hardening with cellular features and H_c up to 10 kOe has also been obtained in melt-spun ribbons after short time annealing [4].

Alternative processing routes for nanostructured magnets with high coercivity are explored in several recent studies. By mechanical milling and subsequent annealing, samples with the 1:7 or

2:17 stoichiometries have been prepared that present coercivities up to 20-25 kOe [5-7]. Rapid solidification by melt spinning and short heat treatment has produced lower H_c values, from 5 to 14 kOe [8-11, 4]. We have obtained much higher coercivities, from 16 to 28 kOe, in B-substituted $Sm(Co_{bal}Fe_{0.1}Cu_{0.12}Zr_{0.04}B_x)_{7.5}$ as-spun ribbons or after short annealing [12,13]. The composition is typical of bulk precipitation hardened Sm-Co magnets, but B-substitution and the heat treatment led to the formation of a different microstructure. Very good results were obtained with 0.04% B at moderate wheel speed, ~40 m/sec. So, with these conditions, we tried to improve the magnetic properties by changing the composition of the alloys. In the present work, we investigate the effect of Cu and Ni on the microstructure and the magnetic properties of these high coercivity nanocomposite materials. Other compositional variations are under study.

Experimental

Four bulk alloys with composition $Sm(Co_{bal}Fe_{0.1}Cu_yZr_{0.04}B_{0.04})_{7.5}$, y = 0, 0.09, 0.12 and $Sm(Co_{0.7}Fe_{0.1}Ni_{0.12}Zr_{0.04}B_{0.04})_{7.5,}$ which are denoted as compositions A, B, C and D respectively, were prepared by arc-melting under argon atmosphere. To compensate for the Sm losses during processing, an excess of 5-10% Sm was added to all samples. Ribbons have been obtained from master alloys by melt-spinning using a quartz tube with an orifice diameter of about 0.5mm, about 2 atm pressure of 99.999 % pure argon at wheel speed of about 40m/sec. The ribbons were wrapped with tantalum foil and sealed in a quartz tube - after three purges with pure argon to avoid oxidization - and then annealed at temperatures 800 - 870°C for different times. The phases in as spun and annealed ribbons were determined by X-ray powder diffraction using FeK_α radiation. Magnetic properties of the samples were measured by vibrating sample magnetometers (VSM) with maximum field of 5T. No correction was made for the demagnetization field effect. The high temperature measurements have been performed in a VSM with a maximum field of 2T up to 600°C. The microstructure of the ribbons was determined by transmission electron microscopy using a Jeol JEM 2000FX microscope.

Results and discussion

The as-spun samples are nano-crystalline, with the hexagonal structure of the $TbCu_7$ -type phase (space group P6/mmm). We have previously reported that melt spinning at a wide range of wheel speed (5-70 m/sec) and boron substitution (x=0.005 to 0.5) in the composition with Cu has not produced amorphous melt-spun samples [13]. For y=0, the broader and lower diffraction peaks that are observed indicate a nanocrystalline microstructure. The lattice parameters of the 1:7 phase are a=b≈4.95 Å, c≈4.07 Å. The diffraction peaks (111), (200) and (220) of fcc-Co may be identified at angles ~56°, ~66° and ~101° respectively. The peaks are scarcely seen in the diffraction patterns of ribbon samples not ground into powder, while they are not distinguished in the diffraction patterns of ribbons ground into powder. After annealing at 810°C for 1 h and subsequent quenching, a phase transformation of the $TbCu_7$-type structure to 2:17 rhombohedral and 1:5 structures is apparent in the X-ray patterns. 1:7 is a metastable phase, commonly considered as a disordered 2:17 rhombohedral phase. The rhombohedral Th_2Zn_{17} type structure can be derived from the $CaCu_5$ type structure if 1/3 of the rare earth atoms are substituted by a dumbbell pair of transition metal atoms. The 1:7 structure is formed if the distribution of dumbell pairs is random. Annealing induces an ordered distribution of dumbell pairs and the formation of 2:17R and 1:5 structures. Traces of fcc-Co are found in short-annealed samples [13]. After long time annealing, 14 hours at 870°C, the fcc-Co phase has become dominant with small amounts of samarium oxide and intermetallic phases.

The values of saturation magnetization Ms, remanence Mr and coercivity Hc are reported in Table 1. All samples are magnetically hard, except for sample A (y=0). The loop of sample A is characteristic of a mixture of non-interacting soft and hard magnetic phases, with a steep soft step at low field and magnetically hard curve with lower slope at higher fields, in both the initial and demagnetization curves. Saturation magnetization is decreasing with increasing Cu content from 72.5 emu/g for the Cu-free sample A to 44.6 emu/g for sample C (y=0.12). Sample D (with Ni) has remanence M_r = 49.3 emu/g, which is 50% higher than the M_r of sample C with an equal amount of Cu (y=0.012, M_r = 32.1 emu/g). Ni substitution for Cu does not deteriorate the coercivity. The as-spun ribbons with Ni and Cu have very high coercivity, 20 and 27 kOe respectively for samples D and C.

Table 1. Magnetization and coercivity of as-spun and annealed ribbons

Composition	As-spun			Annealed at 810°C for 1h		
	M_s (emu/gr)	M_r (emu/gr)	H_c (kOe)	M_s (emu/gr)	M_r (emu/gr)	H_c (kOe)
A	95.8	72.5	1.4	55.7	27.5	9.5
B	66.1	45	10.5	58.8	34.1	4
C	44.6	32.1	27	58.4	32.1	3.2
D	63	49.3	19.9	62.2	44.2	14

The magnetic characteristics of annealed samples are given in Table 1 and Fig.1. Besides phase changes, annealing above 750°C induces growth of the size of nano-crystallites and other microstructural changes, which are reflected as changes of the magnetic properties. For example, in sample C after annealing for 10 min at 800°C, the coercivity is reduced to 14 kOe and the loop shape is modified (fig. 1a). On the contrary, sample A, when annealed at 800°C for 15 min, undergoes microstructural changes that improve significantly the coercivity (17 kOe) and the loop shape (Fig. 1b). Remanence enhancement and suppression of the soft step indicate exchange coupling between the hard and soft phases. The exchange coupling has been confirmed by delta-M plots of remanence values in as-spun samples, which have evidenced positive magnetic interactions [12]. Longer annealing, however, causes a decoupling of hard and soft grains that are detrimental to coercivity and loop squareness (Fig 1d). This effect is not so marked in the composition with Ni (Fig. 1c), which presents a square, single-phase loop shape and Hc=14 kOe even after annealing for 1h. The coercivity of as spun and annealed ribbons has been measured at higher temperatures, in the range of 20°C to 550°C, with maximum applied field of 20 kOe.

Fig. 2 shows the variation of coercivity with temperature. Coercivity is decreasing normally for all samples. At 400°C, values close to 4 kOe are obtained. After heating and cooling back at a temperature close to 100°C, the coercivity was measured again and the same values as on heating were obtained. This behavior with the temperature is important for high temperature applications.

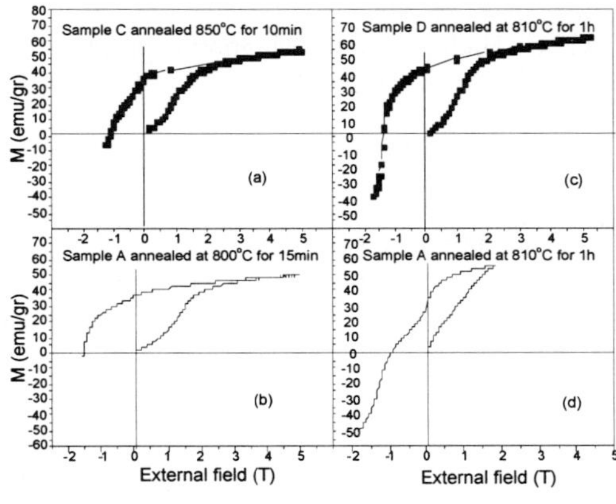

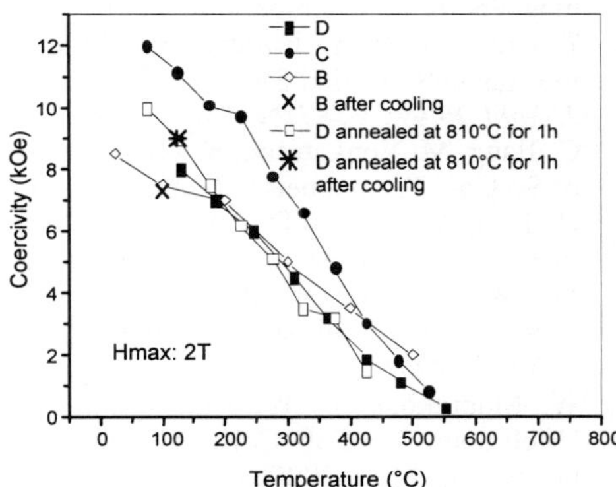

Fig. 1 Hysteresis loops for as-spun ribbons   Fig. 2 Hysteresis loops for annealed ribbons

TEM micrographs display a fine nanoscale structure without precipitation characteristics. The grain size is in the range of 10-100 nm for all samples. As-spun ribbons of sample A (y=0) with grain size of 10-50 nm can be observed in Fig. 3a. After annealing at 800°C for 15 min, the smaller grains have grown larger with narrow size distribution (Fig. 3b). We may conclude that a homogeneous nanograin microstructure is required for high coercivity and stronger interactions between the grains, whereas a non-uniform partition of the grains restrains the exchange coupling and reduces the coercivity. For y=0.12 with Cu or Ni, the microstructure of as-spun ribbons is similar. The only difference observed is that the grain size in ribbons with Ni is about 30% larger than in the sample with Cu, with narrower size distribution.

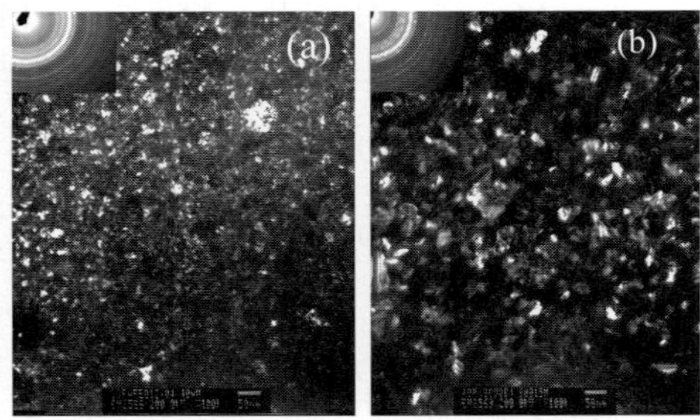

Fig. 3. TEM photos of composition (A): (a) as- spun and (b) annealed at 800°C for 15 min

Acknowledgments

The work is supported by EC project HITEMAG (G5RD-CT2000-00213)

References

[1]   J. F. Liu, Y. Ding, Y. Zhang, D. Dimitrov, F. Zhang, and G. C. Hadjipanayis, "New rare-earth permanent magnets with an intrinsic coercivity of 10 kOe at 500°C," *J. Appl. Phys.*, vol. 85,N. 5, pp. 5660–5662, 1999.

[2]   W. Tang, Y. Zhang and G .C. Hadjipanayis, "Effect of Ni substitution on the microstructure and coercivity of $Sm(Co_{bal}Fe_{0.1}Ni_yZr_{0.04})_{8.5}$ magnets," *J. Appl. Phys.*, Vol. 91, No. 10, pp. 7896–7898, 2002

[3]   A. Yan, A. Bollero, K. H. Muller and O. Gutfleisch, "Fast development of high coercivity in melt-spun Sm(Co,Fe,Cu,Zr)z magnets", *Appl. Phys. Lett.*, V. 80, N. 7, pp. 1243- 1245, 2002

[4]   I. Panagiotopoulos, T. Matthias, D. Niarchos and J. Fidler, "Melt-spun Sm(Co,Fe,Cu,Zr)z magnets for high-temperature applications," *J. Magn. Magn. Mater.*, 2002, in press

[5]   Z. Chen, X. Meng-Burany, H. Okumura and G. C. Hadjipanayis, "Magnetric properties of mechanically milled $Sm_2(Co,M)_{17}$-based powders with M=Zr, Hf, Nd, V, Ti, Cr, Cu and Fe," *J. Appl. Phys.*, V. 87, N.7, pp. 3409-3414, 2000

[6]   C. Jiang, M. Venkatesan, K. Gallagher and J. M. D. Coey, "Magnetic and structural properties of $SmCo_{7-x}Ti_x$ magnets," *J. Magn. Magn. Mater.*, 236, pp. 49-55, 2001

[7]   H. Tang, Y. Liu and D. J. Sellmyer, "Nanocrystalline $Sm_{12.5}(Co,Zr)_{87.5}$ magnets: synthesis and magnetic properties," *J. Magn. Magn. Mater.*, 241, pp. 345-356, 2002

[8]   M. W. Crabbe, H. A. Davies, and R. AA. Buckley, "Remanence enhancement in nanostructured melt spun $Sm(Fe_{0.209}Cu_{0.061}Zr_{0.025}Co_{0.704})_{7.61}$," *IEEE Trans. Magn.*, vol. 30, pp. 696-698, 1994

[9]   W. Manrakhan, L. Withanawasam, X. Meng-Burany, W. Gong, and G. C. Hadjipanayis, "Melt-spun $Sm(CoFeCuZr)_zM_x$ (M= B or C) nanocomposite magnets," *IEEE Trans. Magn.*, vol. 33, pp. 3898-3900, 1997

[10] W. Gong, "Structure and magnetic properties of $Sm(Co_{0.68-x}Fe_{0.25}Cu_{0.07}Zr_x)_{7.7}$ melt spun alloys with Zr additions," *J. Appl. Phys.*, vol. 87, pp. 6713-6715, 2000.

[11] J. Zhang, I. Kleinschroth, F. Kuevas, Z. H. Cheng and H. Kronmüller, "Effect of additives on the structure and magnetic properties of 1:7 type $Sm_2Fe_{15}Ga_2C_3$ permanent magnets," *J. Appl. Phys.*, V. 88, N.11, pp.6618-6622, 2000

[12] S. S. Makridis, G. Litsardakis, I. Panagiotopoulos, D. Niarchos, Y. Zhang, and G. C. Hadjipanayis, "Effects of boron substitution on the structural and magnetic properties of melt-spun $Sm(Co,Fe,Zr)_{7.5}$ and $Sm(Co,Fe,Zr,Cu)_{7.5}$ magnets," *J. Appl. Phys.*, V.91, N.10, pp. 7899-7901, 2002

[13] S. S. Makridis, G. Litsardakis, I. Panagiotopoulos, D. Niarchos, Y. Zhang, and G. C. Hadjipanayis, *IEEE Trans. Magn.*, Mag-38, N5, pp. 2922-2924, 2002

Session F4

Magnetic Composites

Chair: G.D. Li

Contemporary Magnetic Materials and Magnetism:
Research and Applications

Guo-dong Li

Institute of Physics, Chinese Academy of Science,
Beijing, China

Abstract: *This paper discusses several new research directions and applications of contemporary magnetic materials and magnetism, including: (1) high frequency magnetic materials; (2) microwave magnetic materials; (3) nano-magnetic materials and mesocopic magnetism; (4) widespread existence of magnetism and interdisciplinary research.*

Research and applications of magnetic materials and magnetism form an important part of contemporary science and technology. Vast amount of past research work have shown that all materials possess magnetism, from very strong to very weak, and that magnetic fields exist in all space with different strength. Currently, most applications used the strong magnetic materials, or simply magnetic materials. These include ferromagnetic materials (various metals) and ferrimagnetic materials (magnetic oxides). Magnetic materials and magnetism have, therefore, progressed significantly in many aspects, and found many important applications [1] in the modern era, despite its long history and wide domain [2]. In this paper, we introduce several new advances in the research and applications of contemporary magnetic materials and magnetism.

1.   High Frequency Magnetic Materials

In the modern information society, communications, magnetic recording, and electronic computer technology require different high frequency magnetic materials. These include soft magnetic materials that have high magnetic permeability and good frequency behaviour, as well as information magnetic materials with magnetic-pulse properties, good magnetic recording and memory characteristics.

The permeability μ is an important parameter that characterises high frequency magnetic properties. Under AC condition, it is a complex quantity with real part μ' and imaginary part μ". The real part μ' represents the in-phase relationship between magnetic flux density and AC magnetic field, while the imaginary part μ" represents the relation between these two quantities in phase quadrature, and is related to the loss in the magnetic materials under AC magnetic field. The permeability of magnetic materials varies with frequency of the applied magnetic field. This is known as the permeability spectrum. Under normal circumstances, permeability refers to the initial permeability value at very low magnetic field strength. In general, the permeability spectrum is divided into 5 zones, as shown in Figure 1: low frequency spectrum ($< 10^4$ Hz), mid-frequency spectrum ($10^4 - 10^6$ Hz), high frequency spectrum ($10^6 - 10^8$ Hz), ultra high frequency (microwave) spectrum ($10^8 - 10^{10}$ Hz) and extremely high frequency ($> 10^{10}$ Hz). Neglecting the loss due to eddy current and skin effect, the permeability spectrum at each frequency band exhibits different characteristics and dominant mechanism. At low frequency band, μ' is practically constant, while μ" is very small. At mid-frequency band, it is possible to observe magnetic internal friction peak, dimensional resonance or magneto-mechanical coupled resonance. At high frequency band, due to domain wall resonance and relaxation, μ' decreases sharply while μ" increases rapidly. At ultra high frequency band, natural resonance is the dominant mechanism, and gives rise to negative value of (μ' − 1), with a resonant peak observed for μ". At extremely high frequency band, the internal exchange field is the main contributor. A detail study of the permeability spectrum is therefore very important to research on high frequency magnetic materials and its applications.

In high frequency soft magnetic materials, high permeability, high saturation magnetization, low coercivity and low loss are desired. In general, soft magnetic materials can be divided into 3 types: crystal (ferrites, metallic film), amorphous film and nano-crystalline film. The distinct advantages of ferrites are high resistivity ($\rho \sim 10^2 - 10^6$ Ωcm) and good permeability-frequency response curve, thus making it suitable for high and ultrahigh frequency applications. By appropriate selection of its composition and technique to achieve vanishing magnetocrystalline anisotropy constant and magnetostriction coefficient simultaneously, the initial permeability μ_i of the Mn-Zn and Ni-Zn series can rise above 10^3-10^4 [3-5]. The Co-Cr film is a high density vertical magnetic recording material. Fe-Ni (Permalloy) film is widely used as thin film material in magnetic recording head. Amorphous materials can be primarily divided into the metal-metalloid type (Fe-B and Fe-P series), the metal-metal type (Fe-Zr and Co-Zr series), as well as the metal-oxide type (α-$Y_3Fe_5O_{12}$), etc. In the amorphous alloy Co-Zr, $4\pi M_s = 14$ kGs, $H_c < 0.5$ Oe and $\mu_i = 3,500$. For the $Co_{87}Nb_9Zr_4$ film, its permeability can be above 1,000 at 10 MHz [6]. In nano-crystalline film, we have the Fe(Si)-Nb-N series, Fe-Ta-N series, and Fe-Hf-O series, etc. The resistivity of the Fe-Hf-O series lies within 1-100 Ωcm, which, together with its high permeability (see Figure 2) makes it suitable for high frequency applications [7], such as high speed magnetic recording device.

Based upon the foundation of magnetic functional materials, further advancement in research and applications of multi-function magnetic materials are made. Such materials possess simultaneously ferromagnetic properties and other important physical functions. An example is the ferromagnetic-ferroelectric materials [8] ($BiFeO_3$-(Ba, Pb) (Ti, Zr) O_3), which possess both high permeability and permittivity that are potentially useful to the design of miniaturised electronic components. Other examples include ferromagnetic-semiconductor materials [9-11] (such as EuS and $ZnCr_2Se$) which has both high carrier mobility and ferromagnetic properties; magneto-electric materials [12] (e.g. $GaFeO_3$) which possess both strong magnetization and polarization under excitation of the magnetic fields or electric fields; ferromagnetic-superconducting materials (e.g. $ErRh_4B_4$) [13]; organic ferromagnetic materials [14] (e.g. $[C_6H_5(NH_3)]_n$ polymer), such as the weak ferromagnetic properties observed in polybutydine and polytriaminobenzene, with a Curie temperature of about 150 and 400°C.

Magnetic intelligence material is another new development. Besides possessing magnetic properties, the materials possess sensing, feedback and response capabilities to stimulations from external environment. For example, shape memory intelligence has been observed in several magnetic alloys such Ni-Ti, Fe-Pt, Ni-Al, etc, while antenna made of Ni-Ti alloys with shape memory intelligence has been developed for use on spacecrafts. Figure 3 shows wires of Ni-Ti alloys suitably processed by heat to form a large antenna, which is subsequently deformed and shrunk in size before loading onto the spacecraft. Once in space, the antenna recovers its original shape and size for actual applications. Other types of magnetic intelligence materials are also being investigated.

2.   Microwave Magnetic Materials

Due to the advancement in scientific research and new high technology, the frequency in use has been increasing. A key development in modern technology is the shift from high frequency to microwaves, and from microwaves to infra-red and light waves. Similarly, the focus has also shifted from magnetic materials at high frequency to microwave, photo-electronics and photonics magnetic materials, and the corresponding magnetism research and applications.

Among the applications of microwave magnetic materials, non-reciprocal devices are the most outstanding and most widely used. This device exhibits very different transmission, attenuation and other properties in the forward and backward directions of wave propagation. Devices with such properties have been designed, such as isolators with different attenuation in each direction, non-reciprocal phase shifters with different phase delay, circulators with different attenuation for multi-channel transmissions. The working principles of these non-reciprocal magnetic devices are based on non-reciprocal energy absorption and phase shift produced by magnetic materials when the DC magnetic field and AC field satisfy certain conditions. The absorption and shift are related to the tensorial permeability and magnetic resonance of the material. This type of material requires low magnetic and dielectric losses, as well as high gyro-magnetic property. Typical materials are the Li series' and Mg series' spinel ferrites and garnet ferrite YIG ($Y_3Fe_5O_{12}$), with both magnetic and dielectric loss tangents ($tan\delta_m$ and $tan\delta_e$) less than 0.001. For millimetre and sub-millimetre waves, since very high applied magnetic field is required by these devices, the material can only depend on its internal fields, such as the magneto-crystalline anisotropy fields or exchange fields. The hexaferrite series is highly suitable, since most materials in this series possess high uniaxial anisotropy.

Ferromagnetic materials, paramagnetic materials, diamagnetic materials and the system of magnetic nucleus, under a certain constant-magnetic-field and microwave or high frequency magnetic fields, give rise to ferromagnetic resonance, paramagnetic resonance, diamagnetic resonance (also known as cyclotron resonance), and nuclear magnetic resonance, respectively. Figure 4 shows the ferromagnetic resonance linewidth ΔH and curves related to the frequency f [16] for three types of ferrites, namely, $NiFe_2O_4$(1), $(Ni, Zn)Fe_2O_4$(2) and $(Ni, Zn, Cu)Fe_2O_4$(3), at different frequency. It is noted that the ferromagnetic resonance linewidths ΔH (related to microwave loss) are different for different magnetic materials at different frequencies. Figure 5 shows the resonance magnetic field and resonance linewidth of ferrite $Y_{3-3x}Gd_{3x}Fe_5O_{12}$ (which has magnetic sub-lattices) near the compensation temperature. At the wavelength of 3cm, it is observed that the resonance magnetic field H decreases rapidly as the temperature rise to the compensation temperature, and increases rapidly as the temperature increases further. Correspondingly, the resonance linewidth ΔH increases rapidly and then decreases around the compensation temperature. The dashed lines are the corresponding theoretical results [17]. This phenomenon should be given special attention when using ferrimagnetic materials with resonance at microwave frequency.

There are many types of magnetic materials for microwave absorption. The main characteristic of these materials is the strong absorption of electromagnetic waves over a wide bandwidth. Most magnetically lossy materials are based on spinel and hexaferrites. As compared to metal, its resistivity is high and density is low. The complex permeability is due to the interactions of the spin and domain wall with the AC magnetic fields. At microwave bands, 0.1<f<10 GHz, magnetic loss is due primarily to the natural resonance of the spin precession. Currently, most wave absorbers use spinel ferrites, such as NiZn, NiMg, NiCo and LiCd, etc, at lower frequencies. For hexaferrites, due to the poor symmetry in crystal and the high anisotropy fields, the natural resonance frequency is higher than that of spinel. In particular, the c-plane hexaferrites (M-, W-, Z- and Y-) have good microwave magnetic properties. The initial permeability of the Y-type ranges from 4 to 27. The permeability spectrum of $Co_\delta Zn_{2-\delta}$-Z is shown in Figure 6. For magnetic metal powder, carbonyl iron is the main candidate. Others include iron powder, cobalt powder, nickel powder and their alloys. These materials are also use in wave absorbers of the lossy type. However, due to the high dielectric constant, it is not suitable as a matching layer.

The paramagnetic-type maser is an ultra-low noise microwave amplifier, whose early stage research motivated the development and invention of laser. This is also based on the foundation of paramagnetic resonance. In addition, free electron maser, including microwave synchrotron, magnetron, travelling wave tube, klystron, cross-field light-projector, etc, all require magnetic fields of different strengths and distribution, and magnetic materials.

3.   Nano-magnetic materials and Mesoscopic Magnetism

Development in current natural science and new advance technology motivates the appearance of new research areas and applications. Mesocopic science is a new domain bridging the gap between macroscopic and microscopic sciences. In this new domain, mesoscopic magnetism forms an important part. Hence, nano-magnetic materials, which fall within this domain, received much attention in both research and application aspects.

Nano-magnetic materials refer to materials' with one or several dimensions of nanometre (about 10^{-7} - 10^{-9}m) in size, e.g. the nano-magnetic films (one dimension in nm), the nano-magnetic wires (two dimensions in nm) and magnetic dots (three dimensions in nm). These nano-magnetic materials have attracted extensive focus in both research and practical use.

In general, the key features of mesoscopic magnetism are: single domain, as the critical size of the single domain in most materials is of the order of nanometres; the critical sizes appearing in super-paramagnetism and other superferromagnetism are also nanometres in size; magnetic phase change with temperature; change in surface structure; quantum size effects; mesoscopic quantum tunnelling effects; etc. As most mesoscopic magnetism appears in the range of nanometres, this leads to many mesocopic properties observed in nano-magnetic materials, hence, giving rise to some new and unique applications.

Vast amount of investigation have shown that most multi-layer nano-magnetic films, nano-particle magnetic films and nano-oxide magnetic films possess giant magneto-resistance (GMR) effect [19]. Figure 7 shows the relationship between the MR coefficient $\Delta\rho/\rho$ and the film thickness t_{Ag} of Ag, under an external magnetic field of B = 6T, for the multi-layer nano-magnetic film [Fe (1nm)/Ag(t_{ag})]$_{40}$. Curves 1 and 2 are test results at room temperature and ultralow temperature (1.5K). Three oscillations are observed in the magneto-resistivity as the Ag film thickness varies. At room temperature, the highest magneto-resistivity is –6% [20]. As the thickness of the non-magnetic layer increases, ferromagnetic and anti-ferromagnetic exchange coupling alternately occurs between the ferromagnetic layers, and the largest MR coefficient appears at the anti-ferromagnetic coupling [21]. GMR effect was discovered in 1988, and within a few years, integrated GMR components were successfully developed. This class of materials found wide applications in magneto-resistance random storage media, magnetic read head, and magnetic field detector.

The fundamental characteristic of soft magnetic materials is low coercivity. Figure 8 shows the relation between coercivity and grain size. As the grain size reduces to single domain, the coercivity increases linearly. However, when the grain size and spacing between grains are less than the exchange length (few tens of nanometres), the resultant effective magnetocrystalline anisotropy is due to the averaging effect of the local random magnetocrystalline anisotropy over the exchange length, with its value given by the grain size to the power of 6. Hence, nanocrystalline materials have very low coercivity. For example, the nano-alloy FeCuNbSiB has coercivity of the order of 0.01 Oe, with $4\pi M_s \approx$ 10-12 kGs, and initial permeability is 10^5 [22]. Currently, among the permanent magnetic material, Nd-Fe-B has the highest magnetic energy product, which is more than 50 MGOe. This property can be improved by increasing its remanent magnetization. A possible method is to produce a nano-structure alloy with micro crystals of $Nd_2Fe_{14}B$ (20 – 40 nm) separated by α-Fe thin layer (5-10 nm). In 1991, Kneller and Hawig proposed the mechanism of exchange coupling for the two-phase composite permanent magnet [23]. In 1993, Manaf *etal* [24], in their work on the $Nd_2Fe_{14}B/\alpha$-Fe, has shown evidence on this mechanism. Recently, there are reports on permanent magnetic alloy NdFeB/α-Fe multi-layer film with properties far better than that of NdFeB single layer film.

Magnetic liquid is a glue-like liquid obtained by mixing evenly single domain nano-magnetic powders, organic and inorganic liquid, and dispersant with surface activity. The liquid has both ferromagnetic and fluidity properties. Further, it possesses unique properties such as superparamagnetism that does not have hysteresis loop, i.e. both remanence and coercivity vanish; its viscosity, magnetic buoyancy and surface density can be controlled by externally applied magnetic field; it has very good stability to maintain its evenly mixed composition over long time without agglomeration due to influence by magnetic field and gravity; it exhibits different magnetization curves for mixture of different nanoparticles and liquid, which facilitates control and tuning of the magnetic liquid's magnetic properties, magnetic buoyancy, apparent density, and other unique properties. Figure 9 shows the magnetization M as a function of magnetic field H for five types of magnetic liquids comprising different single domain nanoparticles, inorganic or organic fluids and dispersant with surface activity [25]. The single domain nanoparticles and liquid used in the five types of magnetic fluids are: (1) Co/methyl benzene (high concentration); (2) Fe_3O_4/diester; (3) Fe_3O_4/water; (4) Co/methyl benzene (low concentration); (5) Fe_3O_4/oil. The magnetization curves are very different. Magnetic liquids find applications in many areas, including many important applications in electronics, electrical, petroleum-chemical and space technologies, etc. Examples include use in many moving parts for sealing, lubricating and damping; in the selection and separation of materials with different density; as magnetic fuel for use under loss of gravity; in magneto gyrocompass, audio transducer and magnetic fluid dynamo in which the magnetic liquid serves as the working material; in printing and dyeing controlled by magnetic field, etc.

4.   Magnetism, Magnetic technology and Interdisciplinary Magnetism

All materials possess magnetic properties, and magnetic fields exist in all space. Hence, branches of science that are related to magnetic properties and magnetic fields, known as interdisciplinary magnetism or frontier edge magnetism, can be widely found. Here, we will discuss several investigations and applications in biomagnetism, micro-magnetism, geomagnetism and cosmo-magnetism.

4.1   Biomagnetism [26, 27]

Vast amount of scientific observation and experiment cumulated over a long period in time have shown that all living creatures, including human, possess magnetic properties. In addition, magnetic field is produced in life activities, and external magnetic fields do affect living creatures. Hence, many techniques and magnetic methods find important applications in biological and medical studies. In recent years, with the advancement in biomagnetism, research in magneto-cardiogram and magneto-encephalogram has progressed significantly, though electro-cardiogram and electro-encephalogram have been widely used in medical science. Figure 10 shows the magneto-cardiogram (a) and electro-cardiogram (b) of a healthy human. In comparison to electro-cardiogram and electro-encephalogram that measure only AC signals, the magneto-cardiogram and magneto-encephalogram offer the advantages of electrodeless contact (less interference) and measurement of both DC and AC magnetic signals. Further, magneto-cardiogram and magneto-encephalogram can perform three-dimensional measurement, while their corresponding electro-counterparts can only manage two-dimensional measurement.

Experiments have also shown that the resolution obtained from magneto-cardiogram and magneto-encephalogram is better that that of the electro-cardiogram and electro-encephalogram. However, magneto-cardiogram and magneto-encephalogram require the use of very sensitive measurement instrumentation (SQUID), and elimination of interference from the earth's magnetic field, using magnetically shielded chamber and/or magnetic gradiometer. Solutions have been found for these problems. As another example, though X-ray Computerized Tomography (CT) has been an important tool in medical science, the newly developed Nuclear Magnetic Resonance (NMR) CT has more attractive features. It outperforms the X-ray CT in the diagnosis of early stage tumour and encephalic diseases. Biomagnetism is certainly being hotly pursued in current research and applications.

4.2   Micro-magnetism

This mainly includes research and applications of nuclear magnetism and fundamental particles' magnetism. Some examples will be highlighted here. Using the adiabatic demagnetisation of some paramagnetic compounds, it is possible to lower its temperature to about $1K\text{-}10^{-2}K$, while through the nuclear magnetic moment of certain materials, the adiabatic demagnetisation can reduce its temperature to about $10^{-3}K\text{-}10^{-9}K$. Research also shows that some nuclear moment systems are capable of producing magnetic order below certain temperature. For example, the nuclear moment of ^{3}He gives rise to a special anti-ferromagnetic structure of the UUDD type at the ultralow temperature of $10^{-3}K$ [29]. Following Dirac's proposal of the magnetic monopole concept [30] in 1931, past several decades had witnessed various experimental observations and theoretical studies on high-energy accelerator, rocks found on earth and moon, space observation, etc. However, to date, there is no repeatable evidence based on experimental observations.

4.3   Geomagnetism

The existence, effect, change and origin of the earth's magnetic fields have continuously received much attention and investigation, since the earth's magnetic field affects almost everything. Examples include direction and guidance in early sea navigation and aviation, magnetic methods for mine detection, forecast of solar activities, effects on wireless communications and weather, survey of geological structures, determination of geologic ages, evidence on the drift of continents and underwater expansion, effects on radiation in ionosphere and magnetosphere at high altitude, effects on palaeoclimate, understanding of the earth's crust properties, investigation on the interior structure of the earth, effects on the evolution of life and bio-navigation, etc. In addition, it may find possible application in forecasting earthquakes and volcanic eruptions. In early investigation of magnetism in sedimentary rocks, it was discovered that the remanence in these rocks differ greatly and perhaps reverse in direction from the magnetic fields found around the place at the time of discovery. Subsequently, it was observed that in all oceans, the geomagnetic fields on both sides of underwater ridges appeared to display symmetrical oscillatory changes [31], as shown in Figure 11. This shows the reversal in direction of the palaeomagnetic fields. There are more than 10 different theories concerning the origin and reversal in direction of geomagnetic fields. On the former, the fluid dynamo theory of earth's magnetic field is currently widely accepted. However, there is no widely accepted theory on the reversal in direction of the earth's magnetic field.

4.4   Cosmo-magnetism

This is the interdisciplinary investigation on magnetism found in various stellar objects in the universe, the status, purpose, change and origin of the magnetic fields that exist in the vast space between these objects. Its contents are certainly very wide ranging. Only a few examples will be highlighted here. For example, through the Apollo's moon-landing mission, observations were made on the moon's surface and the magnetic fields in the moon's atmosphere, and investigation was done on the magnetism of the moon rocks brought back to earth. From these studies, it was speculated that the interior of the moon is solid. Another example concerns the measurement of Jupiter's magnetic fields by the cosmic spacecraft. The magnetic fields of Jupiter is stronger than that found on most planets, and by combining with the gigantic size of Jupiter and its main chemical constituent (hydrogen), it is speculated that the interior of Jupiter contains the superconducting hydrogenium. Hydrogenium is currently an unsolved important research topic on earth. From the theoretical point of view, it is speculated that hydrogenium is a fuel with high energy density and a superconductor with high transition temperature.

From the above discussion on the research and applications of magnetic materials and magnetism, it is obvious that these areas will have a significant impact on contemporary science and technology, and the development of modern society.

References

[1]   李国栋，《当代磁学》，1990，中国科学技术大学出版社，p1。

[2]   宋德生、李国栋，《电磁学发展史》，1996，广西人民出版社，P1.

[3]   J. Verwell, in "Magnetic Properties of Matterals", ed. J. Smit, (McGraw-Hill, 1971), p64.

[4]   E. Roess, et al, Proc. Int. Conf. Ferrites, Kyoto, eds. T. Hoshion et al, (Univ. Tokyo Press, 1970), p187.

[5]   C. Guillaud et al, in "Solid State Physics", vol. 3, eds. M. Desirant and J. L. Michiels, Academic Press, New York, 1960), p71.

[6]   F. Roozeboom and F. W. A. Dime, J. Appl. Phys. **77**, 5293 (1995).

[7]   J. Huijbregtse et al, J. Appl. Phys. **83**, 1569 (1998).

[8]   黄波（主编），《固体材料及其应用》，广州，华南理工大学出版社，1994，第三章

[9]   李国栋，《磁性材料及器件》，No.2, 584 (1973).

[10]   H. Ohno, Science 281, 951 (1998).

[11]   Y. Matsumoto et al, Science **291**, 854 (1999).

[12]   R. M. Hornreich, IEEE Trans. Magn. **8**, 584 (1972).

[13]   李国栋, 自然杂志 **3**, 29 （1980）。

[14]   李国栋, 世界科技研究与发展 18, 50（1996）。

[15]   李荫远，李国栋，《铁氧体物理学》科学出版社，1962，1978。

[16]   H. Z. Wang et al, Proc. 9th Int. Conf. Microwave Ferrites, Esztergan-Hungary, 1988, p263.

[17]   李国栋等，物理学报 22，119 （1966）；科学通报 22, 1692 （1987）。

[18]   J. Smit and H. P. J. Wijn, in "Ferrite" (Philips Technical Library, Eindhoven, 1959).

[19]   M. N. Baibich et al, Phys. Rev. B **61**, 2472 (1988).

[20]   C. T. Yu et al, Phys. Rev. B **52**, 1123 (1995).

[21]   S. S. P. Parkin et al, Phys. Rev. Lett. **66**, 2152 (1991).

[22]   Y. Yoshizawa et al, J. Appl. Phys. 64, 6044 (1988).

[23]   E. Kneller and R. Hawig, IEEE Trans. Magn. 27, 3588 (1991).

[24]   A. Manaf et al, J. Magn. Magn. Mater. **128**, 302 (1993).

[25]   R. W. Chantrell et al, IEEE Trans. Magn. **14**, 975 (1978).

[26]   李国栋，《生物磁学及其应用》，科学出版社，1983。

[27]   李国栋等，《生物磁学—应用，技术，原理》，北京 国防出版社，1993。

[28]   D. Cohen, Science **156**, 652 (1967).

[29]   A. Abragam et al, in "Nuclar Magnetism: Order and Disorder", (Oxford Clarendon Press, 1982)

[30]   P. A. M. Dirac, Proc. Roy. Soc. **A133**, 60 (1931).

[31]   A. D. Raff et al, Bull. Geol. Soc. Am. **72**, 1267 (1961).

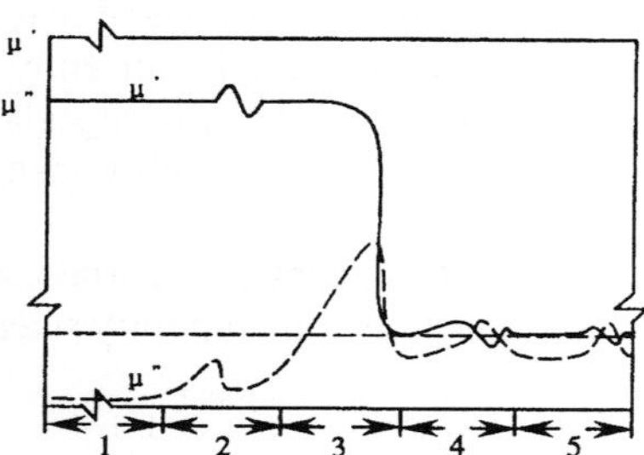

Figure 1. Permeability spectra of magnetic materials: (1) low frequency; (2) mid frequency; (3) high frequency; (4) microwave frequency; (5) extremely high frequency.

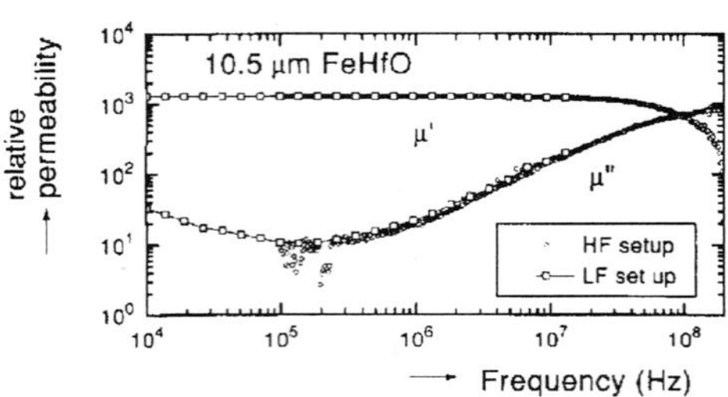

Figure 2. The complex permeability spectrum of Fe-Hf-O thin film. Film thickness is $10.5\mu m$.

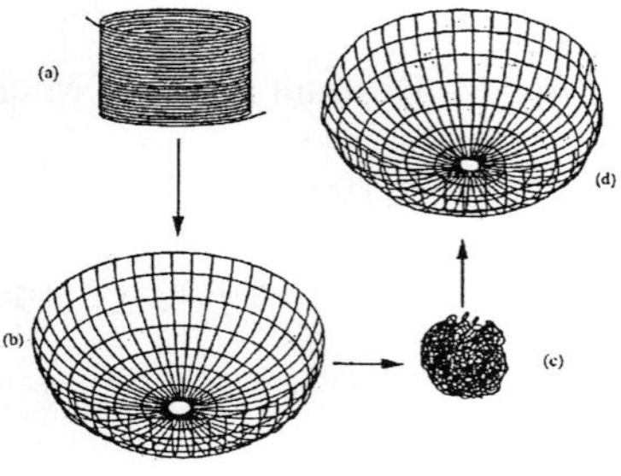

Figure 3. Antenna for use in spacecraft, made of Ni-Ti alloy with shape memory intelligence.

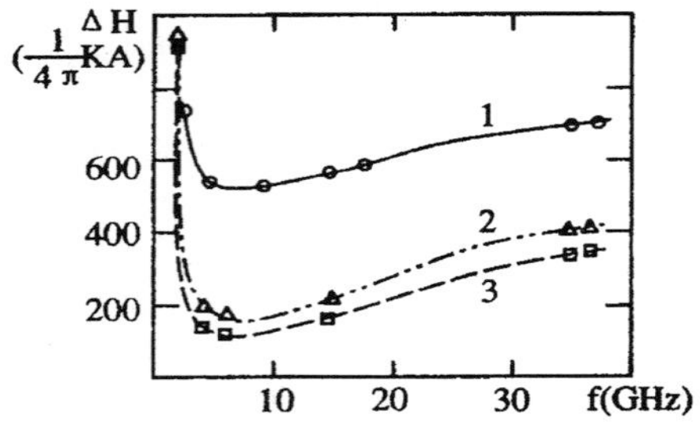

Figure 4. Ferromagnetic resonance linewidths ΔH as a function of frequency: 1. $NiFe_2O_4$; 2. $(Ni,Zn)Fe_2O_4$; and 3. $(Ni, Zn,Cu)Fe_2O_4$.

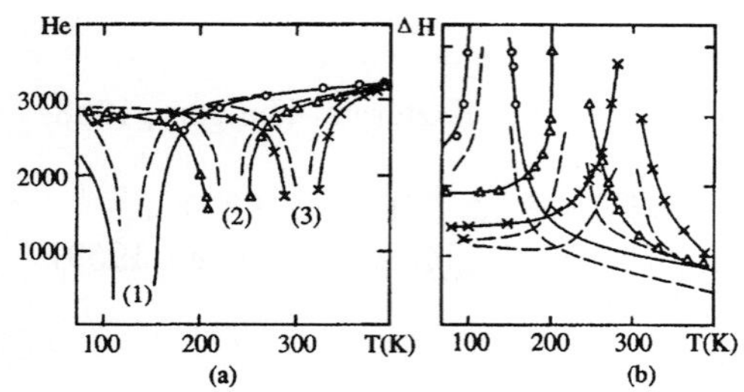

Figure 5. $Y_{3-3x}Gd_{3x}Fe_5O_{12}$ at 3 cm band: (a) ferromagnetic resonance field; (b) resonance linewidth ΔH; as a function of temperature.

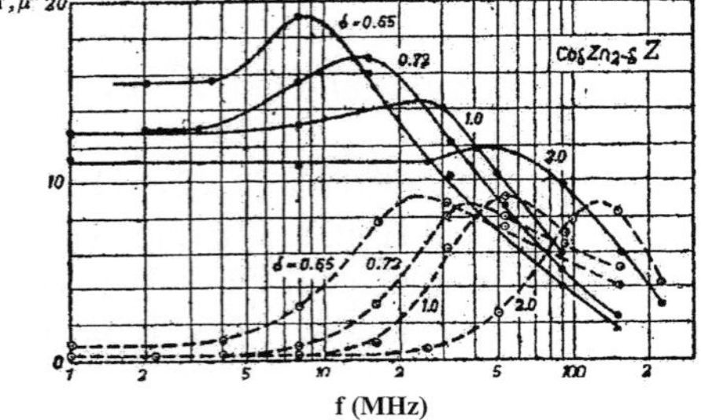

Figure 6. Microwave permeability spectra of the hexaferrite $Co_\delta Zn_{2-2\delta}Z$.

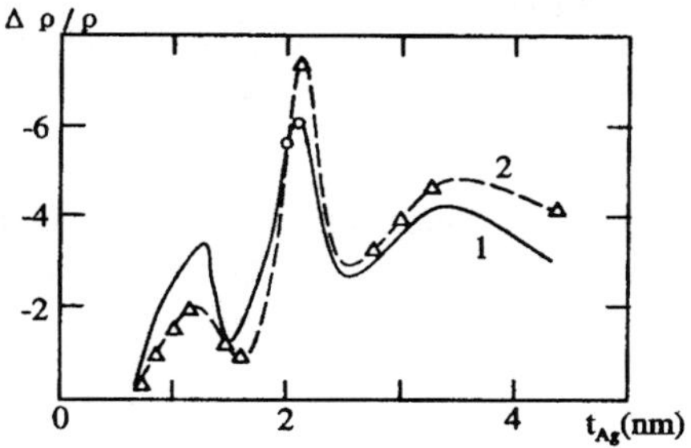

Figure 7. Relationship between the MR coefficient and the layer thickness of Ag for the multi-layer nano-fiilms [Fe(1 nm)/Ag(t$_{ag}$)]$_{40}$.

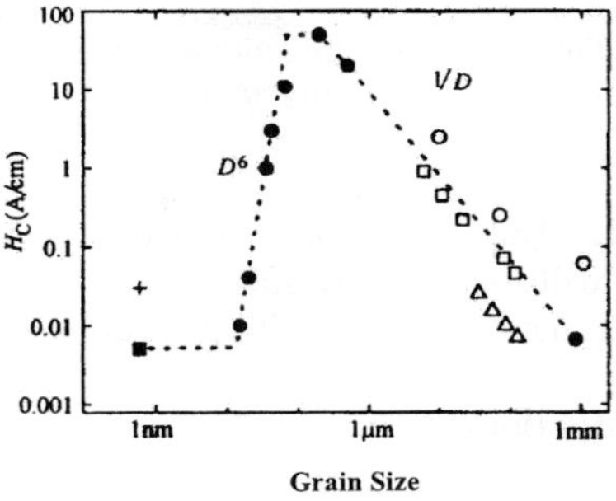

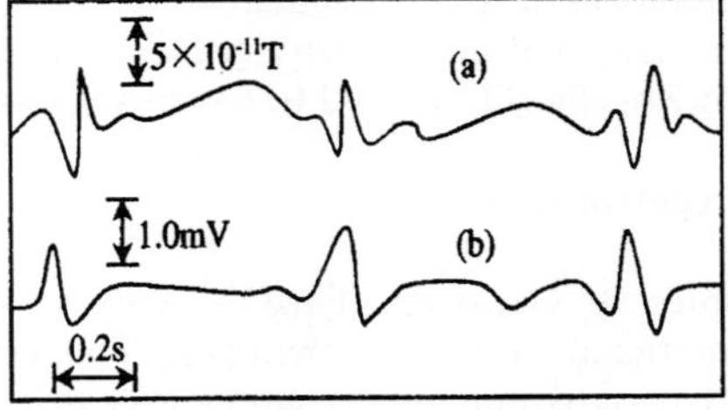

Figure 8. Relationship between Coercivity Hc and grain size.

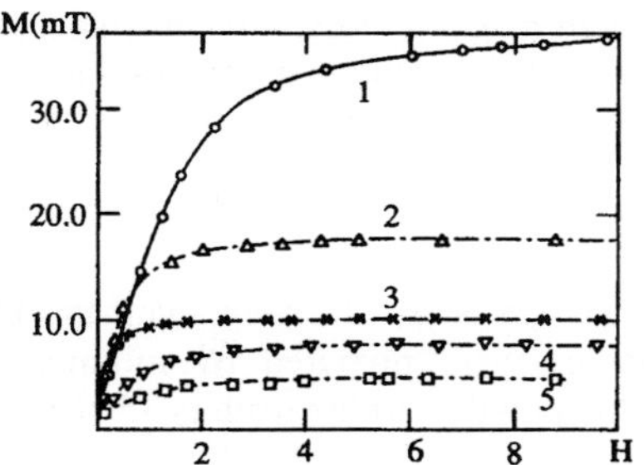

Figure 9. Magnetization curves of several magnetic liquids.

Figure 10. (a) Magneto-cardiogram, and (b) Electro-cardiogram, of a human.

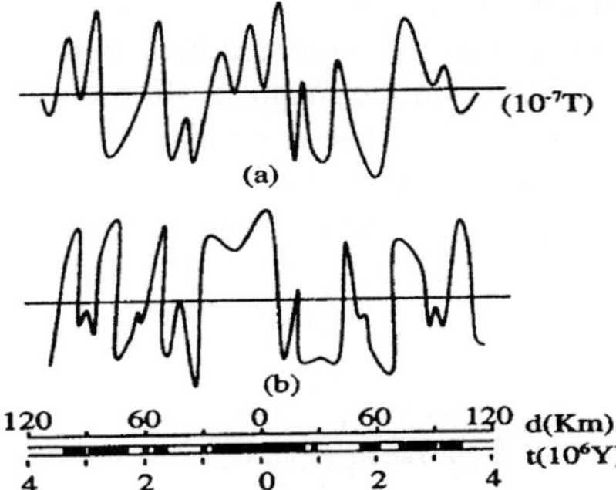

Figure 11. Geomagnetic fields on both sides of the Juan de Fuca ridge in East Pacific Ocean: (a) measured total magnetic field, (b) calculated results based on assumed model. The lower part of the figure shows the distance d from the ridge, and the corresponding expansion time t of rock on underwater.

Magnetic characteristics of $BaCo_xZn_{1-x}Fe_{16}O_{27}$ composites at microwave frequencies

Z. W. Li and Chen Lingfeng

Temasek Laboratories, National University of Singapore, 10 Kent Ridge Crescent, Singapore, 119260

Wu Yuping and C. K. Ong

Centre for Superconducting and magnetic Materials and Department of Physics, National University of Singapore, 10 Kent Ridge Crescent, Singapore, 119260

Abstract: The static and dynamic magnetic properties of $BaCo_xZn_{1-x}Fe_{16}O_{27}$ have been systematically studied. Results show that $BaCo_xZn_{1-x}Fe_{16}O_{27}$ is a potentially good candidate for electromagnetic materials with low reflectivity at microwave frequency

I. Introduction

Electromagnetic (EM) materials have been used extensively in defence, industry and commerce. Due to the strong attenuation on EM waves, the materials can be used in high-speed electronic circuits or other devices to minimize environmental EM interference, to decrease the noise level of signals, and to ensure electromagnetic compatibility. Such EM materials usually comprised of dielectric or magnetic fillers and polymer. Specially designed ferrite powders are ideal fillers for the development of wave attenuation materials, due to their high magnetic loss, high chemical stability, low density and high resistivity [1,2]. In this paper, we report the static and dynamic properties of $BaCo_xZn_{1-x}Fe_{16}O_{27}$ (x=0-2.0) composites for use as EM materials.

II. Experimental

Samples of $BaCo_xZn_{1-x}Fe_{16}O_{27}$ with x=0, 0.5, 0.7, 1.0, 1.5 and 2.0 were synthesized using conventional ceramic techniques. The composite samples were prepared by mixing the fine powders of barium ferrites with epoxy resin; the ratio of the powders to the resin is 35 % by volume.

The magnetization curves and *M-H* loops were measured with applied fields of 0-80 kOe, and between -30 and +30 kOe, respectively, using the VSM made by Oxford Instrument. Saturation magnetizations M_s were deduced from numerical analysis of the magnetization curves based on the law of approach to saturation. Coercivities H_c were obtained from the *M-H* loop. The values of the anisotropy fields were determined based on the magnetization curves parallel and perpendicular to the alignment direction for aligned samples. The complex permeability and permittivity over 0.5-16.5 GHz were measured using a HP8722D Vector Network Analyzer with TRL calibration. In addition, the complex permeability over 0.1-1.8 GHz was measured using the impedance method.

III. Results and discussions
3.1 Static magnetic properties

XRD of aligned samples showed that $BaCo_xZn_{1-x}Fe_{16}O_{27}$ has a c-axis anisotropy for x<0.7 and a c-plane anisotropy (or cone anisotropy) for x 0.7. The values of anisotropy fields H_a (for c-axis anisotropy) and H_θ (for c-plane anisotropy) are shown in Fig.1 (a). With increasing Co substitution, H_a decreases rapidly from 12 kOe at x=0 to 5.0 kOe at x=0.5. Higher Co substitutions modify the anisotropy from c-axis to c-plane. The values of H_θ increase from 4.5 kOe at x=0.7 to 21 kOe at x=2.0.

Saturation magnetization M_s and coercivity H_c, as a function of substitution x, are plotted in Fig. 1(b). With Co substitution, H_c reduces rapidly from 184(5) Oe at x=0 to 15(5) Oe at x=0.7. This result is attributed to a decrease in the anisotropy field H_a, and a change in the anisotropy from c-axis to c-plane. With higher substitution, H_c increases slightly to 76(5) Oe at x=2.0, which is due to an increase in the c-plane anisotropy. On the other hand, when the Co substitution is increased, M_s first increases, reaches a smooth maximum at x=0.7, and then decreases. This appears to be related to the distribution of Co and Zn ions at the seven sites in $BaCo_xZn_{1-x}Fe_{16}O_{27}$ ferrites.

3.2 Dynamic magnetic properties

Dynamic parameters of $BaCo_xZn_{1-x}Fe_{16}O_{27}$ composites are listed in Table I. The real and imaginary values of permittivity, ε' and ε'', are almost constant. The complex permeability spectra are shown in Fig. 2.

For c-axis anisotropy, there exist two resonance absorption peaks; one is at about 1 GHz, and the other is at 15 and above 16.5 GHz for composite with x=0.5 and x=0, respectively. The two absorptions correspond to the domain wall (low frequency) and natural (high frequency) resonances, respectively. On the other hand, for c-plane anisotropy, we observe only one absorption peak that corresponds to natural resonance.

For c-axis anisotropy, although the resonance frequencies are high, the static permeability μ'_0 and the maximum μ''_{max} are rather small. However, when the anisotropy is modified from c-axis to c-plane, the values of μ'_0 and μ''_{max} increase abruptly to 2.4-2.2 and ~0.8, respectively, for x=0.7 and 1.0. With increasing Co substitution, the resonance frequencies f_R are shifted to higher frequency, from 2.5 GHz at x=0.7 to 12 GHz at x=1.5.

3.3 Resonance frequency

It is found that, to a good approximation, the resonance frequency f_R varies linearly with the anisotropy field H_θ for composite with c-plane anisotropy, as shown in Fig. 3.
The natural resonance frequency f_R can be expressed as

$$f_R = \frac{1}{2\pi}\gamma\sqrt{H_\theta \cdot H_\phi} \tag{1}$$

for c-plane anisotropy [3]. H_θ and H_ϕ are the anisotropy fields along the c-axis and in the c-plane, respectively. The gyromagnetic ratio $\gamma/2\pi$ is equal to 2.8 MHz/Oe.

It is assumed that the anisotropy field H_ϕ is proportional to the field H_θ with a proportional coefficient β. Eq. (1) can be rewritten as $f_R=(1/2\pi)\gamma\beta^{1/2}H_\theta$. From the slope of the line, $\beta=0.07$ is obtained. Hence, the values of H_ϕ are estimated to be 0.32, 0.60, 1.2 and 1.9 kOe for samples with x=0.7, 1.0 1.5 and 2.0, respectively. Paoluzi et al [4] have observed that the magnitude of the third order magnetocrystalline anisotropy constant increases with Co substitution, which leads to an increase in the anisotropy field H_ϕ.

3.4 Attenuation properties

Electromagnetic wave attenuation is determined from the reflection loss (RL). In the case of a metal-backed single layer, RL is given by

$$\begin{cases} R.L = 20\log\left|\dfrac{Z_{in} - Z_0}{Z_{in} + Z_0}\right|, \\[2em] Z_{in} = Z_0\sqrt{\dfrac{\mu}{\varepsilon}}\tanh(j\dfrac{2\pi ft}{c}\sqrt{\mu\varepsilon}) \end{cases} \tag{2}$$

where Z_{in} is the impedance of the composites, $Z_0=1$ is the impedance of free space, c is the light velocity, t is the thickness of composite, and μ and ε are complex permeability and permittivity. The optimum reflection loss is related to the thickness of the composites. Fig. 4 shows the frequency response of the reflection loss for $BaCo_xZn_{1-x}Fe_{16}O_{27}$ composites. The maximum RL is -35 dB and the band-widths for 10dB attenuation are 4.8 and 7.0 GHz for x=0.7 and 1.0, respectively.

IV. Conclusions

For $BaCo_xZn_{1-x}Fe_{16}O_{27}$, the optimum soft magnetic properties are obtained at x=0.7 and 1.0, which have c-plane anisotropy. Correspondingly, the maximum μ'_0 and μ''_{max} are 2.4-2.2 and ~0.8, respectively. Co substitution can modify the anisotropy from c-axis to c-plane at x=0.6. With Co substitutions, the anisotropy field H_θ increases and the resonance frequency f_R shifts to higher frequency for c-plane anisotropy. Based on the measurement of dynamic properties and theoretical estimations on the reflection loss, $BaCo_xZn_{1-x}Fe_{16}O_{27}$ is found to be a potentially good candidate for electromagnetic materials with low reflectivity at microwave frequency.

References

[1] T. Nakamura, T. Tsutaoka and K. Hatakeyama, J. Magn. Magn. Mater. **138**, 319 (1994)
[2] Han-Shin Cho and Sung-Soo Kim, IEEE Trans. Magn. **35**, 3151 (1999)
[3] J.L. Snoek, Physica XIV, 245 (1948); G. T. Rado, Rev. Mod. Phys. **25**, 81 (1953)
[4] A. Paoluzi, F. Licci, O. Moze, G. Turilli, A. Deriu, G. Albanese and E. Calabrese, J. Appl. Phys. **63**, 5074 1988.

Table I: Dynamic properties for $BaCo_xZn_{1-x}Fe_{16}O_{27}$ composites

x	Anisotropy	f_R (GHz)	μ_0'	μ_{max}''	ε'	ε''
0	c-axis	0.9, >16.5	1.38	0.20	6.4	0.3
0.5	c-axis	1.2, 13.5	1.64	0.65	6.5	0.4
0.7	c-plane	2.5	2.38	0.80	5.9	0.7
1.0	c-plane	6.5	2.20	0.85	6.2	0.4
1.5	c-plane	12.0	1.75	0.60	6.0	0.5
2.0	c-plane	>16.5	1.42		6.0	0.4

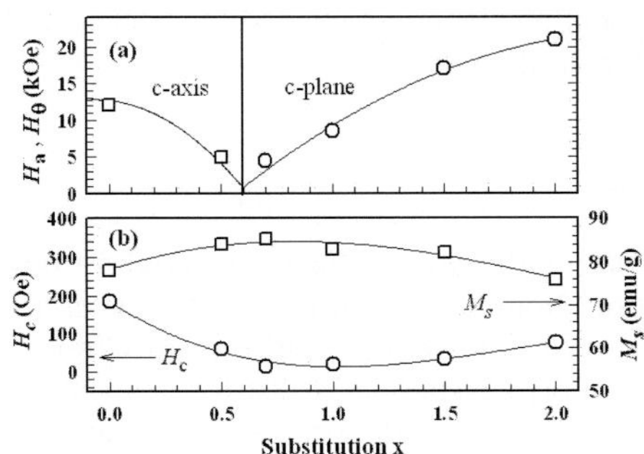

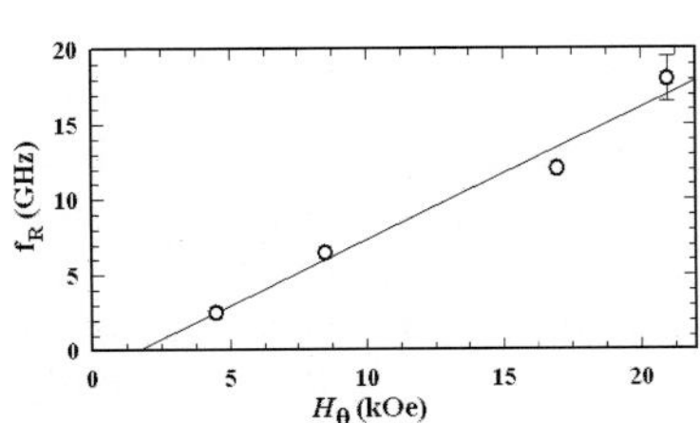

Fig.1 (a) Anisotropy fields, (b) Saturation magnetizations and coercivities for $BaCo_xZn_{1-x}Fe_{16}O_{27}$

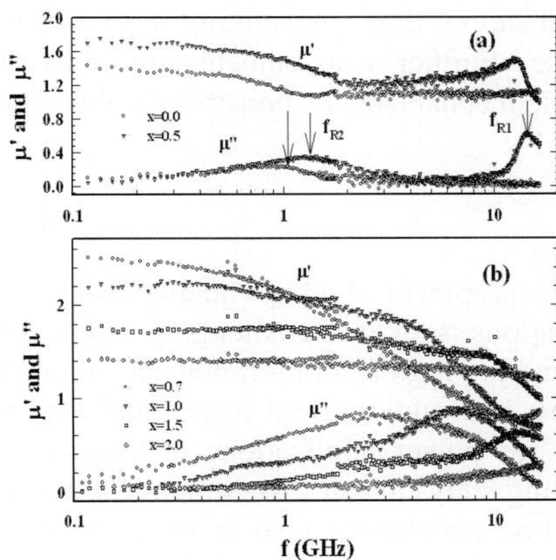

Fig.2 Permeability spectra for $BaCo_xZn_{1-x}Fe_{16}O_{27}$ composites with (a) c-axis and (b) c-plane anisotropies. The volume concentration of ferrite powders is 35%.

Fig.3 Linear relationship between resonances frequencies and anisotropy fields for $BaCo_xZn_{1-x}Fe_{16}O_{27}$ composites.

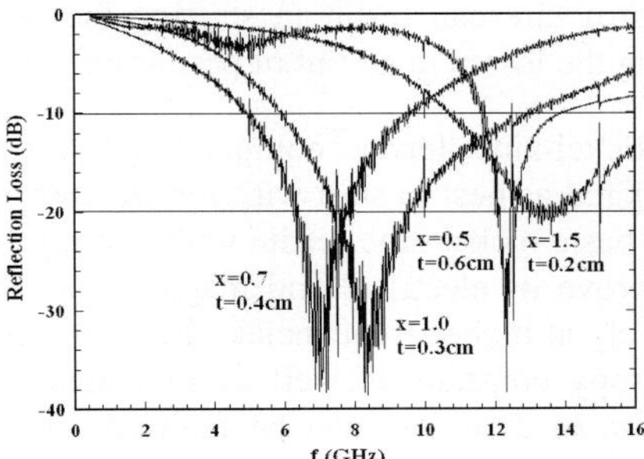

Fig.4 Reflection loss for $BaCo_xZn_{1-x}Fe_{16}O_{27}$ Composites with 35% volume concentration of ferrite powders

Development of Loss-less High Frequency Ni-Zn Ferrites

B.Parvatheeswara Rao, R.V. Rao, P.S.V. Subba Rao and K.H. Rao
Department of Physics, Andhra University, Visakhapatnam 530003, India.
E mail: < bprao25@rediffmail.com > ; Fax: +91-891-2755547.

Abstract- By making careful choice of chemical compositions and preparatory conditions, poly crystalline high saturation magnetisation nickel-zinc ferrites with fixed quantity of titanium and varied quantities of indium have been studied to examine their utility for high frequency power applications. Preliminary results indicate definite improvements in resistivity and saturation magnetisation, thus contributing to minimise core losses- an important requirement for the defined purpose. Further measurements are underway. The present results are explained by discussing the possible mechanisms responsible for the observed positive variations.

I. Introduction

Development of a high quality cost effective loss-less high frequency ferrite material for power applications is an ever challenging aspect for investigation. Important requirements for materials of this kind are high saturation magnetisation and low core losses. Core losses arise mainly from two sources, i.e., eddy current losses and hysteresis losses. Poly crystalline Ni-Zn ferrites find their utility for power applications at frequencies over 1 MHz because they have the advantage of curtailing eddy current loss component due to their higher electrical resistivities. In order to keep the other loss component also at low, one must have higher magnetic permeabilities [1] which can however, partially offset the advantage drawn earlier and set a limit to the driving frequency at which the material can be operated. Therefore, it appears at first difficult to achieve low hysteresis loss and high switching frequency simultaneously. But according to the formulaes derived in the model works [2-4], a careful choice of composition with high saturation magnetisation and fine microstructure can result in shifting the operating frequency relatively to higher limit while not causing the losses to go out of permissible limits.

Nickel-zinc ferrite composition with high room temperature saturation magnetisation is considered as best to start with for the above study. Compositional modifications have been aimed to the basic nickel-zinc ferrite with varying concentrations of indium and fixed quantity of titanium to improve its electrical and magnetic properties, so that the material can be expected to operate relatively at higher frequencies. In^{3+} ions are preferred for substitution by expecting a reduction in anisotropy constant as well as in magnetostriction [5]; thus an increase in permeability and a decrease in core loss can be realised. On the other hand, the choice of titanium ions in small amounts is made because of their ability to improve both electrical and magnetic properties by locking-up pairs of Ti^{4+} - Fe^{2+} and thereby contributing to decrease the total loss component [6]. This paper discusses the results of resistivity and saturation magnetisation of the system under investigation and predicts the utility of these materials for high frequency power applications.

II. Experimental Details

Polycrystalline Ni-Zn ferrites with the formula $Ni_{0.65} Zn_{0.375} Fe_{1.95-x} In_x Ti_{0.025} O_4$ where x values ranging from 0.00 to 0.20 in steps of 0.05 have been prepared by conventional ceramic technique using the procedure described elsewhere [7]. Sintering of the samples has been carried out at 1250° C for 4 hours in air atmosphere followed by natural cooling. X-ray studies confirm single phase spinel structure in all the samples.   Characterisation of the samples was done by measuring the lattice constant and Curie temperature of the composition $Ni_{0.65}\,Zn_{0.35}\,Fe_2\,O_4$ prepared under similar conditions and these are found to be in good agreement with the respective parameters of the same composition reported earlier [8].

DC resistivity and saturation magnetisation measurements have been made on the samples by standard two-probe method and the Ponderometer method respectively. Bulk density measurements have been made by using Archimedes principle and these are approximately 96% to their theoretical limits.

III. Results & Discussion

Fig 1 shows the variations of DC resistivity with indium concentration. Though the variations are small for lower indium concentrations, the substitutions have resulted higher DC resistivities with the increase in x.

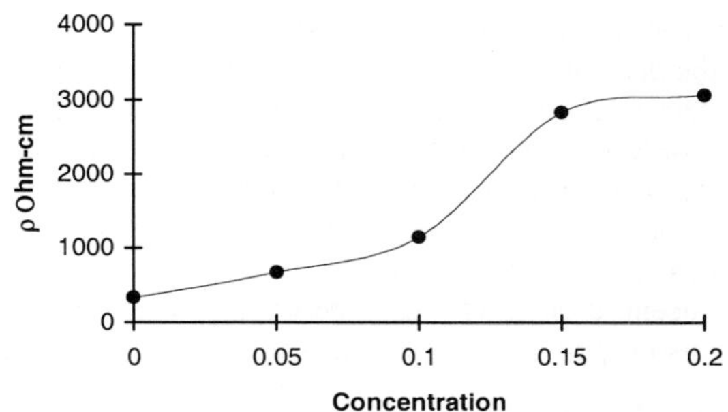

Fig 1. Variation of resistivity with In^{3+} concentration in the system $Ni_{0.65}\,Zn_{0.375}\,Fe_{1.95-x}\,In_x\,Ti_{0.025}\,O_4$

Electronic conduction in ferrites predominantly takes place between Fe^{2+} and Fe^{3+} ions. Since ferrous ions are essential to realise low magnetic losses and high saturation magnetisations [9], the preparation of these ferrites is made in such a way that the final product definitely contains certain quantity of divalent iron. This leads to increase hopping probability between Fe^{2+} - Fe^{3+} ions. However, with increased concentration of indium in place of iron, the available quantity of iron after each substitution becomes lesser and lesser and also the possibility of formation of ferrous content. Even though the introduced fixed quantity of Ti^{4+} ions lock up the Fe^{2+} ions, they may not be able to prevent all ferrous ions from participating in hopping processes with ferric ions at lower concentrations of indium; thus resulting lower resistivities up to $x \leq 0.1$. However, the increase in resistivity with the increase of x may be due to decrease of ferrous content as well as larger inter atomic separations due to successive insertion of larger ionic radii spheres of indium after each substitution. At higher concentrations, it appears that the quantities of Fe^{2+} and Ti^{4+} ions are well-balanced and leading to higher resistivities.

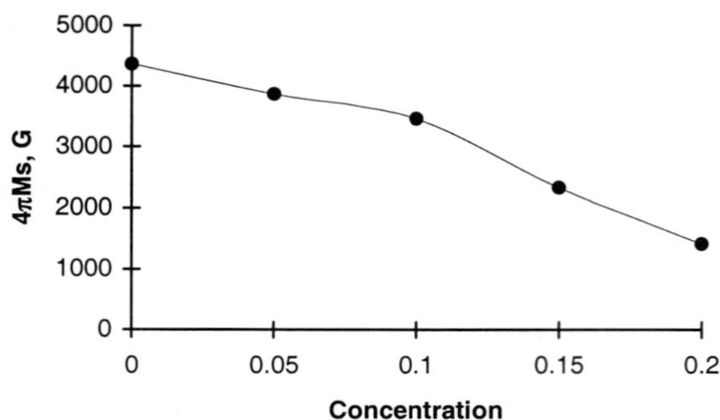

Fig 2. Variation of saturation magnetisation ($4\pi Ms$) with In^{3+} concentration in the system $Ni_{0.65} Zn_{0.375} Fe_{1.95-x} In_x Ti_{0.025} O_4$

Fig 2 shows the variations of saturation magnetisation with x. Though the Ms values seem to be in a decreasing trend, the decrease is not found to be prominent at lower concentrations. Indium ions are known to prefer tetrahedral sites initially. This may lead to forcing the Fe^{3+} ions there to move into octahedral sites only to be filled there in the vacant sites arising out of substitutions. As the tetrahedral Fe^{3+} ions decrease in number with the increase in substitution, the octahedral Fe^{3+} ions line-up in anti parallel and preventing the system to enhance its net magnetisation. However, magnetisation contributions due to Fe^{2+} ions in octahedral sites prevent rapid fall of net magnetisation at lower concentrations. The saturation magnetisation variations are in accordance with these arguments and also support the arguments made earlier regarding resistivity variations.

Hysteresis loops with lesser areas with the increasing concentration of indium have been observed and the results are under analysis. Similarly, permeability measurements and core loss characteristics at higher frequencies and at higher field strengths and scanning electron micrographs for all the samples are underway and final results will be presented as a separate communication.

IV. Conclusions

In conclusion, the preliminary results of the carefully planned and prepared nickel-zinc ferrite system with simultaneous substitutions of titanium and indium have been found to be quiet encouraging and contribute to definite improvements in resistivity and thereby in core losses while retaining the magnetisation relatively at higher levels; thus finding the material its utility for power applications at higher frequencies. However, a complete picture corroborating its usefulness will be made known only after evaluation of all the results for which the measurements are underway. In view of the fact that ferrite materials find their utility in large number of power applications of newer technologies with less expensive packages, the observations made by the authors in this study carry enough significance.

REFERENCES

[1] Smit and H.P.J. Wijn, Ferrites, Philips Technical Library, Eindhoven (1959) 73; R. Becker, Zeit. f. Phys. **33** (1932) 905.

[2] A Globus, Thesis, Paris (1963).

[3] Guyot and V. Cagan, J. Magn. Magn. Materials **27** (1982) 202.

[4] L. Snoek, Physica **14** (1948) 202.

[5] Krishnan and V. Cagan, IEEE Trans. Magn. MAG-7 (1971) 617.

[6] Neamtu, M. I. Toacsen and D. Barb, J. Phys. IV France 7 (1997) Suppl. C1:79.

[7] B.Parvatheeswara Rao and K.H.Rao, J.Appl.Phys. **80** (1996) 6804.

[8] B.Parvatheeswara RAO, P.S.V.Subba Rao and K.H.Rao, IEEE Trans. Magn. **33** (1997) 4454.

[9] Lebourgeois, J. P. Ganne and B. Lloret, J. Phys. IV France 7 (1997) Suppl. C1:105.

Interpretation of microstructure dependent magnetic properties of cobalt substituted lithium ferrites

Watawe S C [*1], Bamne U A[1] and Sarwade B D[2]

1. Lokmanya Tilak Institute of Post Graduate Teaching and Research, Gogate Jogalekar college, Ratnagiri, Maharashtra, India
2. National Chemical Laboratory (N. C. L.) Pune, India

*Corresponding author: Tel 91-02352-226603, Fax 91-02352-241580, email: shrikantwatawe@yahoo.com

Abstract

The magnetic properties of $Li_{0.5-x/2}$ Co_x $Fe_{2.5-x/2}$ O_4 (where x=0.1,0.2,0.3,0.4,0.5,0.6,0.7) ferrites have been studied. The x-ray diffraction studies reveal the single phase formation of the ferrite samples. The SEM micrographs show that the grain size decreases up to x=0.4 and increases for x>0.4. The grain size for all the samples lies in the range 0.2 μm and 6.7μm. The permeability studies at room temperature show that the values lie in the range of 5.5 to 20.6. The AC susceptibility measurements show a single domain structure. Usually the single domain structure is present in case of grain size < 0.1 μm. Both these variations suggest the presence of non- magnetic grain boundaries (NMGB) which involves deposition/ segregation of non magnetic particles over the surface of grains which reduce effective grain size. An attempt has been made to calculate the effective grain size using the data obtained from the measurements of initial permeability with temperature and the grain size obtained from the SEM micrographs.

KEYWORDS: Ferrites; AC susceptibility; SEM; NMGB model

Introduction

Lithium ferrites have been used as cost effective replacements for garnets in the microwave frequency applications. The lithium ferrites have attracted considerable attention due to their squareness of hysteresis loop and higher Curie temperature [1]. The Co^{2+} ions are known to change magnetocrystalline anisotropy because of the non-quenched orbital angular momentum. The Co^{2+} ions being a fast relaxing ion, enhances the microwave properties. The work on lithium ferrites containing various ions such as Cd, Mg, Zn, etc [2-4] has been reported. However, not much has been reported on magnetic properties of cobalt substituted lithium ferrites [5-9]. We report here the microstructure dependence of permeability and AC susceptibility of cobalt substituted lithium ferrites.

Experimental

The Li-Co ferrites having general formula $Li_{0.5-x/2}$ Co_x $Fe_{2.5-x/2}$ O_4 (where x = 0.1, 0.2, 0.3, 0.4, 0.5, 0.6, 0.7) were prepared by standard double sintering ceramic method using AR grade Li_2CO_3, CoO and Fe_2O_3 as starting materials. The oxides were mixed in stoichiometric proportions and the mixtures were presintered at 873 K for 10 hrs. The resultant powder was granulated into fine powder in agate morter and further subjected to final sintering at 1273 K for 24 hr and the samples were furnace cooled at the rate of 80 K/hr. The final sintering temperature was carried out at 1273 K considering the fact that lithium evaporates above this temperature [10-11]. The completion of solid state reaction was confirmed by X-ray diffraction studies carried out on powder samples with Philips PW-1710 diffractometer using Cu-Kα (λ = 1.5418 A°) radiation. and IR spectra recorded using Perkin- Elmar IR spectrometer (model −783). The scanning electron micrographs (SEM) were taken using LICER Stereoscan 440 model. The variation of initial permeability with temperature was measured using LCR-Q meter (Aplab-4912) at a fixed frequency of 1 KHz. The XRD patterns of the samples confirm single phase formation of the ferrites. The low field AC susceptibility measurements were done using a double coil apparatus designed by T.I.F.R. Mumbai operating at 260 Hz in a r.m.s. field of 7Oe.

Results and discussion

Fig.1 shows the variation of lattice constant (a) with increase in cobalt content (x). A linear increase in lattice parameter indicates that the system obeys Vegards law. The increase in lattice constant with cobalt can be attributed to the fact that Co^{2+} (0.82 Å) has larger ionic radii than the radii of Fe^{3+} (0.645 Å) and Li^{1+} (0.74 Å) ions that it replaces. The present values of lattice parameter compare well with the values reported earlier for Li – Co ferrites [5-9].

The SEM micrographs of the finally sintered samples of different composition (reported earlier [6]) indicate the distribution of grains with non-uniform size and accumulation of some small sized particles on the grain boundaries. The features are similar to the one reported for other lithium ferrites [2,4] sintered at lower sintering temperatures used to avoid lithium losses at higher temperature [10,11]. The average grain size of the samples calculated using the line intercept method using the SEM micrographs. The variation of grain size with content of cobalt (x) is shown in Fig 2. The grain size decreases up to x=0.4 and then it increases with further increase in x. This variation in grain size can be attributed to the grain growth mechanism involving diffusion coefficients, sintering temperature and the concentration of dissimilar ions [2,6]. The grain growth mechanism is compromised between driving force for grain boundary movement and retarding force of pores and inclusions during the sintering process. The strength of the driving force, generated due to thermal energy, depends upon sintering temperature and diffusivity of constituent ions.

Fig. 3 shows the variation of initial permeability μ_i with cobalt content room temperature. The values are found to be in the range 5.5 to 20.6 for all the samples. The initial permeability is found to decrease up to x=0.4 and then slight increase is observed for the samples with x > 0.4. Such a variation can be attributed to the dependence of initial permeability on M_s and the grain size according to Globus relation [12].

$$\mu_i = M_s^2\, dm \,/\, K_1$$

where M_s is saturation magnetization, dm the grain size and K_1 the anisotropy constant. The lower values of the initial permeability can be attributed to positive anisotropy constant of Co^{2+} ions and the reduction in effective grain size due to the non-magnetic grain boundary. It is confirmed by some workers [13] that the non-magnetic particles and trace amounts of impurities mostly segregate on the grain boundaries forming a non-magnetic grain boundary. The thickness of the NMGB is estimated by combining the Globus theory [12] and NMGB model proposed by Vissar *et al* [14] using the relation

$$D\,/\,\mu_e = D\,/\,\mu_i + \delta$$

where D is the mean grain diameter, μ_e is the effective permeability, μ_i is the intrinsic permeability of the grains and δ is the thickness of the NMGB. Using the data from the variation of initial permeability with temperature shown in Fig. 4, the $D\,/\,\mu_e$ values of the samples have been estimated at different temperatures, up to the point where μ_i drops off. The values of $D\,/\,\mu_e$ are found to decrease with temperature and the asymptotic limits of each curve i.e. the lowest value is considered as the thickness of NMGB, are shown in Fig.5 and Fig 6. The thickness of NMGB for the samples with x=0.1, 0.2, 03 are found to be in the range of 27 nm to 40 nm, while for the samples with x>0.3 they are in the range 115 nm to 144 nm. In the present samples, the thickness of grain boundary seems to be large in comparison with the small sized magnetic particles. It is assumed that the magnetic ordering can be broken up easily at the NMGB's and it results in short magnetic exchange interaction. The samples of the same composition are found to possess different values of initial permeability because of the difference in the NMGB thickness [2,13]. The values of initial permeability seem to be largely affected by the non-magnetic grain boundary as the effective grain size is decreased.

The variation of normalized AC susceptibility ($\chi_T\,/\,\chi_{RT}$) with temperature is shown in Fig 6. All the samples show an increase in AC susceptibility with temperature. The variation has been used by many researchers to indicate the nature of magnetic particles in ferrites [14-17]. According to these reports the ratio χ_T / χ_{RT} remains independent of temperature for ferrites with MD state of particles, slowly rises to a peak for SD particles and decreases continuously for the SP particles. In the present case it can therefore be concluded that the present nature of χ_T / χ_{RT} curves indicate the presence of SD particles in our samples. Usually SD structure is observed in case of samples with grain size less than 0.1 µm. The higher content of cobalt is further found to favour the formation of SD state in the samples due to its inherent positive anisotropy constant. This behaviour may also be supported by the small effective grain diameter due to the presence of non-magnetic grain boundaries.

Conclusions

The lower values of initial permeability and single domain structure in the large grain size samples can be attributed to the NMGB model which reduces the effective grain size.

References

1. D. Ravinder, J. Appl. Phys. 75 (10) (1994)
2. S.S. Bellad, S.C. Watawe and B.K. Chougule, J. Magn. Magn. Mater. 195, (1999), 57
3. S.S. Bellad, S.C. Watawe, A.M. Shaikh, B.K. Chougule, Bull. Mater. sci. 23 (2) (2000) 83
4. A.M. Shaikh, S.A. Jadhav, S.C. Watawe, B.K. Chougule, Mater. Lett., 44 (2000)192
5. J.M. Song, J.G.Koh, IEEE Trans. Magn. 32 (1996) 411
6. S.C. Watawe, B.D. Sarwade, S.S. Bellad, B.D. Sutar, B.K. Chougule, Mater. Chem. Phys. 65 (2000) 173
7. S.C. Watawe, B.D. Sarwade, S.S. Bellad, B.D. Sutar, B.K. Chougule, J. Magn. Magn. Mater. 214 (2000) 55
8. S.C. Watawe, B.D. Sutar, B.D. Sarwade, B.K. Chougule, Intl J. of Inorganic Materials. 3(2001) 819
9. K Mohan, Y.C. Venudhar, J of Mater. Sci Lett. 18 (1999) 13
10. D.H. Redley, H. Lessoff, J.D. Childless, J. Am. Ceram. Soc. 53 (1970) 304
11. A.J. Pinton, R.C. Saull, J. Am. Ceram. Soc. 52 (1969) 157
12. A. Globus, J. Phys. (Suppl) C1 (1977) 1
13. E.G. Vissar, M.T. Johnson, P.J. Van Der Zaag, Ferrites: Proc. ICF-6, Japan, 1992, p.807
14 G.J. Baldha , R.V. Upadhy and R.G. Kulkarni, Mater. Res. Bull., 21 (1986) 1051
15. A.B. Naik, S.R. Sawant, S.A. Patil and J.I. Powar, Bull. Mater. Sci., II (4), (1988) 315
16. Sawant S. R., Birajdar D. J. and Patil R. N. Indian J. Pure and Appl. Phys. 28 (1990) 424
17. Unnikrishan S. and Chakravorty D. K., Physica Status Solidi (a) 121 (1990) 265

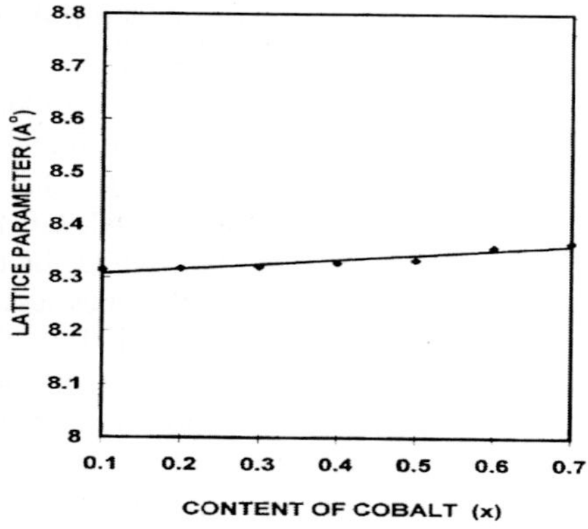

Fig.1: Variation of lattice parameter 'a' with Co content.

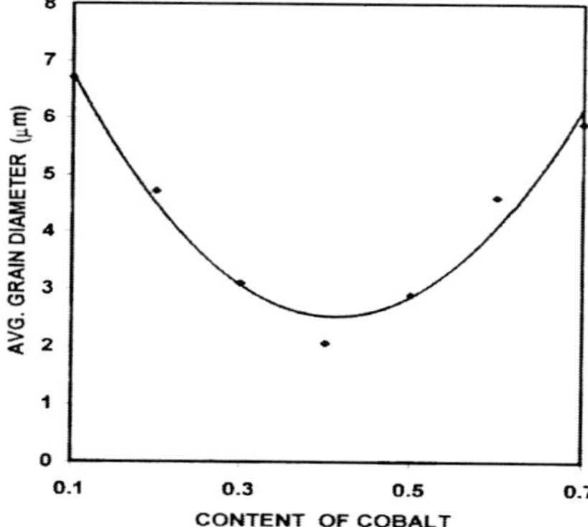

Fig.2: Variation of average grain diameter with Co content.

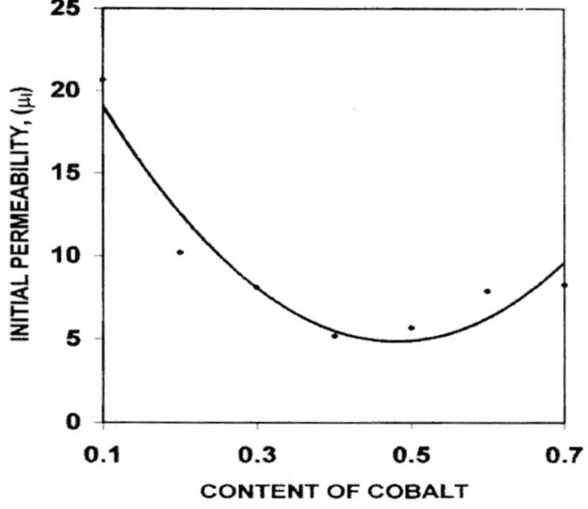

Fig.3: Variation of initial permeability with Cobalt content.

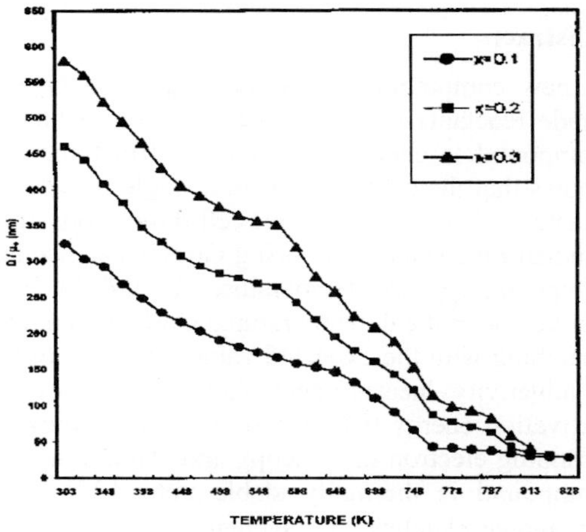

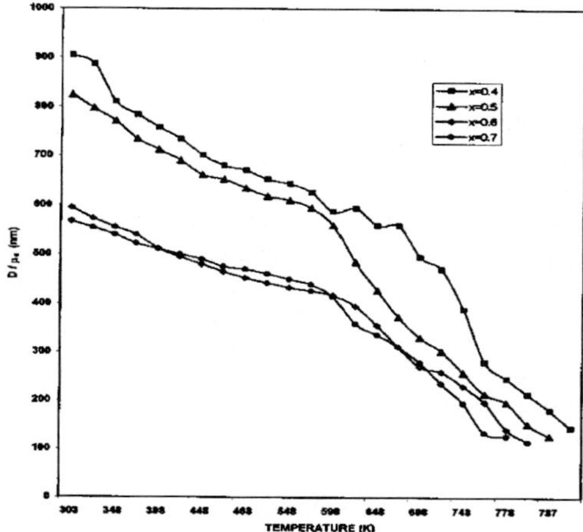

Fig.5: Variation of non-magnetic grain boundary with composition for the samples with x = 0.4, 0.5, 0.6 and 0.7.

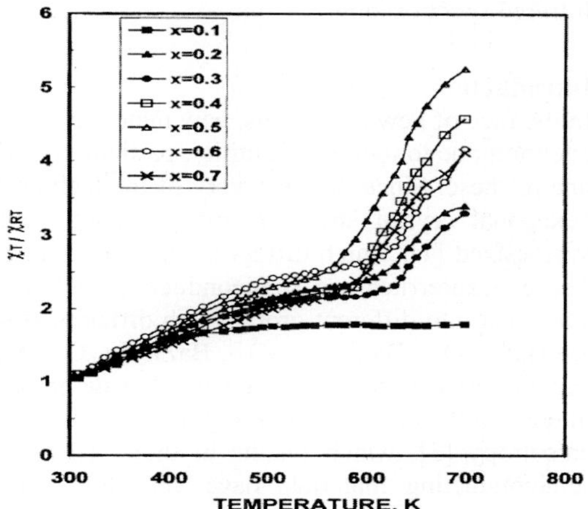

Fig.4: Variation of non-magnetic grain boundary with composition for the samples with x=0.1,0.2 and 0.3.

Fig.6: Variation of normalized AC susceptibility with temperature

TRANSPORT AND MAGNETIC PROPERTIES OF $SrTi_2Zn_2Fe_{10}O_{22}$ HEXAFERRITE

M.Y. Salunkhe, A.S. Potle, S.N. Giradkar,
S.B. Kondawar and **D.K. Kulkarni**
Institute of Science, R.T. Road, Nagpur, India.
Email: nimahsal@yahoo.com

Abstract

A new compound with a chemical formula $SrTi_2Zn_2Fe_{10}O_{22}$ has been synthesized using analar grade reactants by solid-state diffusion technique at $1173^{\circ}K$ for 100 hour. The formation of the compound is checked by X-ray diffractometry with filtered Cu radiation. The structural result shows that the compound is in a single hexagonal phase without traces of unreacting phases of the reactants. The observed unit cell dimensions are a = 5.891 Å and c = 46.064 Å. The compound is studied magnetically by using Gouy's balance and the result shows that it is ferrimagnetic at room temperature (Curie temperature $T_C=334^{\circ}K$). From the paramagnetic behaviour of the compound above the critical temperature, Curie molar constant is worked out and is found to be 42.67, matching with the expected value for stable oxygen states of the cations in the molecule. Electrical conductivity measurement shows slight deviation from linearity near Curie temperature with activation energy 0.44 eV. In addition to these, the compound is analyzed by thermo-gravimetry, scanning electron microscopy and infrared absorption. The thermo-gravimetric analysis shows the compound is thermally stable upto $1273^{\circ}K$; the microstructural study (SEM) shows clearly hexagonal platelets with average particle size 2.263 µm where as the infrared absorption gives two absorption peaks as observed in spinel structures in the range 400 cm^{-1} to 4600 cm^{-1}, which may be due to octahedral and tetrahedral complexes present in the compound.

Keywords: Hexaferrite, XRD, magnetic susceptibility, electrical conductivity, TGA, SEM, Infrared spectroscopy.

Introduction

In the race of new inventions, new materials of technological importance have acquired much more attention due to their direct utility, particularly in the field of materials science. Hexagonal ferrite is one of these. Since the discovery of the technically very important ferrite $BaFe_{12}O_{19}$ in 1951 with hexagonal crystal structure, many related compounds with totally new structures have been synthesized [1], which differs in stacking sequence of blocks, symmetry, size of the unit cell etc. These hexaferrites are semiconducting and form a large family of compounds. These can be classified into different groups with different stoichiometry. Amongst these hexagonal ferrites few are $BaFe_{12}O_{19}$, $Ba_2Zn_2Fe_{12}O_{22}$, $BaZn_2Fe_{16}O_{27}$, $Ba_3Zn_2Fe_{24}O_{41}$, etc [2]. The chemical compositions and the crystal structures of these hexagonal compounds show many similarities. The technical interest in the oxides with hexagonal crystal structures are due to their very high uniaxial magnetic anisotropy [3], which results in their use as permanent magnetic ceramic materials. As these semiconducting materials have very low conductivity at room temperature, they have high frequency applicability in various areas.

Experimental
Sample Preparation:

The oxide with a chemical composition $SrTi_2Zn_2Fe_{10}O_{22}$ was synthesized by a standard ceramic technique. This technique requires a high temperature with a long heating period yet it gives sufficiently homogeneous polycrystalline sample. The analar grade reactants $SrCO_3$, TiO_2, ZnO and Fe_2O_3 were mixed in proper molar ratio and ground in acetone for 6 hours. It is known that the particle size plays an important role in the formation of the compound; the necessary care has been taken by continuous and uniform grinding. The powder was calcined at 1073°K for 24 hours and then ground in agate mortar. After this, the powdered compound is palletized under pressure 8 psi for 10 minutes with PVA as a binder. These pellets then sintered in electrically operated furnace at 1173°K for 100 hours, which then cooled at a rate of 30°C per hour up to 773°K and then furnace cooled to room temperature.

Measurements:

The X ray diffraction spectra was taken at room temperature (RT) using powdered sample on Philips PW – 1710 X-ray diffractometer with Cu-Kα radiation. The structural details viz. d_{obs}, d_{cal}, observed intensity and the reflection indices are included in Table 1. The true density of the compound was performed on micrometrics multi-volume helium Picnometer-1305.

The magnetic susceptibility measurements were performed on Gouy's apparatus in the temperature range 300°K to 800°K. The Curie temperature and Curie molar constant are also estimated. The electrical conductivity measurements were carried out using digital LCR meter by two probes up to 500°K.

The T.G.A. measurements were carried out by STGA/DTA Toledo Switzerland, Mettler-835 apparatus. The scanning electron microscopy was carried out on a Jeol instrument JXA, 840-A operating at 5 KeV. Infrared absorption study of the compound was made according to Mazen's technique [4] by mixing with powdered KBr in the ratio 1:100. The I.R. spectrum was obtained on Infrared spectrophotometer FTIR-8001 in the range 400 cm^{-1} to 4600 cm^{-1}.

Table 1: X-ray diffraction data of $SrTi_2Zn_2Fe_{10}O_{22}$

$d_{obs.}$	$d_{cal.}$	$I_{obs.}$	h	k	L
3.0671	3.0673	4.7	0	1	12
2.9455	2.9455	50.3	1	1	0
2.7651	2.7651	94.8	0	1	14
2.6230	2.6223	100.0	1	1	8
2.5471	2.5469	17.9	0	2	1
2.5082	2.5073	13.7	1	0	16
2.4206	2.4207	39.6	2	0	6
2.3361	2.3368	8.5	1	1	12
2.2332	2.2315	19.8	0	2	10
2.1265	2.1257	14.9	1	1	15
1.6963	1.6937	8.2	2	1	13
1.6644	1.6632	38.3	2	0	21
1.6601	1.6603	22.2	3	0	6
1.6318	1.6330	20.2	1	2	15
1.6191	1.6184	37.7	2	0	22
1.4732	1.4728	42.4	2	2	0
1.4694	1.4642	26.9	3	0	16
1.3003	1.3046	12.6	2	1	26

Lattice Parameters: a = 5.891 Å, c = 46.064 Å.

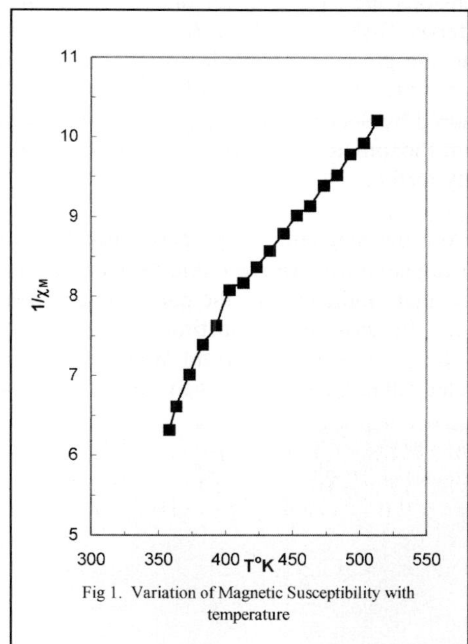

Fig 1. Variation of Magnetic Susceptibility with temperature

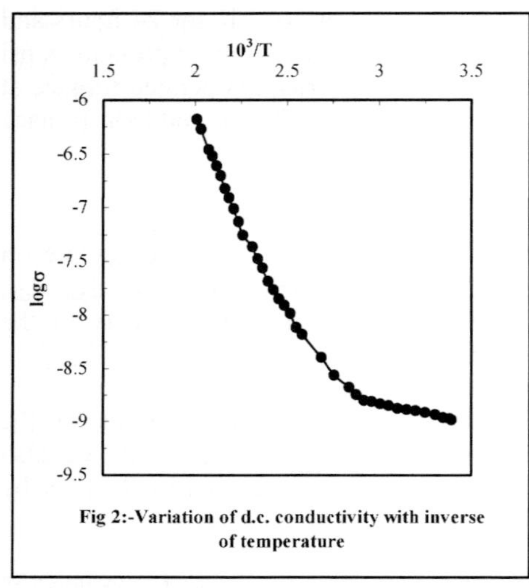

Fig 2:-Variation of d.c. conductivity with inverse of temperature

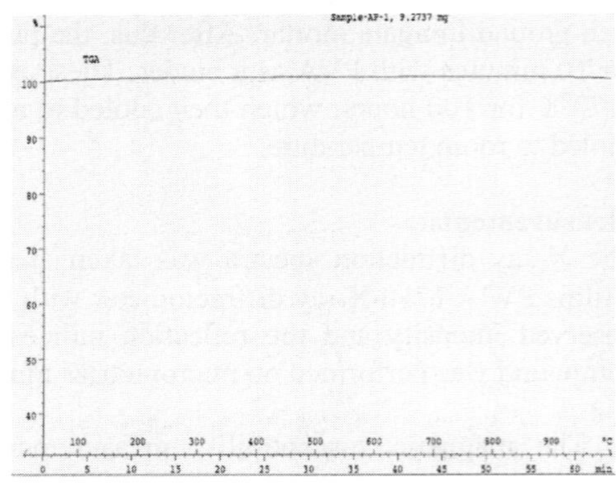

Fig. 3: Thermo-gravimetric analysis of $SrTi_2Zn_2Fe_{10}O_{22}$

Result and Discussions

In the present work, the substituted ions have been chosen to keep the electrical neutrality of the compound. From the crystallographic analysis [5], it is observed that the compound is in a single hexagonal phase without traces of the unreacting phases of the reactants. The hexagonal unit cell dimensions are: a = 5.891Å and c = 46.064 Å, which are closer to the parameters of Y type hexaferrite $Sr_2Fe_2^{2+}Fe_{12}^{3+}O_{22}$ (a = 6.0 Å, c = 41.9 Å) [6]. The decrease in parameter a is obviously due to the presence of single strontium cation in the lattice, whereas the increase in c parameter may be due to substitution of Ti^{4+} and Zn^{2+} cations in the lattice. The average crystallite size measured by Scherrer equation [7] on 100 % peak at $2\theta = 34.2$ was 0.1042 µm. The true density by helium Picnometer was found to be 6.5471 gm/cc; the higher value indicates the formation of better quality ferrite.

From the magnetic study, performed on Gouy's balance [8], the compound is ferrimagnetic at room temperature. Above Curie temperature ($T_C=334^oK$), the plot of $1/\chi_M$ versus T (figure 1) is nearly linear indicative of the paramagnetic behaviour of the compound. The low value of T_C is as expected because of the substitution of non-magnetic Ti, Zn cations. From the linear portion of the plot, the Curie molar constant was worked out and was found to be 42.67, matching with the expected value for stable oxygen states of the cations in the molecule.

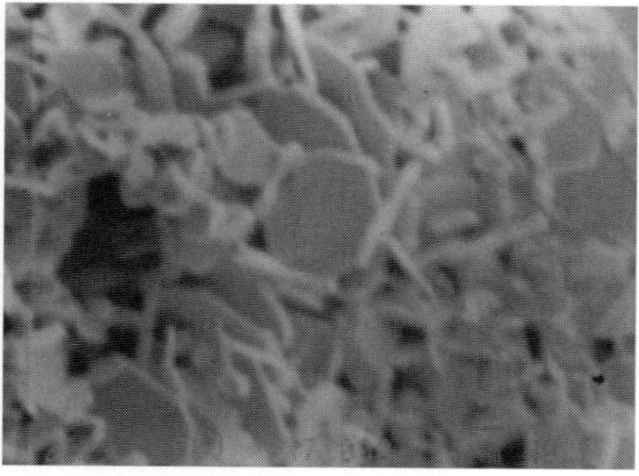

Fig. 4: SEM of $SrTi_2Zn_2Fe_{10}O_{22}$

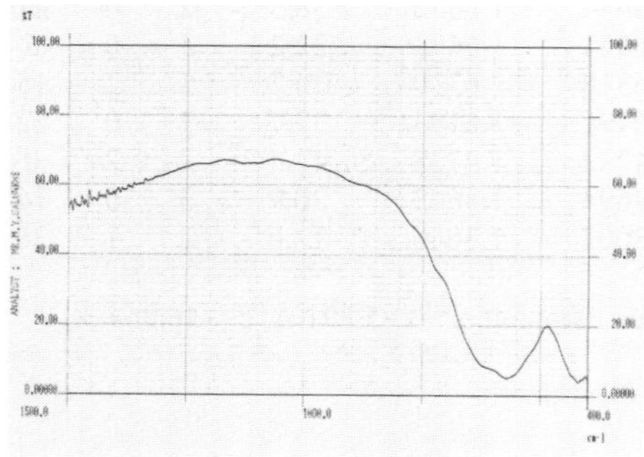

Fig. 5: Infrared absorption spectra

The d. c. electrical conductivity measurements (Figure 2) show that the change of slope near the Curie temperature is due to change in the magnetic ordering from order state to disordered state. Semiconducting nature of the prepared compound is proved with room temperature resistivity 9.548×10^8 Ωcm. The activation energies are found to be 0.44 eV and 0.0758 eV in para and ferri-region respectively.

The thermo-gravimetric analysis (Figure 3) shows no change in weight with increase in temperature that is the compound is thermally stable from room temperature to 1273^oK. The surface morphology was studied by SEM. Clearly hexagonal platelet shaped particles are observed with random orientation (Figure 4). The average particle size observed from microstructural study is 2.263 μm. The particles observed from SEM may contain number of crystallites calculated from x-ray diffractometry.

The infrared absorption spectra show two absorption peaks at 421 cm^{-1} and 575 cm^{-1} with a shoulder at 625 cm^{-1} (Figure 5). This is the common feature of spinels (9,10) and also observed in hexagonal $Sr_2Zn_2Al_{10}Fe_2O_{22}$ [11] structure with non-magnetic trivalent substitution. This may be due to vibrations of octahedral and tetrahedral cations present in the compound. The absence of high frequency (above 700 cm^{-1}) bands indicates the reduced vibrations of Fe-O complexes. Thus the substitutions of Ti^{4+} and Zn^{2+} plays major role in the compound at octahedral and tetrahedral sites respectively.

Acknowledgement

We express our gratitude to Dr. K. V. Ramanarao and Shri M. T. Nimje, Scientist, J. N. A. R.D.D.C., Wadi, Nagpur, for providing the facilities and useful discussion in carrying out SEM and density measurements.

References

1. Kohn J.A. and Eckart D.W., J. of Applied Physics 35, 3 (1964) 968.
2. Smit J. and Wijn H.P.J., Ferrites, Philips Tech. Library, Eindhoven, (1959).
3. Sugimoto M. in: E.P. Wohlfarth (Ed.) Ferromagnetic Materials, Vol. 3, North Holland Physics Publishing, Amsterdam, (1982).
4. Mazen S.A., Abdallah M.H., Sabrals B.A. and Hashem H.A.M., Phys. Stat. Solidi (a) 134 (1992) 263.
5. Klug H. and Alexander I., X-ray diffraction technique, McGraw Hill, New York, (1974).
6. Ram S., J. Magn. Magn. Mater. 22 (1985) 315
7. Puller R.C., Taylor M.D. and Bhattacharya A.K., J. Mater. Sci. 32 (1997) 365.
8. Bates L.F., Modern Magnetism, Cambridge University Press, Cambridge, (1939).
9. Waldron R.D., Phys. Rev. 99 (1955) 1727.
10. Patil R.S., Kakatkar S. V., Sankpal A.M., Sawant S.R., Suryavanshi S.S., Ghodke U.R. and Kamat R.R., Ind. J. Pure and Applied Phys., 32 (1994) 193.
11. Salunkhe M.Y., Ph.D. Thesis, 2002.

A new method to characterize the soft magnetic material

O. Caltun

Faculty of Physics, Al. I. Cuza University, Iasi 6600, Romania, Caltun@uaic.ro

Current developments in the field of power electronic devices are leading to miniaturization. However, if storage inductors or small transformer are needed, the final dimension of a device is determined by these components. A size reduction is usually possible by increasing the operating frequency and the initial permeability, if ferrite with high magnetic performances is available. The ferrites producers and user are more and more interested in quickly and complex characterization of the ferrite cores. To obtain the characteristics of soft ferrites operating in the domain of the high frequency and having low hysteresis and power losses a new and complete experimental set-up was designed and used. Many papers report the dependence of the magnetization processes in soft ferrite cores on excitation frequency [1], [2]. Especially the microstructure, the resistivity and the permittivity could influence the distortion in the RL circuits with MnZn or Ni-Zn ferrite cores [3]. In some ferrites the pores and impurities can impede the movement of domain walls and the coherent rotations leading in magnetization processes. In this case the magnetization processes will be influenced by the frequency above the resonance ferrimagnetic frequency [4]. In ferrites with low porosity in static process or at low frequency the movement of domain walls will lead the magnetization process [5]. At high frequencies the wall can't fallow the field and the movements of the wall are irreversible due to the damping [6]. By using the Fourier techniques for the spectral analysis we have studied the digitized output signal in a secondary coil of a transformer with a Ni-Zn-Cu ferrite core. The cores were samples with different microstructure sintered at various temperatures [7]. The differences observed between the spectral coefficients at various excitation frequencies is discussed in relation with the two main magnetization processes. The behaviour is related to the microstructure and to the experimental hysteresis loops of the samples. In the same time different waveforms for the magnetic field strength were used to obtain minor and major hysteresis loops.

1. EXPERIMENTAL RESULTS

1.1 Experimental set-up and FFT analysis

An experimental set-up presented in fig 1 comprising a measuring system assisted by a personal computer analyses the magnetic properties of the cores. The hysteresis curves can be plotted. The initial permeability, the hysteresis losses and the power losses of soft magnetic materials can be calculated. The measurement system designed by us few time ago [1] derived advantages by the measurement instrument and personal computer upward performances. The generator is a Standford Research System Model DS 345. Direct Digital Synthesis (DDS) is a method of generating very pure waveform with extraordinary frequency resolution, low frequency switching time, crystal clock-like phase noise, and flexible modulation. DDS works by generating addresses to a waveform RAM to produce data for a DAC. The clock is a fixed frequency reference. Instead of using a counter to generate addresses, an adder is used. On each clock cycles, the content of a phase Increment register are added to the contents of the phase Accumulator. The Phase Accumulator output is the address to the waveform RAM. By changing the phase Increment the number of clock cycles needed to step trough the entire waveform RAM changes, thus changing the output frequency. The DS 345 uses a custom Application Specific Integrated Circuit to implement the address generation in a single component. The System is interconnected with the computer and assisted by a special program generating all the waveforms desired by the user. By applying the waveform amplified by the Power Amplifier in the primary (excitation) coil different magnetization process can be studied. To study the influence of the frequency on the spectral coefficients of the induced signal, a RL circuit with a soft ferrite core, toroidal in shape was used.

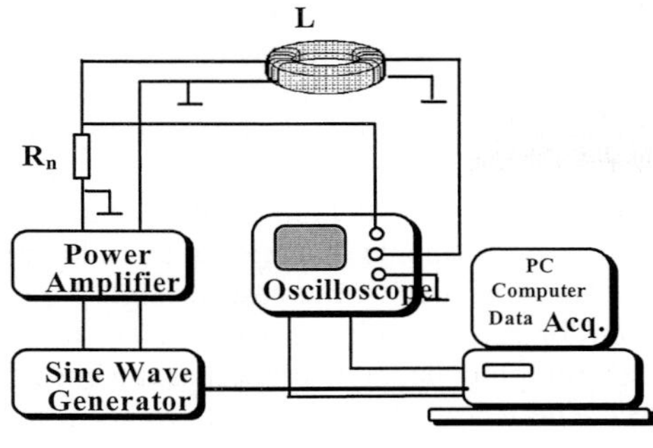

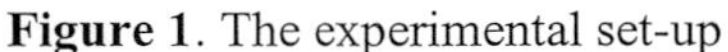

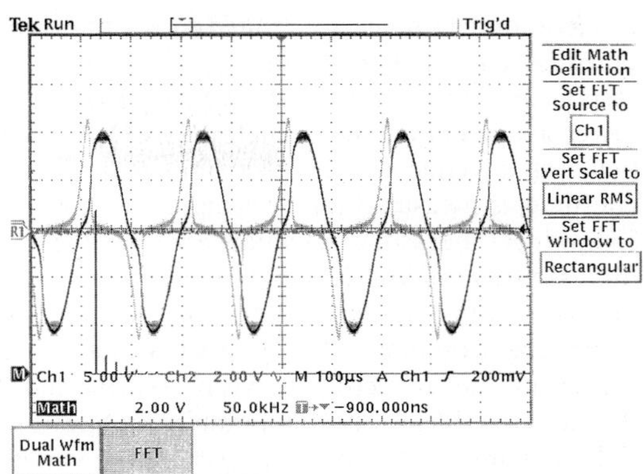

Figure 1. The experimental set-up

Figure 2 The signals memorized using the TDS oscilloscope and the FFT of excitation current

To obtain the required degree of certainty over the computed results we compared these results with data collected using a STANDFORD RESEARCH SYSTEMS SR830 DSP Lock-In Amplifier and the results of FFT analysis performed by the Ossciloscope. The TDS 3000B Series Digital Phosphor Oscilloscope permits the communication via Internet and the storage of the data in the same time in different operating posts. The input signals for channel I is the waveform of the excitation current carried by the primary coil. The input signals for the channel II are the signals induced in secondary coil.

1.2 The samples and experimental data

Ni-Zn ferrites doped with CuO were prepared using the usual ceramic technique [4]. The samples presintered in different conditions with different doping levels and sintered at different temperatures are characterized by different microstructures [7]. Consequently the magnetic properties are different. By increasing the copper substitution the average grain size decreases and the microstructure becomes more uniform without pores. By increasing the sintering temperature the grain size becomes larger. The figure 3 presents a comparison between the experimental and calculated data for samples N4 (for detailed data see the reference [4]). As one can see, the differences are minor, so the computer-based method, designed by us, could be used successfully for the analysis of the magnetization processes.

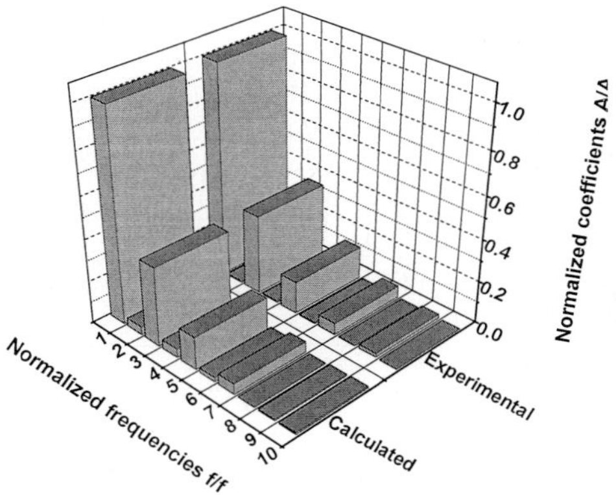

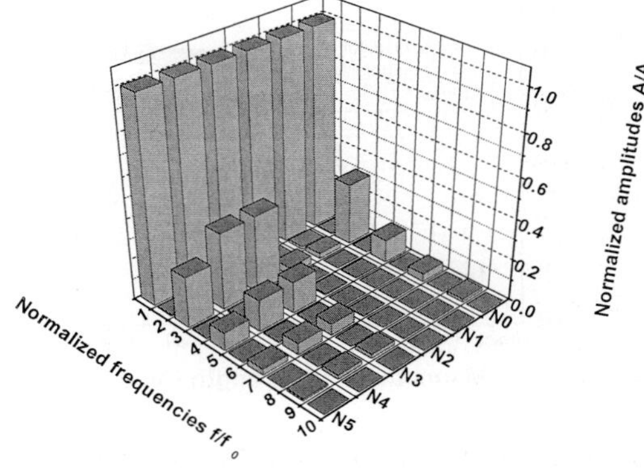

Figure 3. Calculated and experimental Fourier coefficients N4 at 10 KHz

Figure 4. Fourier coefficients calculated for all the samples at 10 KHz and 65 A/m field strength

There are two different behaviors of the samples: N0, N3, N4, and N5 that have high coercive magnetic fields, while the samples N1 and N2 have smaller coercive magnetic fields. This observation can be easily related to the microstructure of samples [8]. The N4 sample has reduced porosity and higher average grain-size that allows magnetic walls movements. These are responsible for the high coercive magnetic field strength. The N1 sample has greater porosity and lower average grain-size. The movement of domain walls is restricted and the permeability of the material is reduced. These conclusions are supported by a subsequent analysis of the Fourier coefficients measured for all samples at 10 KHz, as shown in fig. 4. By comparing the spectrum for samples N1 and N4, one can observe that the amplitudes of the harmonics superior to 2 are lower in the case of sample N1. The motion of the magnetic walls is damped in the case of high porosity materials with small grain-size, and the main mechanism for magnetization is the coherent rotation of the magnetic moments [9]. Small variations of the macroscopic magnetic properties are observed increasing the frequency between 10 KHz and 20 KHz. The recorded hysteresis loops seem to be the same in shape. The saturation magnetic flux density drops (6.25%) and the coercive magnetic field increases (8.61%). By analyzing the variation of calculated normalized Fourier coefficients at 10 KHz and 20 KHz, one could observe the decrease of the A_3/A_0 ratio (25%) for the N4 sample (fig. 7, and 9). In conclusion the spectral analysis becomes an appropriate tool for monitoring the magnetization processes rather than the classical hysteresis loops.

1.3 The influence of the waveform of the applied magnetic field on the hysteresis loop

When a square, triangular or arbitrary waveform of excitation voltage is impressed on the test circuit the shape of the waveform of the magnetic field applied changes. Consequently, the waveform of the magnetic flux density rate and the waveforms of the magnetic flux density changed. A special problem in recording dynamic hysteresis loops is the exact phase relationship of the primary current referred to the induce voltage. The secondary voltage is the derivative of the magnetic flux density, the passing of the voltage trough zero and the maximum of the current should coincide. The applicability of this criterion is facilitated by the saturation of the material. Our program operates the control of the hysteresis loop also for the non-saturation region.

In figure 5 are presented the hysteresis loops obtained for sample N4 exciting the primary coil by sinusoidal and respectively triunghiular waveform at 25kHz. Even if the rate of the magnetic is different to shape of the hysteresis loops are similar. The differences are significant for the coercive magnetic field strength and for the saturation magnetic flux density. The magnetization processes are influenced by the frequency and by the rate of magnetic field as result the amplitude of the superior harmonic coefficient are slowly different.

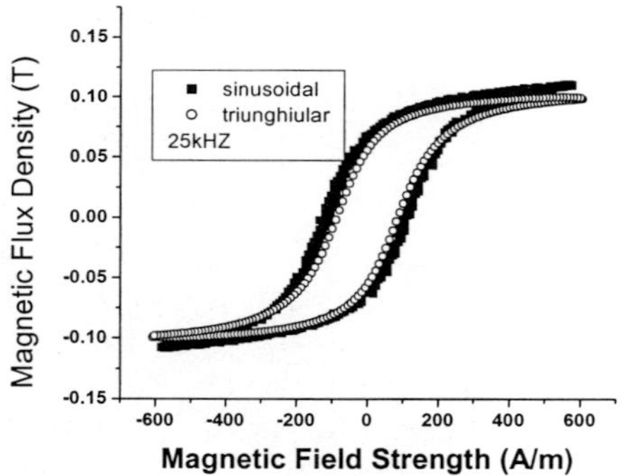

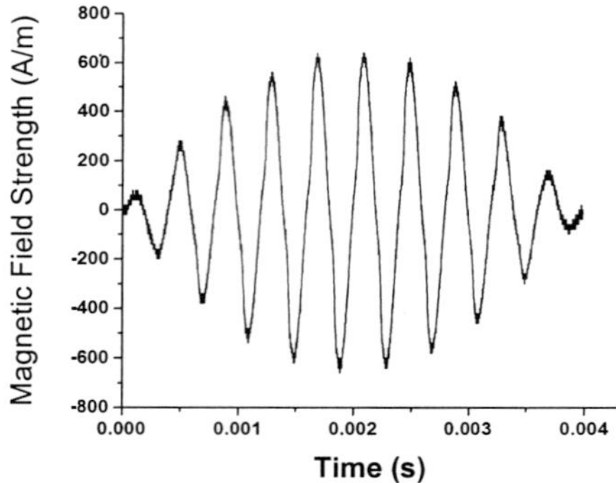

Figure 5: Hysteresis loop at 25kHz excitation frequency for sinusoidal and triangular waveform

Figure 6: Magnetic field strength vs. time Arbitrary waveform

Another operation mode of our experimental set-up is designed to obtain the minor and major hysteresis loops. An arbitrary wave is generated by the Synthesized Function Generator and applied in the primary coil. The waveform is presented in Fig. 6

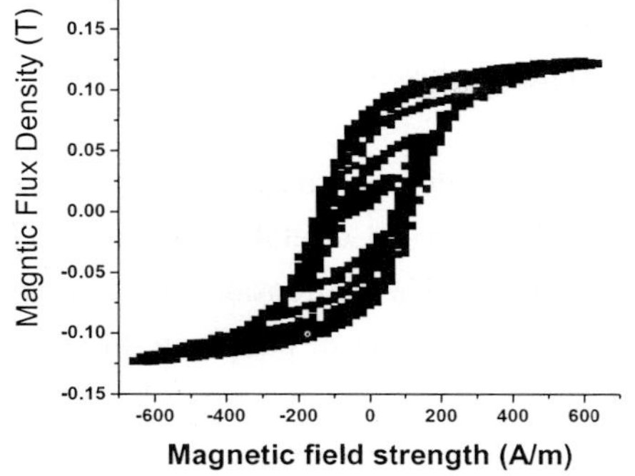

Figure 7: Minor and major hysteresis loops obtained by applying in the primary coil of the arbitrary waveform of the magnetic field strength

2. CONCLUSIONS

The experimental set-up presented in this paper allows visualization of the hysteresis loops in different conditions. The waveform and the frequency of the applied magnetic field can be changed and different magnetization processes can be studied. Minor and major hysteresis loops were obtained by applying in the primary coil of the transformer an arbitrary waveform in which the magnetic field strength sinusoidal in shape increase and decrease to the maximum value.

An others important utility of the experimental set-up is the possibility to make numerical Fast Fourier Transformer by oscilloscope or by our program designed to represent and to calculate magnetic properties of the samples. Even though the hysteresis loop features small changes in shape, the spectral analysis allows discriminating between the magnetization processes (between the domain wall displacement and coherent rotations).

3. REFERENCES

[1] Al. Stancu, O. Caltun, P. Andrei, "Models of hysteresis in magnetic cores", J.Phys. IV France 7, Collq. C1, Suppl. J.Phys. III, p. 209 –210, 1997., O. Caltun, P. Andrei, J. Phys. IV France, Collq. C1, Suppl. J. Phys. III, (1997) 209.

[2] Al. O. Caltun, C. Papusoi, A. Stancu, P. Andrei, W. Kappel, IOS Series "Studies in Applied Electromagnetics and Mechanics", editors V. Kose and J. Sievert, (1998), ISBN 905199 381 1, 594.

[3] N. Schmidt, H. Guldner, IEEE Trans. on Magn., 32 (2) (1996) 489

[4] O.Caltun, L. Spinu, Al. Stancu, L. D. Tung, W. L. Zhou, "Study of the microstructure and of the permeability spectra of Ni-Zn-Cu ferrites", J.Magn. Magn. Mater. 242 (20020 160

[5] T. Nakamura, "Snoek's limit in high frequency permeability of polycrystalline Ni-Zn, Mg-Zn, and Ni-Zn-Cu spinel ferrites", J. Appl. Phys., Vol. 88 (1), pp. 348-353, 2000.

[6] O. Inoue et al, "Low Loss Mn-Zn ferrites", IEEE Trans. on Magn., 29, pp. 3532-3534, 1993.

[7] P. S. Anil Kumar et al, "Particle size dependence of rotational responses in Ni-Zn ferrite", J. Appl. Phys., Vol.83 (11), pp. 6864-6866, 1998.

[8] O.Caltun, A.Stancu, C. Papusoi, P. Andrei, "On the distortion in a RL Circuit with Fine-grained Mn-Zn Ferrite Cores", IOS Series "Studies in Applied Electromagnetics and Mechanics 18", editors P. Di Barba and A. Savini, ISSN: 1383-7281, p. 31-34.

[9] M. Guyot et al., "Mobility and/or Damping of the Domain Wall", Phys. Stat. Sol. (a) 106, 595 (1988).

Effects on electronic and magnetic properties of intermixing at Cr/Cu interface

Byung Sub Kang[1] and Suhk Kun Oh

Institute for Basic Science, Dept. of Physics, Chungbuk National University,

Cheongju, 361-763, South Korea

E-mail address: kangbs@nscience.chungbuk.ac.kr

We study the effects of interfacial intermixing on electronic structure and magnetic properties of Cr/Cu by using the method of *ab initio* full potential linear muffin-tin orbital (FP-LMTO). The formation of CrCu in the subsurface layer is energetically favorable. The antiferromagnetic ordering of in-plane in Cr/Cu is found to be energetically favorable more than that of out-plane. In order to simulate the interfacial intermixing in Cr/Cu, we use the system of 50:50 CrCu alloy layer on the surface or at the interface, as well as the sandwich of $Cr/(Cr_xCu_{1-x})_n/Cu$ (where $x = 0.25$, 0.50, and 0.75 for $n=1$, 2, and 3, respectively). Our result for the energetics of surface alloying shows clearly that the Cr-Cu intermixing has a strong effect on the Cr magnetization.

Keywords: *Surface alloy; Magnetism; Cr/Cu; Ab initio calculation*

1. Introduction

The stability of thinfilm and effect of interfacial intermixing during metal-on-metal growth process is a general subject. The intermixing of ultrathin 3d transition metals layers grown on a Cu surface is a common phenomenon. For instance, it has been shown that Fe[1,2], Co[3,4,5], and Ni[6,7,8] intermixing with Cu atoms at their interfaces and form alloy layers. X-ray magnetic circular dichroism (XMCD) measurement[9] reported that the average magnetic moment of a 4 monolayers (ML's) Ni film on Cu(001) is a half of that in the bulk Ni. Extended X-ray absorption fine structure (EXAFS) and low energy electron diffraction (LEED) measurements also reported that Mn mixes with Cu at Mn/Cu(100) interface[10]. And then, what is the effect of intermixing in thin Cr/Cu film?

Cr and Mn are characterized by a more complicated magnetic structure than the other bulk 3d series. The c(2x2) in-plane antiferromagnetic configuration is the most stable one for an isolated Cr monolayer and a Cr monolayer on Ag[11,12,13], Cu[13,14], and Pd(001)[15]. The c(2x2) in-plane antiferromagnetic configuration at the surface of a Cr crystal leads to frustrated spins in the subsurface layer. A topological antiferromagnetism of Cr (001) surface, which consists a ferromagnetically ordered terraces gives the most natural explanation of the experimental results[17]. This was confirmed by means of the scanning tunneling microscope (STM)[18]. The interfacial intermixing in the magnetic thinfilm is strong interrelation between magnetic and structural properties of ultrathin films and alloy. Thus, the purpose of our study is to investigate the influence of the interdiffusion at the Cr/Cu interface on the electronic structures and magnetic properties.

The method used in the numerical calculations is presented in Section 2. The calculated results and the comparison with other theoretical data are described in Section 3. In the final section, a summary is given.

2. Computational details

The calculations are performed by using a self-consistent full-potential linear muffin-tin orbital (FPLMTO) method[19-22] with the exchange-correlation potential proposed by Perdew and Wang[23], within the framework of density functional theory and the local-density approximation (LDA). The Cu substrate is modeled by eleven atomic layers with a (2x2) surface cell, which separated by six layers of vacuum. The atomic structure of CrCu surface alloy on the Cu substrate includes a 50:50 CrCu alloy layers on the surface and at the interface.

The crystal is divided into regions inside atomic sphere and interstitial region. The LMTO basis in the vacuum regions consists of sets of appropriate s, p, and d states. The ``full'' charge density including all nonspherical terms is evaluated by Fourier transformed in the interstitial region. The basis functions in the interstitial region are expanded by the smoothed Hankel functions[20]. The basis functions of Cr and Cu for the s, p, and d electrons are generated with cut-off energy of E_{cut}=167.28eV, 243.44eV, and 356.32eV, respectively. Spin-orbit interactions are not included. The linear tetrahedron method including the corrections of Blochl et al[24] is chosen to improve the convergence of the electronic structure and total energy with respect to the number of k points. Using 49 k points corresponding to a 12x12x1 grid insures that the total energy and magnetic moment are converged to better than 1meV/cell and $0.01\mu_B$/atom, respectively.

3. Results and discussion

3.1. The atomic structures

Cr and Cu atoms are placed pseudomorphically on each side of the slab (substrate) to model the systems of Cr_2Cu/Cu, $CuCr_2$/Cu, $CrCuCr$/Cu, $(Cr_{0.5}Cu_{0.5})_2Cr$/Cu, and $Cr(Cr_{0.5}Cu_{0.5})_2$/Cu, which are used to simulate the interdiffusion in Cr/Cu. The purpose of this paper is to study the effects of intermixing in Cr/Cu on the electronic and magnetic properties rather than to find a lowest-energy configuration. The in-plane lattice constant in our model is fixed at the optimized lattice constant for the bulk Cu through generalized gradient approximation (GGA)[23] calculations (6.88 a.u.). The vertical positions of all atoms in the unit cell are determined according to their atomic forces. The equilibrium atomic structures are determined through total energy and atomic force calculations with Pulay corrections for the change of atomic positions [20]. The calculated equilibrium z coordinates for each system are listed in Table 1.

For the alloy layer of $(Cr_{0.5}Cu_{0.5})_2$, a buckling between Cr and Cu atoms take place. In $(Cr_{0.5}Cu_{0.5})_2Cr$/Cu(001), Cr atom in the surface layer is higher than Cu atom by 0.283 a.u.. The relaxation of Cr and Cu atoms in the second layer are -0.541 and -0.449A, respectively. The roughness, defined as, |z(Cr)-z(Cu)|/d(Cu bulk) is 4.12%, where z(Cr) and z(Cu) are equilibrium z coordinates for the Cr and Cu atoms in the $Cr_{0.5}Cu_{0.5}$ alloy layer, respectively. d(Cu bulk) (3.639 a.u.) is the interlayer spacing in bulk Cu. The average distance between Cu(3) and the surface $Cr_{0.5}Cu_{0.5}$ is larger than the average of the corresponding interlayer distances of bulk Cr and bulk Cu (d=2.40 a.u.)

For $Cr(Cr_{0.5}Cu_{0.5})_2$/Cu(001), the roughness of interface $Cr_{0.5}Cu_{0.5}$ subsurface layer (2nd layer) is about 3.62% with Cr atom more inward than Cu atom by 0.134 a.u.. The Cr-Cr interlayer distance from the surface to the interior region in Cr film are 3.398 and 3.331 a.u., respectively. They correspond to -1.21% and -3.18% relaxations from the interlayer spacing in the bulk Cu.

The interdiffusion and surface segregation of Cu atom are very strong[8]. The system of $CuCr_2/Cu(001)$ is found to be energetically more favorable by as much as 0.631eV compared to the $(Cr_{0.5}Cu_{0.5})_2Cr/Cu(001)$ system. However, the growth of Cr film on Cu is ruled by kinetics. It depends on growth condition such as a temperature and deposition rate. According to these growth conditions, some metastable structure can be formed.

Table 1. The magnetic moments (M, in mB) and calculated equilibrium vertical positions (z, in a.u.) for each CrCu surface alloys on the Cu (001) and (111) surfaces, and the magnetism for sandwich structure of $Cr/(CrxCu1-x)n/Cu(001)$ (where x= 0.25, 0.50, and 0.75 for n=1, 2, and 3, respectively). Cr(s) and Cu(s) indicate surface Cr and Cu atoms, respectively. z position of Cu(1) is 0.0, the data in parentheses are those for Cu atom in the Cr0.5Cu0.5 alloy layer.

Structure		Cr(s) / Cu(s)	Cr(2) / Cu(2)	Cr(1) / Cu(1)	Cu(3)
		2 monolayers Cr on Cu(001)			
Cr_2Cu/Cu,	M	2.738, -2.738	2.240, -2.240		
	M			0.001, -0.001	0.001, -0.001
	z	16.990	13.525	10.367	6.728
$CuCr_2/Cu$,	M		2.560, -2.560	1.982, -1.982	
	M	0.051, -0.051			0.019, -0.019
$CrCuCr/Cu$,	M	3.126, -3.126		2.668, -2.668	
	M		0.0		0.0
$(Cr_{0.5}Cu_{0.5})_2Cr/Cu$,	M	3.390	-2.525	2.266, -2.488	
	M	0.118	-0.026		-0.017, -0.017
	Z	16.351(16.177)	12.911(12.737)	10.243	6.799
$Cr(Cr_{0.5}Cu_{0.5})_2/Cu$,	M	3.143, -2.833	2.561	-2.870	
	M		0.035	-0.058	-0.058, -0.058
	Z	17.014	13.482(13.616)	10.285(10.316)	6.828
		Sandwich of Cu/Cr/Cu(001)			
$Cr/(Cr_xCu_{1-x})_n/Cu$,	M	-0.488[Cr]	1.098[$Cr_{0.75}$]	-2.347[$Cr_{0.5}$]	2.857[$Cr_{0.25}$]
	M	-0.006[$Cu_{0.25}$]	-0.017[$Cu_{0.5}$]	0.002[$Cu_{0.75}$]	0.029[Cu]
		2 monolayers Cr on Cu(111)			
$CrCuCr/Cu$,	M	2.663, -2.663		1.780, -1.780	
	M		0.021, -0.021		0.009, -0.009
$(Cr_{0.5}Cu_{0.5})_2Cr/Cu$,	M	2.585	-1.230	0.169, 0.169	
	M	0.021	0.009		0.002, 0.002

3.2. The magnetic and electronic properties

The Cr magnetism for thin Cr/Cu(001) film is very sensitive to change in their lattice parameter and the number of k vector as well as their atomic structure. As known in other previous result[13], the magnetic moment is found to be energetically favorable in the antiferromagnetic state.

The magnetism tends to stabilize Cr-deposited atoms on the Cu surface. The Cr magnetic moment at the surface reduces the surface free energy of Cr metal[25,26]. That is, it provides an additional contribution to the surface free energy of Cr film, due to large magnetic moments by additional gain of magnetic energy of Cr film on the surface. For 50:50 CrCu alloy of $(Cr_{0.5}Cu_{0.5})_2Cr/Cu(001)$, the magnetic moments (in μ_B) of Cr atoms from the surface to interior region are 3.390, -2.505, and 2.266(-2.488), respectively. Here, the Cr moments in the interface (the third layer) show in-plane antiferromagnetic ordering. For $Cr(Cr_{0.5}Cu_{0.5})_2/Cu(001)$, the magnetism is very similar to above system. Interestingly, the Cr magnetic moments of $Cr_{0.5}Cu_{0.5}$ alloy layers by intermixing between Cr and Cu atoms is very high due to the proximity effect of Cu atoms. From the result for magnetism of sandwich system, we can see clearly that the Cr-Cu intermixing has a strong effect on the Cr magnetization. These calculated magnetic moments are listed in Table 1. $(Cr_{0.5}Cu_{0.5})_2Cr$ surface alloy is energetically favorable structure more than $Cr(Cr_{0.5}Cu_{0.5})_2$ alloy on the Cu (001) or (111) surface. The atomic structure of Cr/Cu sandwich, which consists of 4 atoms per in-plane cell is as follow:

$$Cr/Cr/Cr/Cr_{0.75}Cu_{0.25}/Cr_{0.5}Cu_{0.5}/Cr_{0.25}Cu_{0.75}/Cu/Cu/Cu.$$

The Cr magnetism in thin $Cr_1/Cu(001)$ and $Cr_1/Cu(111)$[13] films shows in-plane antiferromagnetic ordering. Compared to total energies between ferromagnetic and antiferromagnetic states of these two systems, the differences of total energies are very small by about 0.01eV. For thin $Cr_2Cu/Cu(001)$ and $Cr_2Cu/Cu(111)$ films of 2-ML Cr, the Cr magnetic moment also aligns antiferromagnetically. That is, the in-plane antiferromagnetic ordering is energetically favorable more than to out-plane magnetic ordering. The difference in total energy between for $Cr_2Cu/Cu(001)$ ferromagnetic state and antiferromagnetic state is about 0.51eV. For the 2-ML $Cr_{0.5}Cu_{0.5}$ surface alloy structure, the Cr magnetic moment is very high. However, in $CuCr_2/Cu(001)$ and $CuCr_2/Cu(111)$, the Cr magnetic moments are small. For $CuCr_2/Cu(001)$, the value of density of states (DOS) at E_F for Cr is drastically reduced. Strong hybridization between Cr and Cu is clearly shown in the DOS curves. The DOS is bulk-like for the central layer in the Cu substrate due to an efficient screening. Almost no Stoner exchange splitting between the two spin channels can be found in the Cu substrate, showing the weakness of the induced magnetization in the Cu substrate. The induced magnetic moment in surface Cu atoms is about by $\pm0.05\mu_B$/atom, which aligns with parallel to the Cr atoms.

Figure 1 shows the local density of states (LDOS) for $(Cr_{0.5}Cu_{0.5})_2Cr/Cu(001)$. Resonant satellite peaks appear right below E_F in both the interfacial Cu layer and the Cu in the $Cr_{0.5}Cu_{0.5}$ plane. The LDOS for the Cr- and Cu-component in the surface alloy layer is also shown by different panels in Fig.1(a). The DOS curves of the surface $(Cr_{0.5}Cu_{0.5})$ alloy layer are significantly narrower with a higher DOS peak at below E_F compared to those of the interfacial Cr layer in $(Cr_{0.5}Cu_{0.5})_2Cr/Cu(001)$. The Stoner exchange splitting, therefore, the magnetic moments are large. The electronic structure of $(Cr_{0.5}Cu_{0.5})_2Cr/Cu(111)$ is similar to the system of $(Cr_{0.5}Cu_{0.5})_2Cr/Cu(001)$. The interfacial Cr layer [Cr(1)] has a smaller Stoner exchange splitting, therefore, a smaller magnetic moments as compared with Cr atom of other layers [Cr(2), Cr(s)](see Table 1). The Cu atoms of the $Cr_{0.5}Cu_{0.5}$ alloy layer induce a high magnetic moment of neighboring Cr atoms.

Unlike the cases in $CuCr_2/Cu(001)$ and $CuCr_2/Cu(111)$, the LDOS's of Cr- and Cu-component in the $Cr_{0.5}Cu_{0.5}$ alloy layer of $(Cr_{0.5}Cu_{0.5})_2Cr/Cu(001)$ and $(Cr_{0.5}Cu_{0.5})_2Cr/Cu(111)$ take place separated with each other without obvious hybridization. Compared with the interior Cr layer, the surface Cr-component has a narrower bandwidth with a larger exchange splitting, and therefore it has a larger magnetic moments. As known, this is mainly attributed its lower coordination numbers on surface and a strong effect on the magnetization of the Cr-Cu intermixing.

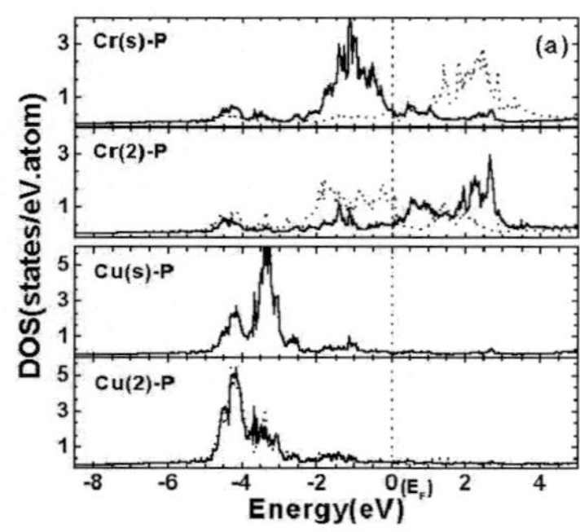
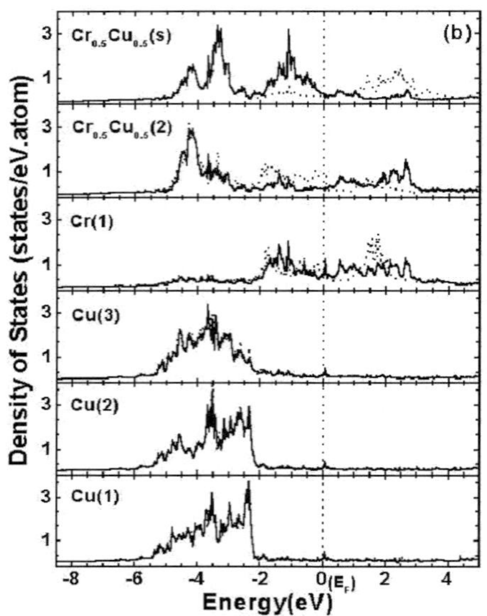

Fig.1 (a) and (b):

The layer- and spin-projected density of states (DOS) for atomic structure of $(Cr_{0.5}Cu_{0.5})_2Cr/Cu(001)$. (a) the DOS for the Cr- and Cu-site in the $Cr_{0.5}Cu_{0.5}$ alloy layer are shown in the panels of Cr-P and Cu-P, respectively. Cr(s) and Cu(s) indicate surface Cr and Cu atoms. (b) the DOS for each layers of slab structure. The majority and minority spin DOS are shown in solid and dotted lines, respectively. The vertical line at E=0 indicates the position of Fermi level.

4. Concluding remarks

We have studied the electronic structures and magnetic properties of CrCu alloy on the surface or in the interface resulting from the interdiffusion in the Cr/Cu system and sandwich alloy by using the FPLMTO method. In order to simulate the intermixing in Cr/Cu, we have used six systems which includes 50:50 CrCu surface alloy. The electronic structure and magnetic properties of the $Cr_{0.5}Cu_{0.5}$ surface alloy are sensitive to its surrounding environment. The Cr atoms of Cr_2Cu/Cu and $CrCuCr/Cu$ thinfilms are shown in-plane antiferromagnetic ordering. The in-plane magnetic ordering (antiferromagnetic) of Cr atoms is energetically favorable more than out-plane magnetic ordering. The magnetism of these systems is weaker than that of 50:50 CrCu surface alloy on the Cu substrate. The magnetic moments of Cr atoms in the $Cr_{0.5}Cu_{0.5}$ surface layer of $(Cr_{0.5}Cu_{0.5})_2Cr/Cu(001)$ are significantly high ($3.390\mu_B$/atom). The nearest neighbors Cr-Cr magnetic moments prefer an antiferromagnetic coupling, the nearest neighbors Cu-Cu moments are coupled to parallel with Cr atom.

Acknowledgements

This work was supported by the Korea Research Foundation, Grant No. KRF-2001-005-D20009.

References

[1] J.H. Kim, K.H. Lee, G. Yang, A.R. Koymen, and A.H. Weiss, Appl. Surf. Sci. 173 (2001) 203.

[2] Christian Pflitsch, Rudolf David, Laurens K. Verheij, Rene Franchy, Surf. Sci. 468 (2000) 137.

[3] J. Fassbender, R. Allenspach, and U. Durig, Surf. Sci. 383 (1997) L742.

[4] U. Kohl, M. Dippel, G. Filleboeck, K. Jacobs, B.-U. Runge, and G. Schatz, Surf. Sci. 407 (1998) 104.

[5] F. Nouvertne, U. May. Bamming, A. Rampe, U. Korte, G. Guntherodt, R. Pentcheva, and M. Scheffler, Phys. Rev. B 60 (1999) 14382.

[6] W. Platow, U. Bovensiepen, P. Poulopoulos, M. Farle, and K. Baberschke, Phys. Rev. B 59 (1999) 12641.

[7] S.H. Kim, K.S.Lee, H.G. Min, Jikeun Seo, S.C. Hong, T.H. Rho, and Jae-Sung Kim, Phys. Rev. B 55 (1997) 7904.

[8] Byung-Sub Kang, Jean-Soo Chung, Suhk-Kun Oh, and Hee-Jae Kang, J. Magn. Magn. Mater. 241 (2002) 415.

[9] P. Srivastava, F. Wilhelm, A. Ney, M. Farle, H. Wende, N. Haack, G. Ceballos, and K. Baberschke, Phys. Rev. B 58 (1998) 5701.

[10] S. D'Addato and P. Finetti, Surf. Sci. 471 (2001) 203.

[11] C. Krembel, M.C. Hanf, J.C. Peruchetti, D. Bolmont, and G. Gewinner, Phys. Rev. B 44 (1991) 11472.

[12] O. Speder, P. Rennert, and A. Chasse, Surf. Sci. 331 (1995) 1383.

[13] {Kru} P. Kruger, M. Taguchi, and S. Meza-Aguilar, Phys. Rev. B 61 (2000) 15277.

[14] S. Blugel, Appl. Phys. A 63 (1996) 595.

[15] Jamil Khalifeh, J. Magn. Magn. Mater. 185 (1998) 213.

[16] F. Meier, D. Pescia, and T. Schriber, Phys. Rev. Lett. 48 (1982) 645.

[17] S. Blugel, D. Pescia, and P.H. Dederichs, Phys. Rev. B 39 (1989) 1392.

[18] R. Wiesendanger, H.-J. Guntherodt, G. Guntherodt, R.J. Gambino, and R. Ruf, Phys. Rev. Lett. 65 (1990) 247.

[19] Bal K. Agrawal, Savitri Agrawal, and Pankaj Srivastava, Surf. Sci. 408 (1998) 275.

[20] S. Yu. Savrasov and D. Yu. Savrasov, Phys. Rev. B 46 (1992) 12181;S. Yu. Savrasov, Phys. Rev. B 54 (1996) 16470; Phys. Rev. Lett. 72 (1994) 372.

[21] O. K. Andersen and O. Jepsen, Phys. Rev. Lett. 53 (1984) 2571;O. K. Andersen, Z. Pawlowska, and O. Jepsen, Phys. Rev. B 34 (1986) 5253.

[22] Byung Sub Kang, Suhk Kun Oh, Jean Soo Chung, and Ki Soo Sohn, Physica B 324 (2002) 279.

[23] J. P. Perdew and Y. Wang, Phys. Rev. B 45, 13244 (1992).

[24] P. E. Bloechl, O. Jepsen, and O. K. Andersen, Phys. Rev. B 49, 16233 (1994).

[25] M. Alden, H. L. Skriver, S. Mirbt, and B. Johansson, Phys. Rev. Lett. 69 (1992) 2296.

[26] T. Hoshino, R. Zeller, P. H. Dederichs, and M. Weinert, Europhys. Lett. 24 (1993) 495.

Session F6

Metamaterials (1)

Chair: V.G. Veselago

Electrodynamics of materials with negative index of refraction

Veselago V.G.
Moscow Institute of Physics and Technology, Moscow, Russia
(v.veselago@relcom.ru)

In the last several years, many efforts were spent in developing the new section of electrodynamics -electrodynamics of materials with negative index of refraction. The first experiments in this area [1, 2] were carried out by a group of physicists at the University of California, San Diego (UCSD, USA). These works demonstrated unusual electrodynamic properties of some composite materials. These characteristics can be formally explained if it is assumed that the given materials possess the negative index of refraction n. This material presents itself as a combination of small metallic elements, disposed in space in some geometric order, forming structure resembling a sort of crystal. It is possible to consider such a type of structure at wavelengths noticeably larger than the size of its elements and the distance between them. The experiments by the authors of the specified work were made in the centimeter range of wavelengths, but the sizes of the elements explored in the composites and the distances between elements are typically of the order of 7-10 mm.

The main experimental result in [1, 2] was a demonstration, for such material, of the rather unusual realization of the Snellius law of refraction. Figure 1 shows the crossing of light rays through the planar boundary between two media, with refraction indices of n_1 and n_2 respectively. If it is assumed, without loss of generality, that $n_1 > 0$ and $n_2 > 0$, the refracted ray follows the path indicated as 1-4. In the UCSD experiments, the ray follows the path 1-3. Such refracted ray will satisfy the Snellius law with $n_2 < 0$. Herewith, the equation for the Snellius law is given by

$$\frac{\sin\varphi}{\sin\psi} = \frac{n_2}{n_1} = n_{21} \tag{0.1}$$

which does not notice the change.

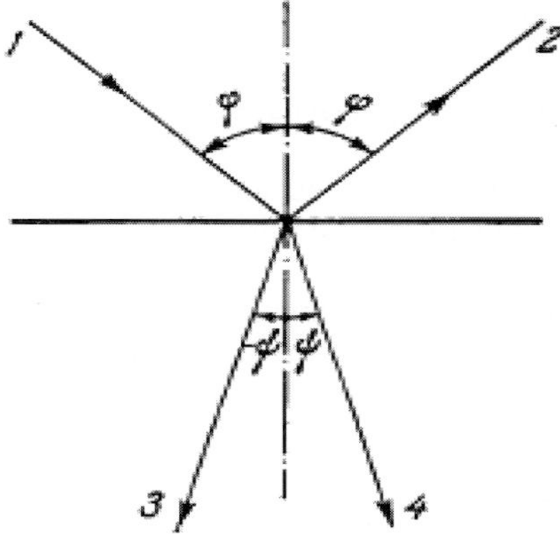

Fig1. The refraction of light on planar boundary between two media. Path 1-4 corresponds to refracted rays for the case of $n_1 > 0$ and $n_2 > 0$, but path 1-3 corresponds to the case $n_1 > 0, n_2 < 0$.

For material with $n < 0$, we have observed the unusual realization not only in the Snellius law, but also in many other phenomena in electrodynamics and optics, in particular, the Doppler effect, Cherenkov radiation, Fresnel equations, and the Fermat's principle. The main principles of electrodynamics of material with negative index of refraction could be found, in particular, in [3, 4, 5, 6]. In these works, it was shown that materials with negative refraction index are also characterized by negative values of dielectric permittivity ε and magnetic permeability μ. It is very important to note that all these statements are correct only for isotropic materials, in which n, ε and μ are scalars.

Negative value of n also corresponds to the fact that in such a material, the directions of the wave vector k and the Poynting vector S are anti-parallel. For the same reason, the directions of the phase and group velocities are also anti-parallel.

To make sure of this fact, it is sufficient to write the Maxwell equations and the expression for Poynting vector in the case of a uniform plane waves in isotropic media:

$$[kE] = \frac{\omega}{c}\mu H$$

$$[kH] = -\frac{\omega}{c}\varepsilon E \qquad (0.2)$$

$$S = [EH]$$

It is obvious that the simultaneous change in signs for ε and μ translates the right three-tuple of vectors k, E and H, into a left one. It is for this reason that in the English literature, such materials are identified as Left-Handed Materials, or, if abbreviate - LHM.

Therefore, it is possible to confirm that isotropic media with both negative values of ε and μ possess negative refraction, or for that matter, negative value of n, while the phase and group velocities are anti-parallel. It is also correct for the inverse statement - if the isotropic material possesses negative refraction index n, it must have simultaneously negative values of ε and μ, with the phase and group velocities having opposite directions.

It should be noted that the fact on the phase and group velocities in opposite directions is not something new. This question was, in particular, discussed by L.I. Mandelstam [7] many years ago. Besides, it was known long ago that in certain electronic devices (for instance, lamps of backward wave, LBW), the phase velocity is in opposite direction to the flow of energy. Recently, intensive discussions are focused on the characteristics of the so-called photonic crystals [8], in which it is also observed that vectors k and S are in opposite directions. However, photonic crystals are mostly anisotropic materials and in general, cannot be characterized by a scalar refractive index n. This pertains also to devices of the LBW type.

The appearance of materials with negative value of n raises very important question - to what measure, for the case of $n < 0$, do all the laws and formulas of electrodynamics, optics and adjacent technical sciences in which enters the value of refraction index n, remain valid? Can we always have the correct result with the direct change of $n \to -n$, as in the case of the Snellius law? In general, the answer to this question is negative.

It is known that most of the laws and equations in electrodynamics and optics correspond to the case when one or the other material is non-magnetic, with the magnetic permeability $\mu = 1$. Using

Table I

Physical law	Equation for nonmagnetic approach	Correct equation
Snellius, Doppler, Cherenkov $n = \sqrt{\varepsilon} \rightarrow n = \sqrt{\varepsilon\mu}$ if $\varepsilon, \mu < 0$, than $n < 0$	$\sin\varphi / \sin\psi = n_{21} = \sqrt{\varepsilon_2 / \varepsilon_1}$	$\sin\varphi / \sin\psi = n_{21} = \sqrt{\varepsilon_2\mu_2 / \varepsilon_1\mu_1}$
Fresnel $n = \sqrt{\varepsilon} \rightarrow 1/z = \sqrt{\varepsilon / \mu}$	$r_{\perp} = \dfrac{n_1\cos\varphi - n_2\cos\psi}{n_1\cos\varphi + n_2\cos\psi}$	$r_{\perp} = \dfrac{Z_2\cos\varphi - Z_1\cos\psi}{Z_2\cos\varphi + Z_1\cos\psi}$
Reflection coefficient for normal fall of light on the border between two media	$r = (n_1 - n_2)/(n_1 + n_2)$	$r = (Z_2 - Z_1)(Z_2 + Z_1)$
Condition for full matching	$n_1 = n_2$	$Z_1 = Z_2$
Brewster angle	$tg\varphi = n$	$tg\varphi = \sqrt{\dfrac{\varepsilon_2}{\varepsilon_1}\dfrac{\varepsilon_2\mu_1 - \varepsilon_1\mu_2}{\varepsilon_2\mu_2 - \varepsilon_1\mu_1}}$

such "non-magnetic approach" leads to many formulas which are drastically changed when μ enters, replacing $\mu = 1$. These formulas turn out to be valid only in this non-magnetic approach. Table I illustrates this situation.

From Table I, it is seen that there exists three groups of physical laws and effects, the wordings of which have different meanings when moving from the "non-magnetic approach" equations to the exact expressions.

The Snellius law and effects of Doppler and Cherenkov are in the first group. In these formulas, usually applicable in the "non-magnetic approach", the simple expression $n = \sqrt{\varepsilon}$ must be replaced by $n = \sqrt{\varepsilon\mu}$. If ε and μ are both negative, then a "minus" sign must be placed before n.

The laws of reflection and refractions of light, and in particular, the Fresnel's formulas, belonged to the second group. In these formulas, when moving from the "non-magnetic approach" to the exact formulas, the value $n = \sqrt{\varepsilon}$ should be changed, not to $n = \sqrt{\varepsilon\mu}$, but to $\sqrt{\dfrac{\varepsilon}{\mu}} = \dfrac{1}{z}$, where the Z is the wave impedance of the medium, i.e. $Z = \sqrt{\dfrac{\mu}{\varepsilon}}$. The wave impedance has the dimension of resistance (Ohm) and is a unique feature of each medium, as its speed of light. From Table I, it is seen that moving from the "nonmagnetic approach", the condition for the absence of reflection of light on planar boundary between two media, in particular, is greatly changed. This condition is now defined not by the equality of the refractive indices of the two media, but by the equality of their wave impedances. It is important to show that with negative values of ε and μ, the wave impedance Z, unlike n, remains positive.

Finally, the third group of correlations, which depends strongly on n and changes greatly when moving from the "non-magnetic approach" to the exact formulas, pertains in particular, to the formula for Brewster angle $tg\varphi = n$. The exact expression for Brewster angle is given in the last line in Table I. It is important to note that the expression under the square root in this exact formula is not changed by the simultaneous change in the signs of ε and μ of one medium. It is necessary to remember that the formula for Brewster angle, given in Table I, corresponds to a determined polarization of light. For the other polarization, perpendicular to this case, the formula can be obtained from that given in Table I by changing $\varepsilon \rightarrow \mu$ and $\mu \rightarrow \varepsilon$ in the expression under the square root. Therefore, reflection under Brewster angle always exists, for any values of permeability, but only for one of the two possible polarizations of the incident light.

Introduction into scientific discussions on the notion of "negative index of refraction" also changes the definition of a fundamental principle such as the Fermat's principle. This question is considered in detail in recent publications [9], where it is shown that the correct definition of the Fermat's principle, applicable to the propagation of electromagnetic wave through material with any sign for the index of refraction n, is that the total length of the optical path is an extremum:

$$\delta L = \delta \int n \partial l = 0 \qquad (0.3)$$

The integration in this expression (which is, in essence, the eikonal) is over the real path traveled by the ray of light. Such an approach shows that the length of the optical path traveled by the electromagnetic wave in medium with negative value of n, is also negative. Hence, it follows, in particular, that in some cases the total length of the optical path can be negative or even zero, though, certainly, the geometric path length on which the light travels and the time taken are not zero.

Such a situation actually exists when light propagates through a planar slab made of material with $\varepsilon = \mu = n = -1$. Such a slab, as shown in Figure 2, is capable of focusing the radiation coming from a point source on one side of the slab, into a point disposed on the other side of slab.

From Figure 2, it is seen that the sum of the path Am, travel by the light from the point source before the slab, and the path nB from the slab to the image, is equal to the path mn, which the light travels inside the slab, i.e.

$$Am + nB = mn \qquad (0.4)$$

Such a correlation is valid for any other possible paths travel by light from A to B, for instance $AcgB$ or $AdfB$. However, since the index of refraction is $n = -1$ inside the slab, and is $n = +1$ outside, the full optical length for light going from point A to B will be, in accordance to the expression (0.3), zero for any possible propagation path. However, as was mentioned earlier, the time taken by the light to travel from point A to B differs greatly from zero.

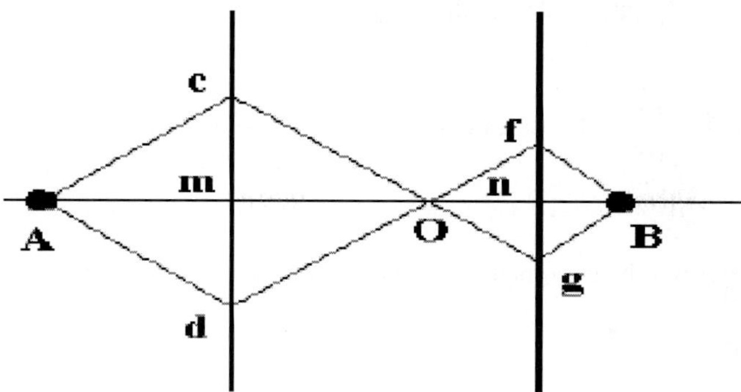

Fig.2 Propagation of light from source A to image B through a planar slab of material with $\varepsilon = \mu = n = -1$. Outside the slab is vacuum, with $\varepsilon = \mu = n = 1$.

The fact that light from a point source is focused into a point disposed on the other side of slab does not mean that this slab is a lens. Such a slab is an ideal optical instrument, which carries the space of object to the space of image without any distortion. But such effect is possible only for objects located at a distance, from the slab, not greater than the thickness of slab. The slab undoubtedly cannot focus parallel cone of rays, coming from infinity, into a point. However, the characteristics of such a slab are undoubtedly interesting and could have practical value.

In the general consideration on the characteristics of material with negative index of refraction, it is necessary to keep in view that these material must possess frequency dispersion. Actually, if ε and μ are both negative, in the absence of dispersion, the full energy in the material based on the equation

$$W = \varepsilon E^2 + \mu H^2 \tag{0.5}$$

will be negative.

However, if frequency dispersion exists, the expression (0.5) must be rewritten in the form

$$W = \frac{\partial(\varepsilon\omega)}{\partial\omega} E^2 + \frac{\partial(\mu\omega)}{\partial\omega} H^2 \tag{0.6}$$

It is clear that both $\dfrac{\partial(\mu\omega)}{\partial\omega}$ and $\dfrac{\partial(\varepsilon\omega)}{\partial\omega}$ are positive if the frequency dispersion for ε and μ are chosen to be of the sufficiently general type

$$\mu = 1 - \frac{A_m^2}{\omega^2} \tag{0.7}$$

$$\varepsilon = 1 - \frac{A_e^2}{\omega^2} \tag{0.8}$$

If it is assumed that

$$A_e^2 = A_m^2 = A^2 > \omega^2 \tag{0.9}$$

the index of refraction $n = 1 - \dfrac{A^2}{\omega^2}$ becomes negative. The phase velocity $V_{\delta} = \dfrac{\tilde{n}}{1 - \dfrac{A^2}{\omega^2}}$ and

group velocity $V_{\tilde{a}\delta} = \dfrac{\tilde{n}}{1 + \dfrac{A^2}{\omega^2}}$ will be governed by the valid equation

$$\frac{\tilde{n}}{V_{\delta}} + \frac{\tilde{n}}{V_{\tilde{a}\delta}} = 2 \tag{0.10}$$

For the waves in medium with negative n, we must choose a "minus" sign before the wave vector k. However, in medium with absorption, the vector k has not only real, but also imaginary part. The appearance of this imaginary part is the result of the presence of imaginary parts in the expressions for ε and μ. So, the question appears - should the sign before the imaginary part of the wave vector be changed, if the sign before its real part is changed?

Let us write the expressions for ε and μ as

$$\varepsilon = \varepsilon' + j\varepsilon'', \mu = \mu' + j\mu'' \tag{0.11}$$

It is not difficult to see that for the case of small dissipation, an expression for k will be

$$k = k' + jk'' = \frac{\omega}{c}\sqrt{(\varepsilon' + j\varepsilon'')(\mu' + j\mu'')} = \frac{\omega}{c}\sqrt{\varepsilon'\mu'}\left[1 + \frac{j}{2}\left(\frac{\varepsilon''}{\varepsilon'} + \frac{\mu''}{\mu'}\right)\right] \tag{0.12}$$

From (0.12), it is obvious that the change in the sign before the real part does not entail automatic change in the sign before the imaginary part of the wave vector. For a change in the sign of the imaginary part of the wave vector, it is necessary to change the sign before the imaginary parts of ε and μ that corresponds to transition from material with positive absorption to material with negative absorption, as for instance, in the case of quantum amplifiers. Such a transition, in general, is not connected to the possible transition from normal materials with positive refraction to materials with negative refraction.

The importance of the new notion - "materials with negative refraction" - depends greatly on whether we can have such materials in reality. This question has appeared since the publications of the work in [3, 4]. We have devoted significant efforts in the preparation of material with negative refraction based on magnetic semiconductor $CdCr_2Se_4$. However, these efforts ended in failure due to essential technological difficulties, which characterize the syntheses of this material. Probably, it is also not appropriate, at this point in time, to discuss the exotic mixture of electric and magnetic charges, which was considered in [5].

A critical turning point appeared when it was reported in [1, 2] on the making of composite material which could be characterized by negative values of ε and μ, and therefore, negative value of n.

In [2] was shown the result on direct measurement of angle of refraction for prism, prepared from this composite. This experiment has been shown to be fully valid for a given material of correlation (0.1) with negative n.

Since these experiments were conducted, they were repeated elsewhere by at least two independent groups of scientist [10, 11], with the same positive results.

The appearance of the new class of materials with several unusual electrodynamics phenomena has brought about, in the literature, motivated objections on its correctness. Hence, in [12], it was firmly established that negative refraction exists for phase velocity only, but the group velocity follows the usual law of refraction with positive n. The authors of this work were not upset that difference in the directions of the phase and group velocities is typical, in particular, of optically anisotropic medium, which undoubtedly cannot be characterized by a scalar refraction index. The mistake made by the authors of [12] is due to the fact that they confused the direction of the group velocity with the direction perpendicular to surfaces of constant amplitude for amplitude-modulated waves. This mistake is discussed in detail and explained in [13].

There is one more problem that appears to be in close relation with the appearance of materials with negative refraction. This is the problem of overcoming the diffraction limit, or for that matter, depending on terminology, the problem of amplification of the so-called evanescent modes. This problem was first discussed in Pendry's work [14], where it was firmly established that in material with negative refraction index, one could successfully propagate waves for which the component of wave vector k_z along the direction of propagation is purely imaginary,

$$k_z^2 = \frac{\omega^2}{c^2} - k_x^2 < 0 \tag{0.13}$$

This inequality is valid for very large k_x, or for very short waves.

In material with positive value of n, the amplitude of such waves (the evanescent modes), in accordance to (0.13), decreases exponentially rapidly along the z axis. It is exactly this circumstance that explains the impossibility of achieving images, by optical systems of objects, with sizes noticeably smaller than wavelength. However, in [14] and in a series of following articles, it became firmly established that in material with negative refraction, wave with greater values of k_x do not weaken, but amplifies. This statement is equivalent to choosing in the following equation

$$k_z = \pm j \sqrt{k_x^2 - \frac{\omega^2}{c^2}} \tag{0.14}$$

the "minus" sign before the imaginary root, rather than the "plus" sign as was done in the usual case.

The author [14] introduced the notion of "superlens" for the device shown in Figure 2, and confirmed that for this devices, the classical restriction on diffraction limit vanished.

The most conclusive evidence on the fallacy of this sort of statements can be found in [15], where by electromagnetic modeling, it is shown that the propagation of evanescent modes in the case of negative n is possible only at distances much smaller than the wavelength, as in the usual case.

However, this does not preclude the need to clarify fully the pecularities on the propagation of such modes in material with negative n.

We are now at the beginning of a long way that leads us to a new, pretty interesting, and perspective area of electrodynamics. The number of researchers, groups and organizations, involved in the discussed problems grows rapidly. Correspondingly, the number of publications in this area also grows. Interested parties can refer to very detailed selection on this area of work posted in INTERNET on http://physics.ucsd.edu/~drs/left_home. The works [3-6,9] by the author are posted in English and Russian on http://zhurnal.ape.relarn.ru/~vgv.

REFERENCES

1. Smith, D.R., Padilla, W., Vier, D.C., Nemat-Nasser, S.C., Shultz, S., (2000), *Phys.Rev.Lett.*, **84**, 4184
2. Shelby, R.A., Smith, D.R., Shultz, S., (2001), *Science*, **292,** 77
3. Veselago, V.G., *Soviet Phys.Solid State,* **8**, 2853 (1967)
4. Veselago, V.G., *Soviet Phys.USPEKHI*, **10**, 509 (1968)
5. Veselago, V.G., *Soviet Phys.JETP*, **25**, 680 (1967)
6. Veselago, V.G., in E.Burstein and Francesco de Martini (eds), POLARITONS, Proceedings of first Taormina research conference on the structure of matter, 2-6 October 1972, Taormina, Italy, Pergamon Press (1972)
7. Mandelstam, L.I., (in Russian) *JETP*, **15**, 475 (1945)
8. Notomi, M., *Optical and Quantum Electronics,* **34,** 133 (2002)
9. Veselago, V.G., *Soviet Phys.USPEKHI*, **172**, 1215 (2002)
10. Parazzoli, C.G., Greegor, R.B., Li K., Koltenbah, B.E.C., Tanielian, M., *Phys.Rev.Lett.,* **90,** 107401 (2003)
11. Houck, A.A., Brock, J.B., Chuang, I.L., *Phys.Rev.Lett,* **90**, 137401 (2003)
12. Valanju, P.M., Walser, R.M., Valanju, A.P., *Phys.Rev.Lett,* **88,** 187401 (2002)
13. Pendry, J.B., Smith, D.R., cond-mat/0206563
14. Pendry, J.B., *Phys.Rev.Lett.,* **85,** 3966 (2000)
15. Rao, X.S., Ong, C.K., cond-mat/0304474

Resolution Enhancement of a Left-Handed Material Superlens

C. K. Ong [1] and X. S. Rao [2]

[1] *Centre for Superconducting and Magnetic Materials and Department of Physics, National University of Singapore, Singapore 117542*
[2] *Temasek Laboratories, National University of Singapore, Singapore 119260*

1. Introduction

In a paper published in 1968, Veselago [1] predicted that a planar slab of left-handed materials (LHM), which possess both negative permittivity and negative permeability, could refocus the waves from a point source. Recently, Pendry [2] made a revolutionary extension and pointed out that not only the propagating waves can be refocused by a LHM slab but also the evanescent waves can be amplified inside the slab and reconstructed at the image plane. Since the evanescent waves carry structural details of the object finer than the wavelength, subwavelength resolution of the image can be achieved using the LHM slab. Due to its superior imaging performance, the LHM slab is also known as a superlens. It is soon realized that the superlens can have extensive technological applications. For example, if the superlens can be applied in photolithography, it may greatly reduce the linewidth of the ULSI circuits and improve the performance of microchips.

Pendry's superlens has initiated great interest in research and also disputes. It is the fact that all the LHM are necessarily dispersive and dissipative. Therefore, absorption, no matter how small it is, is always present in the realistic LHM slab. Garcia and Nieto-Vesperinas [3] studied analytically the effect of this unavoidable absorption and concluded that this unavoidable absorption in LHM would suppress the amplification of evanescent waves and turn it into decay. This casted shadows on the superlens concept. In a recent Letter [4], using a full-wave finite-difference time-domain (FDTD) method incorporating a specially-designed boundary condition, we studied the effect of the absorption on the behavior of evanescent waves inside the LHM slab and we showed that the amplification is a realistic effect even when finite absorption is present. In this paper, we will apply the same scheme to study the resolution limit of the LHM slabs of practically achievable parameters. We will show that great resolution enhancement can be achieved by the realistic LHM slabs and its dependence on different factors.

2. Model and Method

In the FDTD simulation, we apply the standard Yee's algorithm in which the leapfrog staggered electric field and magnetic field are updated in time using the two coupled Maxwell's curl equations [5]. For simplicity, we assume that ε and μ of the LHM have the identical plasma-like causal frequency dependence,

$$\varepsilon(\omega) = \mu(\omega) = 1 - \frac{\omega_p^2}{\omega^2 - j\nu_c\omega}$$

where ω_p is the plasma frequency and ν_c is the collision frequency. By choosing ω_p and ν_c appropriately, we have $\varepsilon(\omega)=\mu(\omega)=-1-j\gamma$ (γ denotes the loss term) in all our simulations. Figure 1 illustrates the geometry of the imaging system we are interested in. A planar LHM slab extends from $x=0$ to L with surfaces normal to the x-axis and is infinite in the yz-plane. It is surrounded by vacuum. The z-polarized electric field source is excited in the $x=-d=-L/2$ plane, which corresponds to the transverse electric (TE) case. The plane wave solution of the system is of the general form

$$E_z(x, y, t) = E_{z0} \exp(\omega t - k_x x - k_y y)$$

where k_x and k_y are wave vectors along the x- and y-directions, respectively, and they satisfy the dispersion relation

$$k_x^2 + k_y^2 = \frac{\omega^2}{c^2}$$

The sign of k_x^2 determines the propagation property of the wave along x-axis. Positive values of k_x^2 correspond to propagating waves and negative k_x^2 correspond to evanescent waves in the x-direction. Pure evanescent waves can be generated and simulated in the scheme of FDTD by applying a periodic-like boundary condition in the y-direction [4],

$$E_z(x, y + \Delta y) = E_z(x, y)\exp(-k_y\Delta y)$$

with an explicitly-assigned large transverse wave vector $k_y>k_0$ ($k_0 = \omega/c$).

In the simulation, the computational space is taken to be 4000×1 cells with $\Delta x=\Delta y=0.3$mm. At both ends in the x-direction, 10-cell uniaxial anisotropic perfectly-matched layer (UPML) [5] is added to truncate the computational space. The time step used is $\Delta t=\Delta x/(2c)=0.5$ps. A sinusoidal source with a smooth turn-on of 30 periods [6] is employed to excite a cw wave at the angular frequency $\omega\approx11$GHz. The corresponding wavelength in vacuum λ_0 (=$2\pi/k_0$) is about 566Δx.

3. Results and Discussions

Figure 2 shows the transfer function (TF), defined by the ratio of the field at the image plane to that of the source, for two LHM slabs as a function of the normalized transverse wave number k_y/k_0. The two slabs have the same absorption $\gamma=0.01$ but different thickness $L=80\Delta x$ and $L=160\Delta x$. The numerical results clearly show that evanescent waves of certain k_y can be reconstructed at the image plane. The TF is close to unity for small k_y and decreases slowly with increasing k_y until it passes through a peak and decays exponentially beyond. The peak for the thinner slab is located at at k_y =4.7 k_0 while that for the thicker one is at at k_y =2.7k_0. The emergence of the peak results from the resonance of the coupled SPs. Due to finite absorption, the resonant divergence [7] is removed. The location of the peak can be used as an estimate for the resolution enhancement (R). For the thinner slab R=4.7 and for the thicker slab R=2.7. Both of them clearly show superior performance compared to conventional imaging systems. It is clear that with constant absorption, R is dramatically reduced for the thicker slab, which is consistent with the conclusion drawn in [4] that large L tends to suppress the amplification of the evanescent waves inside the LHM slab. Since absorption also has negative effects on the absorption, it is expected that the resolution enhancement factor will be reduced with increasing γ.

We noted that two recent FDTD studies [8,9] on the imaging properties of the LHM slabs showed controversial results. In [9], Cummer showed that the superlensing effect really happens. In the numerical example he simulated, a resolution enhancement $R\approx2.5$ was achieved for a LHM slab (γ=0.0006 and L=0.44λ_0), which is consistent with our results. However, in [8], it was shown that the image of a line source focused by a LHM slab (n=-1.001+j0.013 and $L\approx3\lambda_0$) is of the order of the wavelength showing no resolution enhancement and there is no evidence of the amplification of evanescent waves. As we have shown above, increasing L dramatically degrades the performance of the superlens and the superlensing effect only happens for near-field imaging. Based on this, it is now clear that the confusing results shown in [8] were obtained with an inapproporiate numerical example where $L\approx3\lambda_0$ is too large for the superlens concept to be applicable.

4. Conclusions

We have carried out full-wave numerical simulations to examine the superlensing effect of the LHM slabs. Our results have shown that the resolution of the image can be greatly enhanced above the resolution limit of conventional imaging systems. However, the resolution enhancement factor is limited by the absorption and the physical dimension of the LHM slabs.

References

[1] V. G. Veselago, *Sov. Phys. Usp.* **10**, 509 (1968).

[2] J. B. Pendry, *Phys. Rev. Lett.* **85**, 3966 (2000).

[3] N. Garcia and M. Nieto-Vesperinas, *Phys. Rev. Lett.* **88**, 207403 (2002).

[4] X. S. Rao and C. K. Ong, *Amplification of evanescent waves in a lossy left-handed material slab*, submitted to *Phys. Rev. Lett.*.

[5] A. Taflove and S. C. Hagness, *Computational Electrodynamics: The Finite-Difference Time-Domain Method* (Artech House, Norwood, Mass., 2000).

[6] R. W. Ziolkowski and E. Heyman, *Phys. Rev. E* **64**, 056625 (2001).

[7] D. R. Smith, D. Schurig, M. Rosenbluth, S. Schultz, S. A. Ramakrishna, and J. B. Pendry, *Appl. Phys. Lett.* **82**, 1506 (2003).

[8] P. F. Loschialpo, D. L. Smith, D. W. Forester, F. J. Rachford, and J. Schelleng, *Phys. Rev. E* **67**, 025602 (2003).

[9] S. A. Cummer, *Appl. Phys. Lett.* **82**, 1503 (2003).

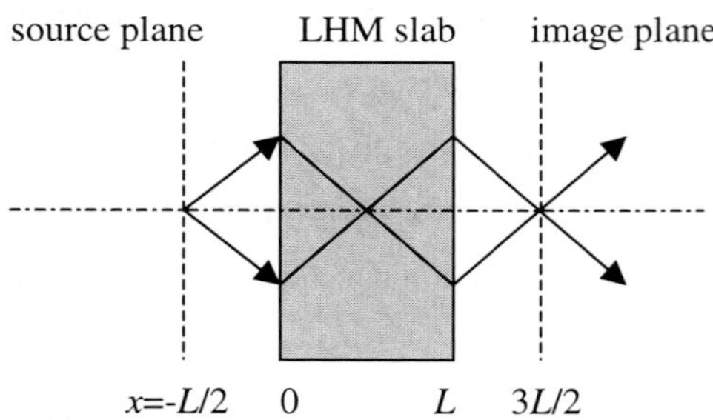

Figure 1. Schematic illustration of the geometry of the imaging system.

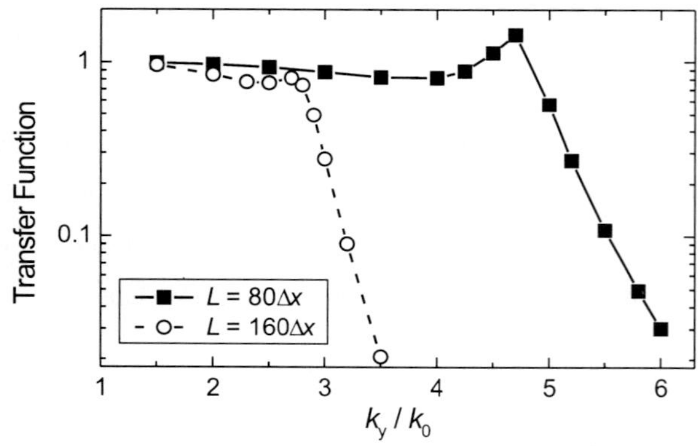

Figure 2. Transfer function for two LHM slabs (γ=0.01 for both) as a function of normalized transverse wave number k_y/k_0. The two slabs have different thickness L=80Δx ≈0.14λ_0 (solid square) and L=160Δx≈0.28λ_0 (open circle).

The Perfect Lens: Possible Alternative Realizations

Stanislav Maslovski, Sergei Tretyakov
Radio Laboratory, SMARAD
Helsinki University of Technology, Finland

Abstract

Alternative possibilities of how to create an electromagnetic device being able to reconstruct near-field images of a source with sub-wavelength resolution (so-called perfect lens) are considered. It is shown that there is a variety of such means not involving backward-wave (double-negative, left-handed, or *Veselago*) materials or periodical backward-wave structures.

1. Introduction

It is known that using materials with simultaneously negative permittivity and permeability (at a given frequency) focusing of divergent homocentric electromagnetic beams by "planar lenses" becomes possible because of the negative refraction, see [1]. As it was found in [2], such lenses are also able to "amplify" evanescent fields carrying information about sub-wavelength details of the near field of a source. There were also a few experimental papers (e.g., [3], [4]) demonstrating negative refraction effects in microwave composite materials. These facts received a lot of attention in the recent literature.

Despite the fact that there is a way to realize of new lenses by means of backward-wave materials, we will make a step aside in this paper. Starting from analysis of an ideal Veselago slab lens performance we will formulate a couple of equivalent problems and show that the special electromagnetic property of the material filling, namely, existence of backward-waves, is not, in fact, crucial for a planar device operating as a lens or, moreover, as a perfect lens. What makes a Veselago lens to behave as such, are the properties of the two slab interfaces. Devices working in a similar manner can be constructed using artificial plane grids or sheets imposing necessary boundary conditions at two parallel planes in air.

2. The ideal Veselago lens and an equivalent problem

Consider an ideal Veselago slab lens positioned in free space. The slab relative permittivity and permeability are both equal to -1 at the operating frequency (we work in the frequency domain and use the time dependence of the form $\exp(+j\omega t)$). The boundary conditions at two slab interfaces are the usual Maxwellian boundary conditions.

It is easy to notice that the field equations in free space regions and in the Veselago slab differ only by complex conjugation. Substitution

$$\mathbf{E}_{(old)}, \mathbf{H}_{(old)} = \mathbf{E}^{*}_{(new)}, \mathbf{H}^{*}_{(new)} \tag{1}$$

(here and thereafter * denotes the complex conjugation operation) into the field equations in Veselago slab results in a new but equivalent problem. In the new formulation the field equations are the same in all three regions, and they are simply the Maxwell equations in free space. The boundary conditions on the two interfaces, however, are no more the standard continuity conditions, but they involve the complex conjugation operation:

$$\mathbf{E}_{t(1)} = \mathbf{E}^{*}_{t(2)}, \quad \mathbf{H}_{t(1)} = \mathbf{H}^{*}_{t(2)} \tag{2}$$

We discuss the physics of these boundary conditions involving complex conjugation later. At this stage we see that an ideal Veselago slab refraction problem is mathematically equivalent to the refraction at two conjugating planes in free space. Hence, for the system of two such planes in free space the field solutions are the same as for a Veselago material slab: propagating plane waves are refracted negatively, and the evanescent modes are "amplified", which are the conditions for a perfect lens.

An illustrative plane wave incidence problem for a single conjugating interface placed in

"

free space can be considered. It can be shown that the wave vector tangential component k_t changes sign across the interface due to complex conjugation in the boundary conditions (2). The wave vector normal component k_n does not change across the interface. The same holds for the Poynting vector components. Clearly, negative refraction takes place. Further, if A, B, and C denote the incident, transmitted, and reflected wave electric field tangential components, respectively, then for the TM incidence one can write:

$$A + C = B^*$$
$$\frac{A - C}{\eta} = \frac{B^*}{\eta^*} \tag{3}$$

Here $\eta(k_t)$ is the wave impedance connecting tangential components of electric and magnetic fields of a wave. It is obviously a function of k_t. For the propagating modes, η is purely real, for the evanescent ones it is purely imaginary. The solution of (3) is

$$C = \frac{1 - \eta/\eta^*}{1 + \eta/\eta^*} A, \quad B = \frac{2A^*}{1 + \eta^*/\eta} \tag{4}$$

It is seen that for propagating modes (real wave impedance) the interface is perfectly matched: reflected field amplitude $C = 0$, and the transmitted field amplitude $B = A^*$. For evanescent modes (imaginary wave impedance) the denominator of (4) is zero, and a surface wave (surface polariton) resonance occurs: $B, C \to \infty$.

An important conclusion from this analysis is that all the phenomena necessary for perfect focusing of the entire wave spectrum take place *at the two interfaces* of the Veselago slab. If one can realize a sheet such that travelling waves refract negatively when crossing this sheet, a system of two such sheets *in free space* will focus propagating modes of a point source just like a Veselago slab. Also, if a sheet supports surface waves for any $k_t > k_0 = \omega\sqrt{\varepsilon_0\mu_0}$, then two such sheets in free space will also reconstruct the entire evanescent spectrum.

3. Physical restrictions

Complex conjugation in the frequency domain corresponds to time reversal in the time domain. If the boundary condition (2) were true for every spectral frequency then this condition would obviously violate the causality principle. Speaking in simple words, one could in this case say that transformation (1) forces the time to go in the opposite directions inside sub-regions of a single physical system. But if we are interested only in steady-state operation at a given frequency, then the complex conjugation becomes possible, at least using non-linear or active devices, e.g., mixers.

Applying transformation (1) and concluding that the middle region has become a true free-space region we have silently assumed that *all* electromagnetic quantities and relations have been kept in their original free-space form after such a transformation. The physical reason of this is the time reversal invariance of a reciprocal electromagnetic system.

4. Realizations not involving complex conjugation of the fields

In the previous section we have considered a potential device that is able to ideally imitate the operation of a Veselago slab lens. Now we will concentrate on other possibilities providing additional freedom in making sub-wavelength resolution lenses and, in the same time, not involving complex conjugation operator. We will see that under certain restrictions devices that "amplify" evanescent fields can be designed using only passive elements.

Complete reconstruction of the field distribution in the source plane at a distant image plane must involve phase compensation for the propagating space harmonics and "amplification" for the evanescent ones. In other words, we need to synthesize a device that somehow inverts the action of a free space layer. The microwave circuit theory has a powerful synthesis method based on so-called transmission matrices. In short, a transmission T-matrix is a matrix defined by

$$\begin{pmatrix} E_2^- \\ E_2^+ \end{pmatrix} = \begin{pmatrix} t_{11} & t_{12} \\ t_{21} & t_{22} \end{pmatrix} \cdot \begin{pmatrix} E_1^- \\ E_1^+ \end{pmatrix} \tag{5}$$

where $E_1^\pm$ and $E_2^\pm$ denote tangential components of the electric field complex amplitudes of partial waves at the first (input) and the second (output) interfaces of a device, respectively (we restrict ourselves by plane structures and plane waves). The signs $\pm$ correspond to the signs in the propagator exponents $\exp(\pm jk_n z)$ of the partial waves, z is the axis orthogonal to the interfaces. It is known that the T-matrix of a serial connection of several devices described by their T-matrices is simply a multiplication of the matrices in the order determined by the connection.

If we apply this approach to our problem then in terms of T-matrices we, obviously, need

$$T_{\text{total}} = T_{\text{after}} \cdot T_{\text{device}} \cdot T_{\text{before}} = \begin{pmatrix} 1 & 0 \\ 0 & 1 \end{pmatrix} \tag{6}$$

for every spatial harmonic of the source field. Here, T_{before} and T_{after} represent air layers between the source plane and the device, and between the device and the image plane.

Condition (6) is a strict condition requiring not only the device one-way transmission to be such that it reconstructs the source field picture at the image plane, but also the matching to be ideal and the device operation to be symmetric (reversible in the optical sense). There are at least two solutions known for this idealistic case by now: an ideal Veselago slab lens and the conjugating planes described above. Any practical realization by no means is ideal, that is why some less strict conditions can be also considered.

A space layer of thickness $d/2$ has the T-matrix defined by

$$T_{\text{space}} = \begin{pmatrix} \exp(-jk_n d/2) & 0 \\ 0 & \exp(+jk_n d/2) \end{pmatrix} \tag{7}$$

To compensate the action of two such layers before and after the device, the device T-matrix has to be, obviously,

$$T_{\text{device}} = \begin{pmatrix} \exp(+jk_n d) & 0 \\ 0 & \exp(-jk_n d) \end{pmatrix} \tag{8}$$

As it was already mentioned, the Veselago lens and the conjugating planes are particular solutions[i]. But let us consider a more general case given by such simple matrix multiplication:

$$T_{\text{device}} = \begin{pmatrix} a & b \\ c & d \end{pmatrix} \cdot \begin{pmatrix} \exp(-jk_n d) & 0 \\ 0 & \exp(+jk_n d) \end{pmatrix} \cdot \begin{pmatrix} e & f \\ g & h \end{pmatrix} \tag{9}$$

It is easy to show that if $a = d = e = h = 0$, and $bg = cf = 1$, then the total device T-matrix takes form (8), i.e., the necessary matrix of an ideal lens! There is no magic here in obtaining growing exponents in place of decaying ones. The result comes just from simple permutations of the matrix elements accordingly to the matrix multiplication (9).

As it is seen from (9), the device is a combination of two "field transformers" (e.g. thin sheets of certain electromagnetic properties) and an air layer in between, similarly to the considered conjugating planes device.

5. Impedance grids

Let us consider a simple system: a planar lossless isotropic grid, e.g., a conductive wire mesh. If the grid supports only electric current, and there is no effective magnetic current induced in the grid (e.g., when the grid structure is completely planar), then the grid reflection coefficient R and transmission coefficient T at the grid plane are connected as

---

[i] Perhaps, it is not seen directly why (8) holds for a pair of conjugating planes as there is only the normal component of the wave vector in (8) which is kept untouched by (2). But considering conjugations at both planes together taking the inner space partial wave propagators $\exp(\pm jk_n d)$ into an account the result (8) can be easily obtained.

$$T = 1 + R \tag{10}$$

provided that they are defined through the electric field tangential components. The corresponding T-matrix of such a grid is

$$T_g = \begin{pmatrix} \dfrac{1+2R}{1+R} & \dfrac{R}{1+R} \\[2mm] -\dfrac{R}{1+R} & \dfrac{1}{1+R} \end{pmatrix} \tag{11}$$

It is possible to make grids supporting propagation of surface modes (known as *slow waves* in radio engineering). If the tangential component of the wave vector of an incident wave coincides with the propagation factor of a surface mode, the surface mode resonance appears. Obviously, the incident wave should be evanescent in this case to match with the propagation constant of the surface mode. At a surface mode resonance $R \to \infty$ (for evanescent modes R is not bounded by $|R| \leq 1$). Then, the grid T-matrix takes the form

$$T_g = \begin{pmatrix} 2 & 1 \\ -1 & 0 \end{pmatrix} \tag{12}$$

Calculating the total device matrix directly we obtain

$$T_{\text{device}} = T_g \cdot T_{\text{space}} \cdot T_g = \begin{pmatrix} 4\exp(-jk_n d) - \exp(jk_n d) & 2\exp(-jk_n d) \\ -2\exp(-jk_n d) & -\exp(-jk_n d) \end{pmatrix} \tag{13}$$

which seems to be far from the desired, but let us check what is the device S-matrix (scattering matrix). After some algebra involving the known relations connecting the elements of T and S matrices we get

$$S_{\text{device}} = \begin{pmatrix} -t_{21}/t_{22} & 1/t_{22} \\ t_{11} - t_{12}t_{21}/t_{22} & t_{12}/t_{22} \end{pmatrix} = \begin{pmatrix} -2 & -\exp(+jk_n d) \\ -\exp(+jk_n d) & -2 \end{pmatrix} \tag{14}$$

Notice pluses in the exponents for s_{12}, s_{21}. The device can "amplify" the resonating evanescent modes.

6. Conclusions

In this paper, several alternative possibilities for realization of planar lenses being able to reconstruct near-field images of a source with sub-wavelength resolution (so-called perfect lenses) have been considered. It has been shown that one does not necessarily need a composite medium possessing negative permittivity and permeability or another kind of backward-wave medium to produce a perfect lens. An analogous device can be constructed using artificially made surfaces or sheets imposing necessary boundary conditions on the fields in free space.

References

[1] V.G. Veselago, The electrodynamics of substances with simultaneously negative values of ε and μ, *Soviet Physics Uspekhi,* vol. 10, pp. 509–514, 1968.

[2] J.B. Pendry, Negative refraction makes a perfect lens, *Phys. Rev. Lett.,* vol. 85, no. 18, pp. 3966–3969, 2000.

[3] D.R. Smith, W.J. Padilla, D.C. Vier, S.C. Nemat-Nasser, S. Schultz, Composite media with simultaneously negative permeability and permittivity, *Phys. Rev. Lett.,* vol. 84, no. 18, pp. 4184–4187, 2000.

[4] R.A. Shelby, D.R. Smith, S. Schultz, Experimental verification of a negative index of refraction, *Science,* vol. 292, pp. 77–79, 2001.

Strong spatial dispersion in wire media in the very large wavelength limit

P.A. Belov[1,2], R. Marques[3], S.I. Maslovski[1], I.S. Nefedov [1,4],
M. Silveirinha[5], C.R. Simovski[1,2], and S.A. Tretyakov[1]

[1]Radio Laboratory, Helsinki University of Technology, P.O. Box 3000, FIN-02015 HUT, Finland
[2]Physics Department, St. Petersburg Institute of Fine Mechanics and Optics,
Sablinskaya 14, 197101, St. Petersburg, Russia
[3]Department of Electronics and Electromagnetism, Facultad de Fisica,
University of Sevilla, Av. Reina Mercedes s/n, 41012 Sevilla, Spain
[4]Institute of Radio Engineering and Electronics, Russian Academy of Science,
Zelyonaya 38, 410019, Saratov, Russia
[5]Instituto Superior Tecnico, Instituto de Telecomunicacoes,
Av. Rovisco Pais, 1049-001, Lisboa, Portugal

We have found that the wire medium (array of long conducting cylinders), one of the component of the famous realization of Veselago media, exhibits strong spatial dispersion even in the very large wavelength limit. Our analysis reveals that the conventional description of this medium by means of a local dispersive uniaxial dielectric tensor is not complete, leading to unphysical results for the propagation of electromagnetic waves at all frequencies. Since non-local constitutive relations have been usually considered in physics as a second-order approximation, meaningful in the short wavelength limit, the aforementioned result is important in a much more general sense for the theory of electromagnetic materials.

Causality imposes that all material media must be dispersive. In most cases this behavior results in local dispersive constitutive relations – i.e. in frequency-dependent constitutive permittivity and permeability tensors. Non-local dispersive behavior (i.e., spatial dispersion), which results in constitutive operators depending also on the spatial derivatives of the mean fields (or, for plane electromagnetic waves, on the wavevector components) is usually considered as a small effect, meaningful in the short wavelength limit. Specifically, spatial dispersion will always appear when the higher-order terms in the series expansion of the constitutive parameters in power series of the dimensionless parameter a/λ, (a is the lattice constant of the crystal, and λ is the wavelength inside the medium) cannot be neglected [1]. Thus, it is usually assumed that non-local dispersive reations are only meaningful when λ approaches a. The usually weak natural optical activity of chiral materials is a well-known example of the application of this principle [1]. When such principle is translated to the analysis of discrete artificial media, also called metamaterials, it would imply that non-local dispersive constitutive relations are only expected to be a small refinement of the local constitutive relations usually considered. However, there is at least one counter example for this assumption, which we will describe in this presentation.

The parallel wire medium is a medium formed by a regular lattice of ideally conducting wires with small radii compared to the lattice periods and the wavelength, see Fig. 1. It has been known in microwave engineering for a long time [2–4] as an artificial dielectric, also called rodded medium, and quasistatic models of the effective permittivity are available [5]. Recently, some attention to wire media has been paid also in optics (e.g. [6, 7]) and in the realization of Veselago ("left-handed" media) [8, 9] as composite media made from lattices of long conducting wires and split ring resonators [10–12] (see discussions corresponding to that in [13]). The electromagnetic response of the specific wire medium shown in Fig. 1 is analyzed in [3, 7] following different approaches. Both analyses are carried out for the wave propagation perpendicular to the wires and show that, for the electric field polarization along the wires, the medium is characterized (if a/λ, $b/\lambda <<1$) by a frequency-dependent effective dielectric constant

$$\varepsilon = \varepsilon_0\left(1 - \frac{k_0^2}{k^2}\right) = \varepsilon_0\left(1 - \frac{\omega_0^2}{\omega^2}\right). \tag{1}$$

The circular frequency ω_0 (corresponding to wavenumber k_0) in (1) plays the role of an equivalent "plasma frequency". Thus, this medium is often called "artificial plasma", since the ideal (collisionless) electron plasma is described by the same equivalent parameter. If the wires are assumed to be very thin, so that their polarization in the direction orthogonal to the wires can be neglected, the effective permittivity for the electric field polarization orthogonal to the wires is ε_0.

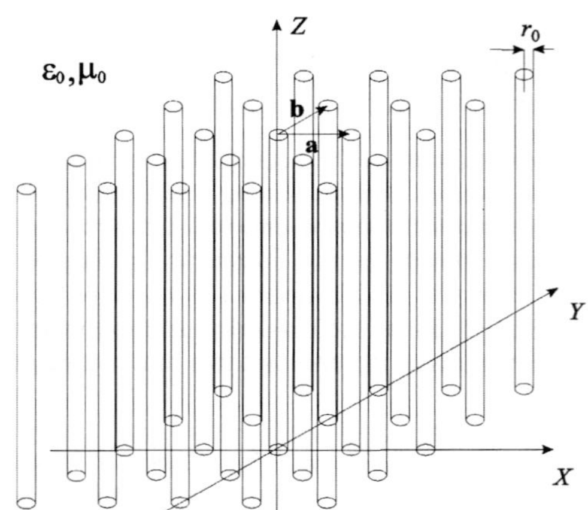

FIG. 1: The geometry of wire media: a lattice of parallel ideally conducting thin wires.

The aforementioned analysis suggests that this wire medium could be modeled as an uniaxial dielectric with the following local permittivity dyadic:

$$\overline{\overline{\varepsilon}} = \varepsilon\, z_0 z_0 + \varepsilon_0 (x_0 x_0 + y_0 y_0), \tag{2}$$

whose permittivity in the axial direction, ε, would be given by (1). However, it will be shown in the following that this naive hypothesis leads to unphysical results and must be substituted by a non-local dispersive relation. In fact, assuming that the medium can be described by the uniaxial dyadic (2), the dispersion equation for extraordinary plane waves with the wavevector $(q_x;\ q_y;\ q_z)^{\mathrm{T}}$ in this uniaxial dielectric reads [14, 15]:

$$\varepsilon_0(q_x^2 + q_y^2) = \varepsilon(k^2 - q_z^2), \tag{3}$$

where k is the phase constant of the host matrix.

We have shown in [16], that the dispersion equation for wire media is

$$q_x^2 + q_y^2 + q_z^2 = k^2 - k_0^2. \tag{4}$$

Comparing (3) and (4), one obtains

$$\varepsilon(k,q_z) = \varepsilon_0\left(1 - \frac{k_0^2}{k^2 - q_z^2}\right). \tag{5}$$

This shows that wire media still can be described by the permittivity dyadic (2), but the axial permittivity ε must be a non-local parameter of form (5). The conventional expression (1) is only a particular case of (6), valid for wave propagation in the x - y plane.

The main difference between the local uniaxial model (1) and the proposed non-local model (5) for the parallel wire medium is that the non-local model predicts a stop band (at frequencies below ω_0) for extraordinary waves propagating along any direction in the medium. On the contrary, (1)–(3) predict propagation of extraordinary waves at any frequency provided $q_z > k$. Thus, both models predict qualitative very different behaviors, even near the cutoff "plasma" frequency. That is, the non-locality of the new constitutive relations affects the electromagnetic response of the medium even in the very large wavelength limit, thus being important for waves with any values of the a/λ ratio inside the medium.

In the following we will describe some relevant effects in the analyzed parallel wire medium, associated with the spatial dispersion. Refraction and reflection of plane waves at a plane interface show strong differences between the local and non-local models. Let us consider an interface between an isotropic dielectric with the permittivity ε_1 and a uniaxial dielectric with ε described by (1). The interface is in the $(y$ - $z)$-plane, and it is illuminated by a plane wave coming from an isotropic dielectric. The wave vector and electric field vector lie in the $(x$ - $z)$ plane ($q_y = 0$, $E_y = 0$). The incidence angle of the plane wave is θ. If $\varepsilon_1 > \varepsilon_0$, $\varepsilon < 0$, and $\sin^2(\theta) < \varepsilon_0/\varepsilon_1$, the wave will be completely reflected, but for $\sin^2(\theta) > \varepsilon_0/\varepsilon_1$ some part of the wave will be transmitted through the interface. This transmitted wave will be an extraordinary wave, as it follows from its polarization state, and can be excited at any frequency. This amazing behavior disappears when the non-local model summarized in (5) is used. Indeed, if the non-local axial permittivity (5) is used for the wire medium, we observe that no transmission inside the wire medium is possible for $k < k_0$. At $k = k_0$ transmission is possible only in the case of the normal incidence. Only if $k > k_0$, a refracted wave appears.

Let us next consider the guidance of electromagnetic waves in a parallel-plate waveguide infinite in the x and y directions and bounded by parallel perfectly conducting planes orthogonal to the z-axis. Separation between the conducting walls is d. We assume that this waveguide is filled with a wire medium with the wires along the z-direction. We will consider eigenwave propagation along the x-axis of the TM_{01} mode. For waveguides filled by a local uniaxial dielectric with anisotropy axis along the z direction we have from (3)

$$\varepsilon_0 q_x^2 = \varepsilon(k^2 - q_z^2), \qquad q_x = \sqrt{\frac{\varepsilon}{\varepsilon_0}\left[k^2 - \left(\frac{\pi}{d}\right)^2\right]}. \tag{6}$$

If $\varepsilon > 0$, Equation (6) gives a cut-off for $k < \pi/d$ and propagation for $k > \pi/d$. In contrast, if $\varepsilon < 0$, propagation is allowed when $k < \pi/d$ (and forbidden for $k > \pi/d$). Within this pass band a backward wave ($dq/d\omega < 0$) propagates, as one can see in Fig. 2 (thin solid lines).

This nonphysical behavior disappears if one fills the waveguide with the non-local wire medium. Using (4), we have in this case

$$q_x^2 + q_z^2 = k^2 - k_0^2, \quad q_x = \sqrt{k^2 - k_0^2 - \left(\frac{\pi}{d}\right)^2}, \tag{7}$$

and we obtain the usual frequency behavior: cutoff for $k^2 < k_0^2 + (\pi/d)^2$ and propagation for $k^2 > k_0^2 + (\pi/d)^2$. An increase of the cut-off frequency is observed compared to the case when there is no filling medium, as one can see in Fig. 2 (thick solid line and dashed line).

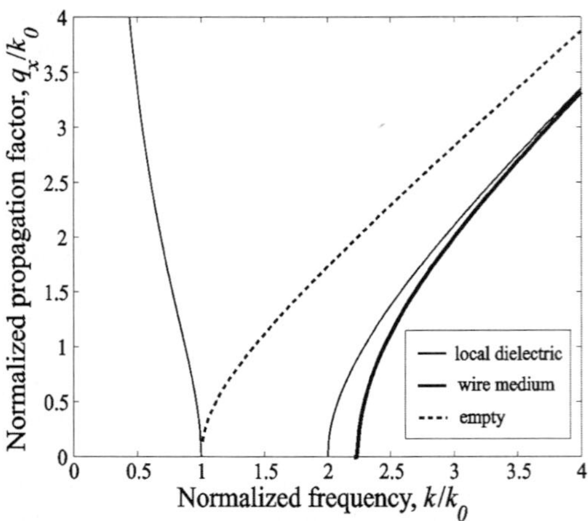

FIG. 2: Normalized propagation factors in a parallel–plate waveguide vs. normalized frequency for different types of waveguide filling: thin solid lines – uniaxial dielectric with a negative permittivity; thick solid line – wire medium; dashed line – empty waveguide. The wire medium and the uniaxial dielectric have the same $k_0 = \pi/(2d)$.

In summary, it has been shown that parallel wire media possess very strong spatial dispersion effects at any frequencies, including the very large wavelength limit. Inconsistency of the local model for parallel wire media with non-vanishing wavevector component along the wires has been shown. Dramatic differences in the predicted behavior of that media, arising from the use of the conventional local and/or the non-local model for the permittivity are shown. We feel that the reported results opens the question of the role of spatial dispersion in the adequate characterization of discrete metamaterials as effective media, at least if arbitrary directions of propagation and/or polarization of the electromagnetic field should be considered in the analysis. Also, this study provides an example of an effective medium in which spatial dispersion is important at any frequency, in contrast with some commonly assumed ideas about the physical relevance of this effect.

[1] L. Landau, E. Lifschitz, L. Pitaevski, *Electrodynamics of continuous media* (Pergamon Press, 1984)
[2] J. Brown, *Progress in dielectrics* **2**, 195 (1960)
[3] W. Rotman, *IRE Trans. Ant. Propag.* **10**, 82 (1962)
[4] R. King, D. Thiel, and K. Park, *IEEE Trans. Antennas and Propagat.* **31**, 471 (1983)
[5] S. Maslovski, S. Tretyakov, and P. Belov, *Microw. and Optical Technol. Lett.* **35**, 47 (2002)
[6] J. Pendry, A. Holden, W. Steward, and I. Youngs, *Phys. Rev. Lett.* **76**, 4773 (1996)
[7] J. Pitarke, F. Garcia-Vidal, and J. Pendry, *Phys. Rev. B* **57**, 15261 (1998)
[8] V. Veselago, *Sov. Phys. Usp.* **10**, 509 (1968)
[9] J. Pendry, *Phys. Rev. Lett.* **85**, 3966 (2000)
[10] D. Smith, W. Padilla, D. Vier, S. Nemat-Nasser, and S. Schultz, Phys. Rev. Lett. **84**, 4184 (2000)
[11] R. Shelby, D. Smith, and S. Schultz, *Science* **292**, 77 (2001)
[12] D. R. Smith and N. Knoll, *Phys. Rev. Lett.* **85**, 2933 (2000)
[13] A. Pokrovsky and A. Efros, *Phys. Rev. Lett.* **89**, 093901 (2002)
[14] V. Ginzburg, *The propagation of electromagnetic waves in plasmas* (Oxford Pergamon, 1964)
[15] I.V. Lindell, S. Tretyakov, K. Nikoskinen, S. Ivonen, *Microw. and Optical Technol. Lett.* **31**, 129 (2001)
[16] P. Belov, S. Tretyakov, and A. Viitanen, *J. Electromagn. Waves Applic.* **16**, 1153 (2002)

Session F7

Metamaterials (2)

Chair: J.A. Kong

Theoretical, numerical, and experimental studies of metamaterials with a negative refractive index

J. A. Kong, T. M. Grzegorczyk, J. Pacheco Jr., C. D. Moss, B.-I. Wu, Y. Zhang
Research Laboratory of Electronics
Massachusetts Institute of Technology
Cambridge, MA 02139, USA.

L. Ran, X. Zhang, K. Chen
Department of Information and Electronic Engineering,
Zhejiang University, Hangzhou 310027, China.

We study negative refractive index metamaterials from theoretical, numerical, and experimental point of views. Theoretically we calculate the time domain power (Poynting vector) transmitted at a boundary between a right-handed medium and a frequency dispersive left-handed medium, both homogenous and isotropic, for case of a multi-frequency signal. Numerically we perform full wave 3D FDTD simulations of the known rod and split-ring lattice in various configurations. Experimentally, we study the shifting of a Gaussian beam and the deflection of power inside a T-junction filled with an L-shaped metamaterial. In all three approaches, we find that materials that exhibit negative refraction not only theoretically exist, but can be built in the form of metamaterials made of rods and split-rings.

I. Introduction

Metamaterials with simultaneously negative permittivity and permeability have recently been the center of much research. Veselago first predicted in 1968 [1] that such materials, which he termed left-handed (LH) due to the LH triad formed by the $\vec{E}$, $\vec{H}$, and $\vec{k}$ vectors, would have negative indices of refraction among other interesting properties. Although no such naturally occurring materials are known, it has been found that a periodic collection of metallic split-ring resonators (SRR) and rods can been constructed with an effective permittivity and permeability that are simultaneously negative over some frequency band [2-3]. Experimentally such metamaterials have been built and measured to have negative indices of refraction [4-6]. In addition to metamaterials composed of SRR and rods, there are also other configurations that exhibit negative refraction such as transmission line networks [7] and photonic crystal structures [8-9].

In this paper, we consider three approaches that demonstrate both the theoretical and practical aspect of constructing LH materials. The first approach demonstrates the theoretical existence of negative refraction under the assumption of an isotropic homogeneous material. Specifically, in the first part of this paper we analyze the transmission of multi-frequency wave from an isotropic RHM into a frequency dispersive isotropic LHM by explicitly calculating the time and space dependent electric and magnetic fields to verify the negative refraction of the power. The other two approaches demonstrate the practicality of successfully constructing a material with LH properties through numerical and physical experiments. In particular, in order to verify the LH behavior of the design given in [4], we have performed a series of full-wave 3D numerical FDTD simulations of the SRR-rod lattice in various macroscopic configurations including a prism shape. Finally, in order to corroborate theoretical and numerical predictions, we have performed various experiments using one specific microscopic metamaterial design in various macroscopic configurations. We shall report here two new experiments, namely the shifting of a Gaussian beam impinging on a slab of metamaterial and the deflection of power inside a T-junction filled with an L-shaped metamaterial. In both cases we find a frequency band where the metamaterial clearly operates in a left-handed regime.

II. Theoretical results

In order to study the effect of dispersion on the direction of power propagation, we consider the transmission of an incident field, composed of many frequencies, from an isotropic non-dispersive region into an isotropic dispersive region. The total TE polarized incident electric field is constructed as a weighted sum of plane waves, and is given by,

$$E_{\ell y}(\bar{r},t) = \int_{-\infty}^{\infty} d\omega A(\omega) e^{i\phi_\ell(\bar{r},\omega) - i\omega t}, \tag{1}$$

where $A(\omega) = \alpha(\omega) + \alpha^*(-\omega)$ is the signal spectrum, $k_x(\omega) = k_0 \sin\theta_i$, $k_{0z}(\omega) = k_0 \cos\theta_i$, $k_0 = \omega\sqrt{\mu_0 \varepsilon_0}$, and $\phi_\ell(\bar{r},\omega) = k_x(\omega)x + k_{\ell z}(\omega)z$, with ℓ referring to region 0 or 1.

In order to understand the effects of dispersion, we first consider an incident field composed of two slightly separated frequencies with the following spectrum,

$$\alpha(\omega) = \frac{E_0}{2}\left[\delta(\omega - \omega_1) + \delta(\omega - \omega_2)\right] \tag{2}$$

where E_0 is the real-valued amplitude, $\omega_2 - \omega_1 = \delta\omega$, and $|\delta\omega/\omega_1| = 1$. From (1), we can find the expression for the time and space dependent transmitted electric field,

$$E_{1y}(\bar{r},t) = E_0\left[\cos\psi_1 + \cos\psi_2\right], \tag{3}$$

where $\psi_j = \phi_1(\bar{r},\omega_j) - \omega_j t$, with $j = 1,2$. Note that here, we consider the lossless case and let $T(\omega,\theta_i) = 1$ for the simplicity of derivation. The more general case of a Gaussian spectrum that includes the transmission coefficient can also be done [10]. To proceed with the Poynting vector calculation, we first determine the magnetic field, which is given by

$$H_{1z}(\bar{r},t) = E_0\left[\frac{k_x(\omega_1)\cos\psi_1}{\omega_1\mu_1(\omega_1)} + \frac{k_x(\omega_2)\cos\psi_2}{\omega_2\mu_1(\omega_2)}\right] \tag{4}$$

where, for brevity, we list only H_{1z}. In order to determine the power propagation direction, we calculate the time average value of the time dependent Poynting vector by integrating $\bar{S}_1(\bar{r},t) = \bar{E}_1(\bar{r},t) \times \bar{H}_1(\bar{r},t)$ over a period T, which is chosen to be the period of the combined signal, i.e., the common period of the frequencies $\omega_1 + \omega_2$ and $\omega_1 - \omega_2$. Integrating, we find that the cross terms average to zero if $\omega_1 \neq \omega_2$ yielding,

$$<\bar{S}_1(\bar{r},t)> = \frac{E_0^2}{2}\left[\frac{\bar{k}_1(\omega_1)}{\omega_1\mu_1(\omega_1)} + \frac{\bar{k}_1(\omega_2)}{\omega_2\mu_1(\omega_2)}\right], \tag{5}$$

where $\bar{k}_1(\omega) = \hat{x}k_x(\omega) + \hat{z}k_{1z}(\omega)$. From this expression, we see that the power flow is in the average direction of the two single frequencies. Thus, depending on the values of $\mu_1(\omega_1)$, $\mu_1(\omega_2)$, $\varepsilon_1(\omega_1)$, and $\varepsilon_1(\omega_2)$, the wave refracts at either a positive or a negative angle. In the case that both the permittivities and permeabilities at each frequency are negative, the wave refracts at a negative angle.

We now consider the cases of an RHM-LHM interface and an LHM-LHM interface for a wave with two frequency components at $f_1 = 10.5\,\text{GHz}$ and $f_2 = 11.5\,\text{GHz}$ $(\omega = 2\pi f)$. Fig. I shows at a specific time, the resulting $\hat{x}$-component of the Poynting vector, overlayed with white colored arrows that indicate the overall direction of the vector. Clearly, in the case of the RHM-RHM

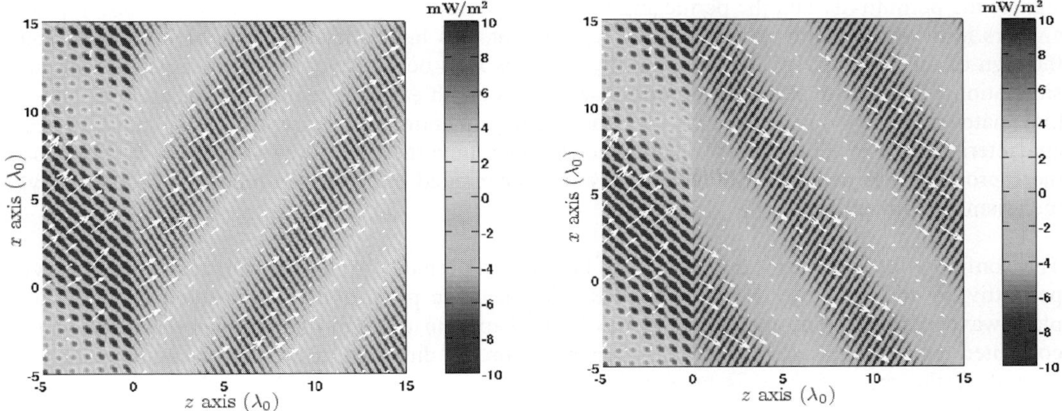

Figure 1: Sx overlayed with arrows indicating the direction and magnitude of the Poynting vector for RHM-RHM (left) and RHM-LHM (right) interface.

interface, the $\hat{x}$-component of the Poynting vector is positive, while in the RHM-LHM case the $\hat{x}$-component is negative. Note, however, that in both cases an interference pattern is formed whose normal vectors are not in the direction of the respective power flows. This interference is a result of the fact that the waves for each frequency are refracted at different angles.

Additionally, it can be seen that negative refraction does not violate causality. Note that at a RHM-RHM interface, points on a single phase front in the incident region can all be mapped to the same phase front in the transmitted region; however, in the RHM-LHM interface case the points along the incident phase fronts are each mapped to different backward propagating phase fronts in the LHM. On the other hand, the interference fronts of a multi-frequency signal are indeed distorted at an RHM-LHM interface since its different frequency components are refracted at different angles. In actuality, each point on the interference front moves in the direction of power flow (downward and to the right). The apparent upward movement is due to the combined effects of semi-infinite extent, the periodicity of the interference pattern and its slanted angle. Indeed, a spatially finite extent signal such as a Gaussian beam, which can be represented by a collection of plane waves [11-12], will also propagate downwards in this case since each plane wave component will refract negatively.

III. Numerical Results

In order to corroborate the theoretical conclusions, we have performed a series of numerical simulations on the metamaterial reported in [4] aiming at the demonstration of the existence of a left-handed behavior. Due to the complexity of these media, one of the main approach we use is based on numerical FDTD simulations [13]. Although we shall report here the results related to the prism experiment only, our purpose is to use a three dimensional FDTD method to study the transmission characteristics, phase propagation, and index of refraction of metamaterials to unambiguously determine their left-handed or right-handed property.

Currently, there exist two main methods used to determine whether a particular medium has left-handed properties. The first is a prism-based experiment, which measures the angle of peak power and then uses Snell's Law to determine the effective index of refraction. The second method is to measure the transmission and reflection coefficients of a slab. A transmission band reflects a similar sign of the permittivity and the permeability, which is taken to be negative if the transmission peak appears in an otherwise transmission-free region. Yet, this last conclusion may be erroneous, and the sign of the permittivity and permeability could as well be positive, as we have verified in our simulations. From an experimental point of view, it is much easier to realize and measure a slab of metamaterial rather than a prism. Yet, as already mentioned before, the sole transmission characteristic of a material cannot unambiguously determine the sign of the index of refraction, and more processing is needed. We have therefore concentrated our work on repeating (numerically) the prism experiment first.

The configurations of the metamaterials are based on a periodic arrangement of rods (for negative permittivity) and split ring resonators (SRRs, for negative permeability) embedded in a parallel plate waveguide [4]. The metamaterial has been shaped in a prism-like geometry, and we have computed the output power in far-field. Depending on the direction of the power compared to the normal of the prism (which is at about 18.4 degrees), we can conclude on the bulk index of refraction exhibited by the metamaterial at the specific working frequency being positive or negative. For the sake of comparison, we have also performed the same simulation with a solid block of Teflon. The results are depicted in Figure II for the Teflon prism and for the metamaterial measured at two distinct frequencies. The results clearly show that the peaks corresponding to the Teflon material and those corresponding to the metamaterial are on opposite side of the normal (solid vertical line in Figure II), showing that the metamaterial operates in a left-handed regime at these frequencies. More exact calculation show that the two frequencies actually correspond to indices of refraction of -1 and -3.4.

IV. Experimental Results

In addition to theoretical and numerical considerations, we have also physically realized various metamaterials and verified their potential left-handed behavior properties. In order to do so, we have performed four different experiments on various realizations of metamaterials: Transmission through a slab-shaped metamaterial, prism experiment on a prism-shaped metamaterial, gaussian beam deflection on a slab-shaped metamaterial, and T-junction measurement, on an L-shaped metatarsal. In this paper, we want to emphasize the last two experiments, as they have not been thoroughly addressed in the literature.

The principle of using the propagation of a Gaussian beam through a slanted slab and measuring the deflection as a way of determining the left-handed or right-handed property of a slab has been first theoretically studied in [11] for transmission and in [12] for reflection. It has been shown that depending on the amount of beam shift obtained, one can conclude on the index of refraction of the material being positive or negative. The limit between the two regions is well defined by a limit beam shift expressed by:

$$d_{\mathrm{lim}} = w \sin \theta \tag{6}$$

where w is the width of the slab measured along the normal direction and θ is the incident angle measured from the normal. We have used the structure reported in [4] to realize a slab of 10 units in the shorter direction and 30 in the longer one, the height being 10 mm. The slab is placed in a parallel plate waveguide with absorbers on the sides, and the incident wave impinges on one of its sides at an angle of about 18.4 degrees. The maximum beam shift obtainable with a right-handed material is therefore $d_{\mathrm{lim}} = 15$ mm, and only a slab of left-handed material would yield a shift beyond this limit. Upon measuring the peak of output power for different materials (air, Teflon and the metamaterial slab) in order to determine the value of the shift, we obtain the results shown in Fig. 1, where the power has been normalized for a better comparison. The curve corresponding to air shows the calibration of our instruments to a shift of about -3 mm at about 9.5 GHz. The result for the Teflon slab indicates a shift of about 7.5 mm, or 4.5 mm away from the calibrated center, corresponding to a typical index of refraction of 1.4. Finally, the beam shift recorded for the metamaterial slab is (effectively) of about 26 mm, corresponding to an index of refraction of -1.27. This experiment proves to be easier to realize than the prism one and can predict as well the region where the metamaterial operates in a left-handed regime.

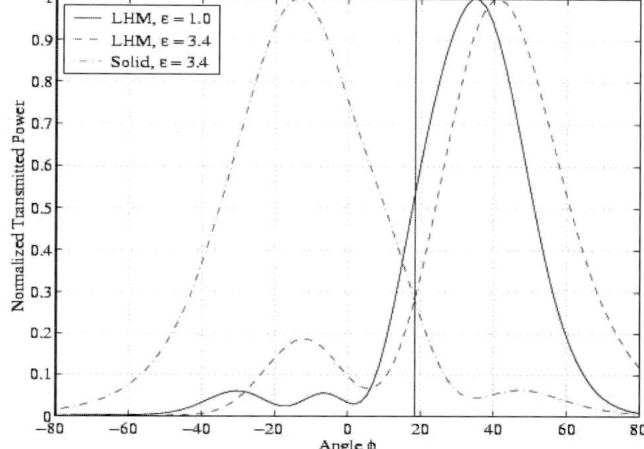

Figure II: Normalized output power simulated in far-field of a prism of metamaterial.

Yet, another experiment is also possible to achieve the same goal, which consists in measuring the power at the output ports of a T-junction waveguide filled with an L-shaped metamaterial. The microscopic structure of the material is the same as the one used for the experiment described above, and only the overall shape of the structure varies (L-shaped as in figure 2 versus slab in the previous experiment). In this experiment, we launch power into port 1 and measure the received power at ports 2 and 3, as shown in Fig. V. For an empty T-junction, we expect the major part of the power to be received at port 2, as the electromagnetic wave tends to propagate directly from 1 to 2. We then introduce the metamaterial inside the T-junction, in such a way that the wave from port 1 impinges on an interface at an angle of 45 degrees. If the metamaterial exhibits a negative index of refraction at the input frequency, we expect most of the power to be deflected to port 3, the latter getting therefore more power than port 2. Upon sweeping the frequency and measuring the output power at both ports 2 and 3, we can determine the frequencies at which the effective index of refraction of the metamaterial is negative. The results obtains for three measurements at 9.73 GHz are summarized in table 1 and show that at this frequency, most of the power is transmitted to port 3, which reflects a left-handed behavior.

	S11 [dB]	*S21 [dB]*	*S31 [dB]*
Empty T-junction	-12.8	-2.18	-4.65
With metamaterial	-1.9	-48.3	-36.6

Table 1: S parameters measured at 9.73 GHz at the three outputs of the T-junction.

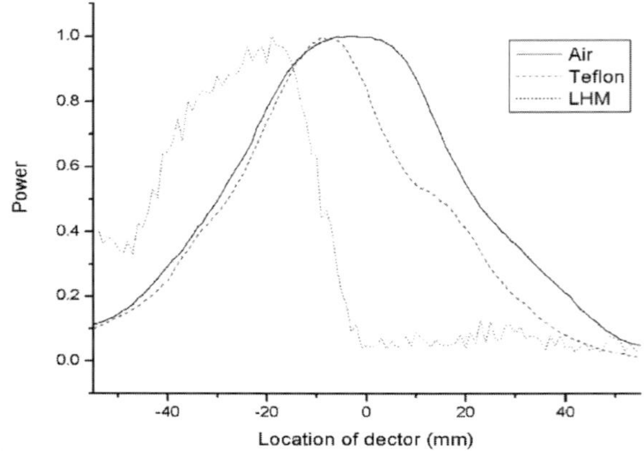

Figure III: Normalized power at about 9.5 GHz at the output of the metamaterial slab for the beam shifting experiment.

Figure IV: Photograph of the metamaterial sample used in the T-junction experiment.

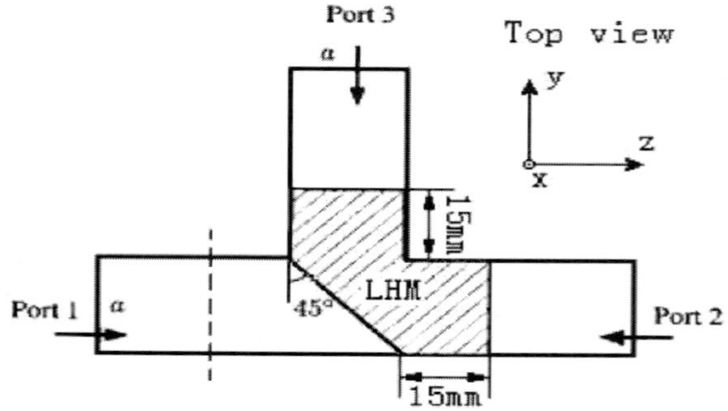

Figure V: Schematic representation of the T-junction experiment.

V. Conclusion

In this paper, we have presented three approaches to demonstrate unambiguously the possibility and the actual existence of a negative index of refraction, which reflects a left-handed behavior. In particular, we have shown that negative refraction is possible for multi-frequency signals by explicit calculation of the Poynting vector in LHM. Using two discrete frequencies, we have shown that the direction of the time-averaged Poynting vector is in the average direction of the time-averaged Poynting vectors for each frequency treated separately, implying that negative refraction is possible, without violating fundamental laws. This conclusion has been corroborated by numerical and experimental results, which show that for a real metamaterial realized as a periodic arrangement of rods and split-rings, we can determine a frequency band where the material behaves in a left-handed regime. This has been shown here for a prism, slab and L-shaped metamaterial in a prism, beam shifting and waveguide T-junction experiments, respectively.

Acknowledgement

This work was supported in part by the MIT Lincoln Laboratory under Contract BX-8133 and the Office of Naval Research under Contract Nos. N00014-10-1-0713, and N00014-99-1-0175.

References

[1] V. G. Veselago, "The electrodynamics of substances with simultaneously negative values of ε and μ", *Sov. Phys. Usp.* **10**, 509 (1968).

[2] J. B. Pendry, A. J. Holden, W. J. Stewart, and I. Youngs, "Extremely Low Frequency Plasmons in Metallic Mesostructures", *Phys. Rev. Lett.* **76**, 4773 (1996).

[3] J. B. Pendry, A. J. Holden, D. J. Robbins, and W. J. Stewart, "Magnetism from Conductors and Enhanced Nonlinear Phenomena", *IEEE Trans. Microwave Theory Tech.* **47**, 2075 (1999).

[4] R. A. Shelby, D. R. Smith, and S. Schultz, "Experimental verification of a negative index of refraction", *Science* **77**, 292 (2001).

[5] C. G. Parazzoli, R. B. Greegor, K. Li, B.E.C. Koltenbah, and M. Tanielian, "Experimental Verification and Simulation of Negative Index of Refraction Using Snell's Law", *Phys. Rev. Lett.* **90**, 107401 (2003).

[6] A. A. Houck, J. B. Brock, and I. L. Chuang, "Experimental Observations of a Left-Handed Material That Obeys Snell's Law", *Phys. Rev. Lett.* **90**, 137401 (2003).

[7] A. K. Iyer and G. V. Eleftheriades, "Negative refractive index metamaterials supporting 2-D waves", *IEEE MTT-S Int. Microwave Symp. Digest* (2002), vol. 2, pp. 1067–1070.

[8] M. Notomi, "Theory of light propagation in strongly modulated photonic crystals: Refractionlike behavior in the vicinity of the photonic band gap", *Phys. Rev. B* **62**, 10696 (2000).

[9] C. Luo, S. G. Johnson, J. D. Joannopoulos and J. B. Pendry, "All-angle negative refraction without negative effective index", *Phys. Rev. B* **65**, 201104 (2002).

[10] J. Pacheco, T. M. Grzegorczyk, B.-I. Wu, Y. Zhang, and J. A. Kong, "Power Propagation in Homogeneous Isotropic Frequency Dispersive Left-Handed Media", *Phys. Rev. Lett.* **89**, 257401 (2002).

[11] J. A. Kong, B.-I. Wu and Y. Zhang, "A Unique Lateral Displacement of a Gaussian Beam Transmitted Through a Slab with Negative Permittivity and Permeability", *Microwave Opt. Tech. Lett.*, vol. 33, n. 2, pp. 136-139, 20 April 2002.

[12] J. A. Kong, B.-I. Wu and Y. Zhang, "Lateral Displacement of a Gaussian Beam Reflected from a Grounded Slab with Negative Permittivity and Permeability", *Appl. Phys. Lett.*, vol. 80, n. 12, pp. 2084-2086, 22 March 2002.

[13] C. D. Moss, T.M Grzegorczyk, Y. Zhang, and J. A. Kong, "Numerical Studies of Left Handed Metamaterials", *PIER*, vol. 35, pp. 315-334, 2002.

A Study into the Possibility of Field Focusing
Using "Left-Handed" Materials

V.N. Kissel[1] and A.N. Lagarkov

Institute for Theoretical and Applied Electromagnetics
Russian Academy of Sciences, Moscow, Russia

The problem of focusing fluxes of electromagnetic field energy in the presence of structures of "left-handed" matter (LHM) assumed a special importance in the light of recent idea of Prof. J.B. Pendry [1] that a parallel-sided slab of LHM under certain conditions ($\varepsilon = \mu = -1$) may serve as a "superlens" with unique focusing properties, whose resolution is not restricted by the well-known diffraction limit. The phenomenon is expected to exist owing to amplification of evanescent modes in an LHM-slab, which exponentially decay in ordinary media. Papers with argumented objections were published (in particular, [2, 3]); however, some researchers did not recognize these objections as being exhaustive and convincing and, for all we know, this scientific discussion is far from being finished [4-6].

In order to solve the problem of the quality limit of focusing of the image of a point source utilizing the realistic finite-width dissipative LHM-slab, we used numerical methods to investigate the spatial distribution of components of electromagnetic field and its major characteristics (in particular, the Poynting vector) under conditions of excitation of a slab by one and two filamentary sources spaced some small distance apart. The solution was obtained by the method of volume integral equations using the Green's function of free space, with the slab parameters $\varepsilon = \mu \approx -1$ included in the expressions for polarization currents. This solution (as distinct, for example, from eigenfunction expansions or surface integral equations) does not at all involve the concept of the propagation coefficient of an LHM-filled space and is not associated with the use of other multiple valued functions whose arguments would include (possibly negative) permittivity and permeability. Moreover, the polarization currents singled out in the Maxwell equations may be identified with actually existing currents in the inclusions of the composite whose model is provided by LHM. Therefore, the method of volume integral equations offers a good tool for checking assumptions and testing solutions obtained by other methods.

The calculation results have demonstrated that, indeed, zones of local concentration of electromagnetic field energy do arise within the parallel-sided slab and in the vicinity of this slab in the neighborhood of the expected points of focusing; in so doing, the direction of energy transport by and large corresponds to the ray representations. However, if the plate thickness is of order of wavelength and the material exhibits even small losses, the size of these zones and the resolution of the system (the extremely small distance between the sources, at which their images are discernible separately) usually provide no advantages over conventional focusing systems.

Given below by way of illustration are the results of calculation for a structure whose cross section (in the $z=0$ plane) is shown in Fig. 1a, with the excitation source provided by one or two electric current filaments arranged symmetrically and in parallel with the slab at a distance $|c\text{-}a|$ from the latter; k_0 is the free space wavenumber. Figure 1 also shows the surface relief and isolines of the modulus of the real part of Poynting vector within the slab (Figs. 1b and 1c, respectively), as well as its vector distribution (Fig. 1c) under excitation by one source (by current filament). One can see that the lines of the Poynting vector are directed mainly along the rays shown in the approximation of geometrical optics in Fig. 1a. Figure 2 shows the images of the surface relief of distribution of the modulus of electric field intensity E_z within and in the vicinity of the slab at different distances between two filamentary sources (($k_0d=0$ (a), $k_0d=4$ (b), $k_0d=5$ (c)); the geometry of the problem is given in Fig. 3a. One can see that the regions of local concentration of the field are quite diffuse, and the images of two sources spaced apart show up separately from each other

---

[1] Corresponding author. E-mail: kis_v@mail.ru

only when the distance between them somewhat exceeds half the wavelength. Note that the increase in the slab length along the Y-axis (parameter b in Fig. 1a) had no significant effect on the calculation results. Finally, Fig. 3b shows the distribution of the phase of E_z for $k_0 d$=4.0, from which one can see that the phase velocity of the wave within the slab has a negative value; the focusing effect is observed, but its quality is restricted by the diffraction limit.

Further investigations were carried out to clarify the conditions in which the use of LHM-plate should result in a higher resolution. As a result of solution of the problem on the excitation of an infinite plane-parallel plate (Fig. 4) by a filament of electric current of amplitude I_0, the vector potential A (the electric field amplitude proportional to this potential) is represented in the form of Fourier integral

$$A = \left(I_0/4\pi\right)\int_{-\infty}^{+\infty}\exp(-i\xi y)F(\xi)d\xi,$$

with $x > a$ (i.e., in the half-space where the focusing point is located),

$$F(\xi) = \frac{-4\mu q\exp(q_0(x_0 - x))\exp(2aq_0)}{\left(\mu q_0 - q\right)^2\exp\left(-2aq\right) - \left(\mu q_0 + q\right)^2\exp\left(2aq\right)},$$

$$q_0 = \begin{bmatrix} i\left(k_0^2 - \xi^2\right)^{1/2}, |\xi| \le \mathrm{Re}(k_0) \\ \left(\xi^2 - k_0^2\right)^{1/2}, |\xi| > \mathrm{Re}(k_0) \end{bmatrix}, \qquad q = \begin{bmatrix} i\left(k^2 - \xi^2\right)^{1/2}, |\xi| \le \mathrm{Re}(k) \\ \left(\xi^2 - k^2\right)^{1/2}, |\xi| > \mathrm{Re}(k) \end{bmatrix},$$

where $k_0 = 2\pi/\lambda$ and $k = k_0\sqrt{\varepsilon}\sqrt{\mu}$ are the wavenumbers of external space and of the plate material, respectively; λ is the wavelength, x_0 is the point of filament location; $2a$ is the plate thickness; the time dependence is selected in the form $\exp(i\omega t)$; and the principal values of radicals are used. Obviously, the accomplishment of "ideal" focusing implies that the function $F(\xi)$ must become equal to the spectral density of the vector potential of the employed linear source $f(\xi) = 1/q_0$ in some neighborhood of the focusing point (at least, within the zone of resolution of conventional systems).

Let a source be located at point $x_0 = -2a$. We will treat the dependence $F(\xi)$ in the plane of intended focusing $x = 2a$ and compare it to $f(\xi) = 1/q_0$ (Fig. 4, curve 1). If the losses in the plate material are vanishingly small (tend to zero), a direct substitution of $\varepsilon = \mu \to -1$ gives $F(\xi) \to f(\xi)$. This implies that an "ideal" image is obtained in the $x = 2a$ plane. However, if the material exhibits even minor losses $\varepsilon'' = -\mathrm{Im}(\varepsilon) > 0$, $\mu'' = -\mathrm{Im}(\mu) > 0$, the equality $F(\xi) \approx f(\xi)$ is valid only approximately, in a limited segment of the infinite region of integration with respect to ξ, outside of which the function $F(\xi)$ reveals rapid decreasing with increasing $|\xi|$. In particular, if we take $\mu = -1$, $\varepsilon = -1 - i\alpha$, where $\alpha << 1$, we can have the asymptotic representation

$$F(\xi) \approx f(\xi)(1 + \Delta)^{-1}, \qquad \text{where } \Delta = \alpha^2\left(2q_0\right)^{-4}\exp(4aq_0).$$

Consequently, for any value of $\alpha > 0$ at sufficiently high values of $|\xi|$, the spectral density $F(\xi)$ exponentially tends to zero. The greater the plate thickness $2a$, the sooner (at smaller values of $|\xi|$) one must assume that $F(\xi) \approx 0$. Needless to say, the parameters $x_0 = -2a$ and $x = 2a$ also vary with the plate thickness.

One can select the criterion of closeness of $F(\xi)$ to $f(\xi)$ (for example, the limiting relative error) and, for the given parameters of the problem, determine the number ξ_0 such that, at

$|\xi| < \xi_0$ $F(\xi) \approx f(\xi)$. With $|\xi| > \xi_0$, the spectral density $F(\xi)$ decreases rapidly; therefore, the numbers $-\xi_0$ and $+\xi_0$ preassign the width of spectrum of spatial harmonics which are involved in image formation. The detailing of image is associated with completeness of the spectrum of modes taken into account; in particular, the size of the focal spot is inversely proportional to the quantity $\widetilde{\xi}_0 = \xi_0 / |k|$. Therefore, the limiting resolution of the system will be defined by the attained value of the parameter $\widetilde{\xi}_0$. The ideal image of a point source is reached at $\widetilde{\xi}_0 \to \infty$. Corresponding to the case of $\widetilde{\xi}_0 = 1$ is the focusing of "standard" quality, when decaying harmonics are lost. This image is formed only by oscillations of propagating types, for which $-1 < \xi/k < 1$. For all intermediate options of $1 < \widetilde{\xi}_0 < \infty$, the focusing properties and resolution are improved compared to the "diffraction limit". It is such values of $\widetilde{\xi}_0$ that are realized in a system with a plane-parallel plate of a "left-handed" material with low losses.

The calculation results demonstrate that, given a plate thickness comparable to the wavelength, very high requirements are placed on the quality of the LH-material. For example, in the case of $\varepsilon'' = \mu'' = 0.01$ and thickness $2k_0 a = 3.0$ (about half the wavelength), the value of $\widetilde{\xi}_0$ is only about 1.5 (Fig. 4, curve 2). In order to attain the value of $\widetilde{\xi}_0 \approx 3$ for the same geometry of the problem, the losses must be reduced to $\varepsilon'' = \mu'' = 0.00005$ (Fig. 4, curve 3), which is unlikely to achieve. With a greater plate thickness, even lower values of $\widetilde{\xi}_0$ are attained. Note that minor losses in dielectric do not result in such a dramatic effect on the amplitudes of propagating harmonics (Fig. 4, $|\xi|/k_0 < 1$); therefore, with the treated plate thickness, it is easier to register the effect of focusing of "standard" quality (as was demonstrated above, see Figs. 2, 3).

On the other hand, the analysis of expressions for $F(\xi)$ and Δ indicates that the use of thin LHM-plates brings about a drastic *reduction of loss requirements* and leads one to hope for the possibility of practical utilization of the effect of superresolution. The results of computer modeling and experiment which demonstrate the possibility of "superresolution" in the system with thin LHM-plates are discussed in the next presentation [7].

References

1. J.B. Pendry, "Negative refraction makes a perfect lens", Phys.Rev.Lett., **85**, 3966 (2000).
2. N. Garcia, M. Nieto-Vesperinas, "Left-handed materials do not make a perfect lens", Phys.Rev.Lett., **88**, 207403 (2002).
3. N. Garcia, M. Nieto-Vesperinas, "Is there an experimental verification of a negative index of refraction yet?", Opt. Lett., **27**, 885 (2002).
4. R. Bansal, "The road not taken?", IEEE Antenna's and Propag. Mag., **44**, No.5, October, 103 (2002)
5. E. Cartlidge, "Negative reaction to negative refraction", Physics World, August 2002 (available online at http://www.physicsweb.org/).
6. J. Minkel, "Left-handed materials debate heats up", Physical Review Focus, May 3, 2002 (available online at http://focus.aps.org/v9/st23.html).
7. A.N. Lagarkov, V.N. Kissel, Numerical and experimental investigation of the superresolution in a focusing system based on a plate of "left-handed" material – to be presented at the International Conference on Materials for Advanced Technologies ICMAT 2003, December 7-12, 2003, Singapore.

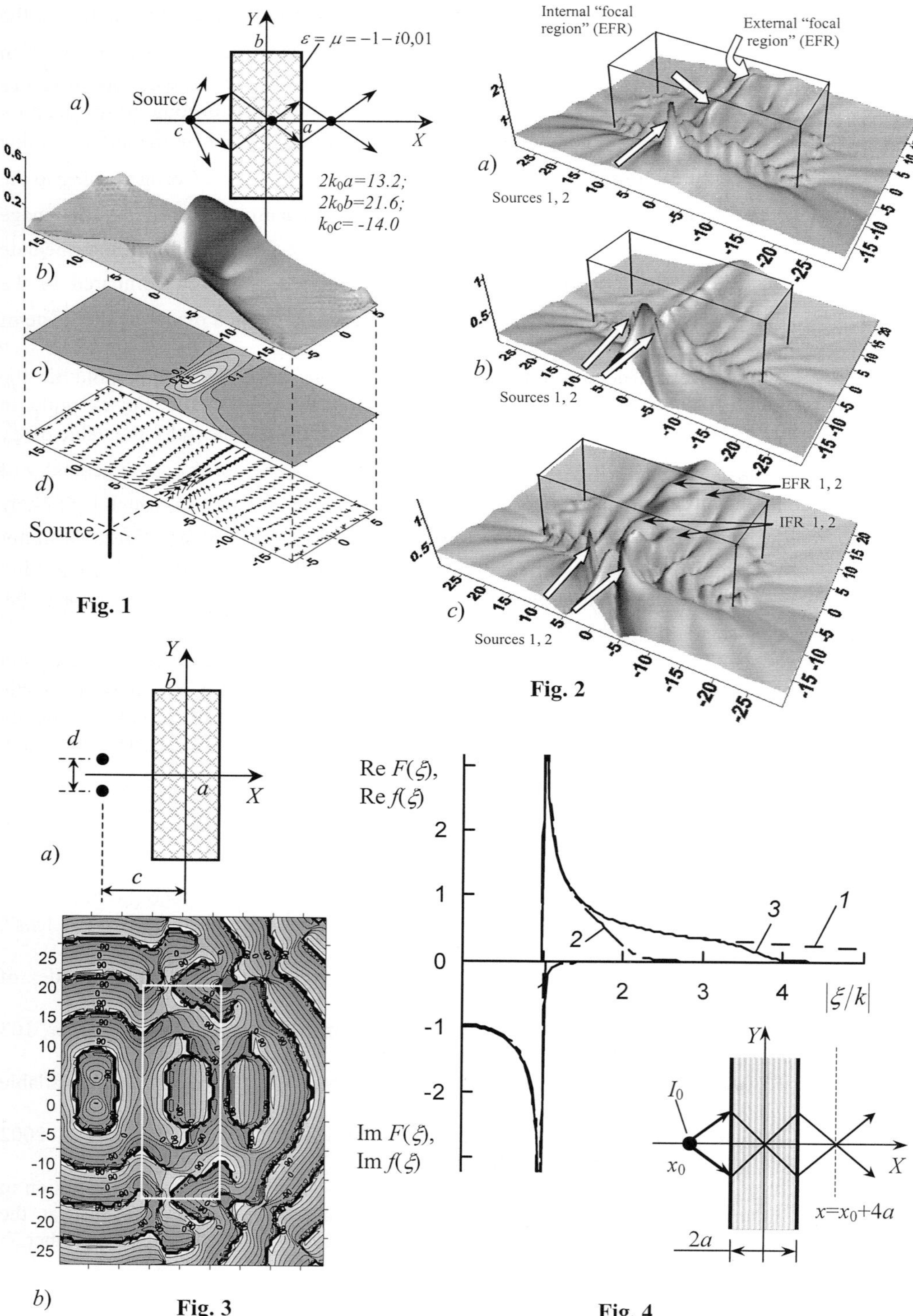

Y
b
ε = μ = −1 − i0,01
Source
a)
c
a
X
2k₀a=13.2;
2k₀b=21.6;
k₀c= −14.0
0.6
0.4
0.2
b)
c)
d)
Source
Fig. 1
Internal "focal region" (EFR)
External "focal region" (EFR)
a)
Sources 1, 2
b)
Sources 1, 2
c)
Sources 1, 2
EFR 1, 2
IFR 1, 2
Fig. 2
Y
b
d
a
X
a)
c
25 20 15 10 5 0 -5 -10 -15 -20 -25
b)
Fig. 3
Re F(ξ),
Re f(ξ)
2
1
0
2   3   4 |ξ/k|
-1
-2
Im F(ξ),
Im f(ξ)
1
2
3
Y
I₀
x₀
X
x=x₀+4a
2a
Fig. 4

Computational study of electromagnetic wave functional materials

C.T. Chan, Lei Zhou, Jensen Li, Ping Sheng

Physics Department, The Hong Kong University of Science and Technology, Clear Water Bay, Hong Kong, China.

With the help of computer simulations, we have studied the properties of some electromagnetic (EM) wave functional materials. Some of these materials have been successfully fabricated in laboratory* with experimentally measured optical properties in good agreement with theoretical predictions.

We show by both experiment and theory that a specific class of planar conducting fractals possesses a series of self-similar resonances, leading to multiple gaps and pass bands for electromagnetic waves over an ultra-wide frequency range [1]. The generator of our fractal is a horizontal H-shaped metallic structure with equal height and breadth. In the subsequent generations, four (generator) elements, scaled in size by a factor of 0.5, are attached to the free ends of the mother element. This procedure is then iterated, with each set of parallel metallic lines defined to be one "level". With a increasing number of levels, the pattern approaches a space filling curve that tiles a 2D square. Figure 1 shows the measured normal transmission and reflection of the fractal plate from 1 to 18 GHz with E vector polarized along the y axis. Three band gaps are clearly seen where EM wave transmission was strongly attenuated. We have performed finite difference time domain (FDTD) simulations to study the observed phenomena. The calculated results are shown as solid lines in Fig. 1. Simulations reproduce all the salient features of the experiments. In particular, the calculated and measured spectra are both log-periodic, an inherent consequence of fractal pattern's self-similarity, which means that the resonances (hence the stop bands) of the fractal cover an ultra-broad frequency range. This unique feature is hard to match with periodic frequency-selective systems, which typically have one single dominant response. FDTD simulations revealed the physics to be governed by geometric resonances.

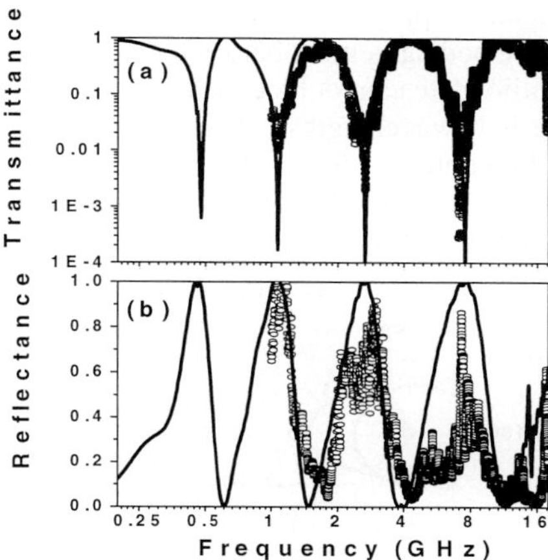

Fig. 1. (a) Measured (circles) and calculated (lines) transmittance; (b) reflectance of a 15 level metallic fractal pattern (first level line length=144.76mm, line width and thickness = 0.1mm) deposited on a 1mm-thick dielectric substrate, for electromagnetic waves polarized with E vector along y axis.

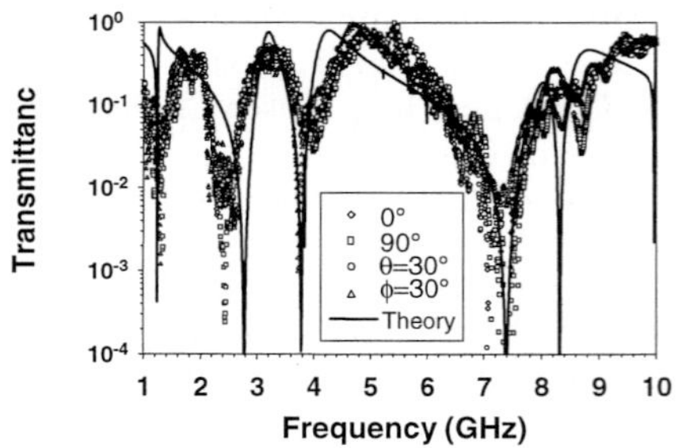

Fig. 2. Transmission through two stacked identical fractal plates with a $90°$ rotation with respect to each other. The diamonds and squares denote measured normal transmission through the double stack with E vector along x and y axes, and the lines are calculated results for normal incidence (identical for both polarizations). The circles and triangles are transmission measured with the plate rotated (θ) and tilted (ϕ) by $30°$, respectively from the normal.

By stacking two identical fractal patterns, one rotated 90 degrees relative to the other; we demonstrate that the double stack structure can simulate the incident-angle and polarization-independent total reflection that is usually characteristics of a photonic crystal slab. Figure 2 shows the transmission of such a double stack when it was rotated and tilted, so as to change the angle of incidence. We found the transmissions to be nearly identical for different angles of incidence; hence the spectral gap frequencies are stable with respect to a large range of incidence angles. In addition, the transmission is independent of polarization since the double stack is rotationally invariant.

We found that a small dielectric plate ($28mm \times 29mm$) covered by a six level metallic fractal pattern can reflect EM waves, at some frequencies with wavelengths that are significantly larger than the lateral dimension of the fractal plate. A metal plate of the same size fails to reflect when its lateral size is smaller than the half of corresponding wavelength [2]. Figure 3 shows the FDTD calculated and experimentally measured H-plane radiation patterns for a monopole antenna, when a fractal plate and a metal (copper) plate placed nearby. Open (fractal) and solid (metal) symbols denote the experiment. Good agreement between theory and experiment is noted. It is clear from Figs. 3 (a-b) that the metallic plate cannot block EM waves at 2.25GHz and 3.85GHz. At these frequencies, the corresponding half-wavelengths (~67mm and ~39mm) are larger than the plate's lateral dimension (~29mm). In contrast, the fractal plate can effectively reflect at these two frequencies (the reflectivity is a bit worse for lower frequency case). For some particular frequencies, the forward radiation for the metallic plate can be stronger than the backward radiation.

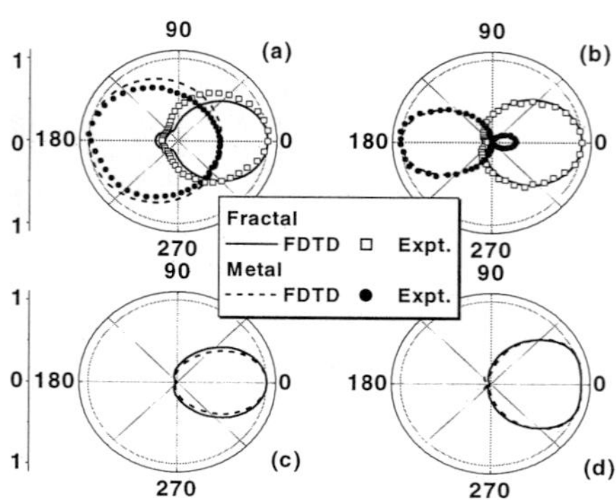

Fig. 3. FDTD calculated (lines) and experimentally measured (symbols) H-plane radiation patterns for a monopole antenna of length L under an operation frequency f, with a fractal plate or a metal plate placed at a distance d at $180°$ from the antenna. (a) f=2.25GHz, L=24mm, d=15mm; (b) f=3.85GHz, L=24mm, d=9mm; (c) f=7.0GHz, L=8mm, d=9mm; (d) f=10.5GHz, L=8mm, d=9mm.

At the two higher frequencies (Fig. 3(c-d)), where half wavelengths are smaller than the plate size, both the fractal and the metallic plates reflect the EM waves effectively, with similar radiation patterns.

We have showed mathematically that layered heterostructures combining ordinary and negative refractive index materials exhibit a new type of photonic band gap corresponding to zero volume-averaged refractive index [3]. Distinct from band gaps induced by Bragg scattering, the zero-$\bar{n}$ gap is invariant upon a change of scale length and is insensitive to disorder. A metallic structure that exhibits such a band gap is explicitly designed, with building block details shown in Fig. 4(a). The building block is replicated to tile the xy plane to form a 3.5mm-thick negative-n slab. We first calculate the transmission spectra through such a slab by FDTD simulation with E vector along the y axis, and then derive its effective $\varepsilon(\omega)$ and $\mu(\omega)$, which are plotted in Fig. 4(a). Between 4.1 and 4.8 GHz, both $\varepsilon_{\mathrm{eff}}$ and μ_{eff} are found to be negative due to resonance, demonstrating that the material has an effectively negative refractive index in this regime. We then create a 1D photonic crystal with unit cell consisting of the negative-n slab and a positive-n one (7mm-thick air). The FDTD calculated transmission spectra through a 16-unit-cell slab is shown as solid line in Fig. 4(c). With $\varepsilon_{\mathrm{eff}}(\omega)$ and $\mu_{\mathrm{eff}}(\omega)$ of the negative-n slab determined, we calculated the band structure (Fig. 4(b)) of the 1D photonic crystal, and the transmittance through a 16-unit-cell slab (shown as open circles in Fig. 4(c)). Both the band structure and the transmission spectra clearly show two band gaps. We note in Fig. 4(c) the excellent agreement between the rigorous FDTD (solid line) result and those calculated using $\varepsilon_{\mathrm{eff}}(\omega)$ and $\mu_{\mathrm{eff}}(\omega)$ (open circles). We also calculated $\bar{n}(\omega)$ for the combined system, and found that $\bar{n}=0$ at about 4.5 GHz, which corresponds to the gap at 4.5 GHz. We emphasize that the $\bar{n}=0$ gap is found not just from effective $\varepsilon_{\mathrm{eff}}$ and μ_{eff} calculations, but also from the FDTD simulations.

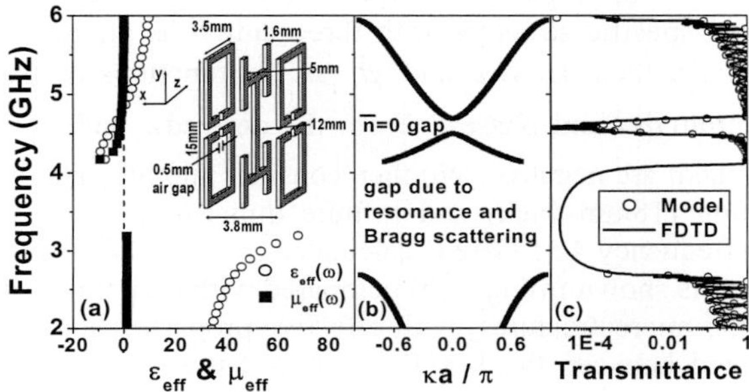

Fig. 4. (a) $\varepsilon_{\mathrm{eff}}$ and μ_{eff} as the functions of frequency of a designed negative-n material, with structural details shown in the inset; (b) Band structure for a photonic crystal with alternating layers of air (7mm thick) and the designed negative-n material (thickness = 3.5mm, $\varepsilon_{\mathrm{eff}}$, μ_{eff} shown in (a)); (c) Transmittance through a slab consisting of 16 unit cells with details described above, through direct FDTD simulation (solid line) and material properties represented by $\varepsilon_{\mathrm{eff}}$ and μ_{eff} (open circles).

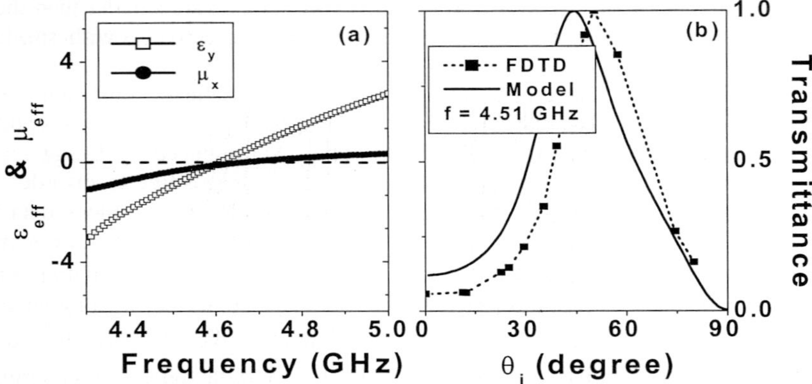

Fig. 5. (a) Effective ε_y and μ_x as the function of frequency for our designed material, with building block structure shown in the inset. (b) Transmittance at frequency 4.51 GHz as the function of incidence angle through a slab of our designed material with three unit cells in the x direction, calculated by FDTD simulations (solid squares) and with effective ε_y and μ_x shown in (a) (solid line).

We also find that anisotropic negative index materials have interesting angle dependent optical properties. The effective refraction index can be positive, negative, or purely imaginary under different conditions. With dispersion taken into account, we show that the material can allow frequency selective total oblique light transmission (and the mechanism is different from that of Brewster angle). Structures that can realize such effects are proposed, which have basically the same building block shown in Fig. 4(a) but with a different lattice structure and substrate material (for details, see [4]). We first employed FDTD simulation to calculate the normal transmission spectra of a slab of the specific structure with three unit cells along z direction (22.5mm-thick) replicated in the xy plane, then derive ε_y and μ_x as the functions of frequency from the FDTD results (using both amplitudes and phases). From the derived ε_y and μ_x, we do find a frequency regime where both of them are negative. We then consider another slab of such material with three unit cells in x direction (18mm-thick) and infinite dimensions in yz plane. The transmittance through such a slab at frequency 4.51 GHz is calculated by the FDTD simulations as the function of the incidence angle, and is shown in Fig. 5(b) compared with that obtained for a homogeneous slab (without considering the microstructures) with effective properties shown in Fig. 5(a). Reasonably good agreement is noted between the FDTD result and the model calculation, and both results clearly show the total transmissions at a particular off-normal incidence angle. This serves to demonstrate that the total oblique transmission does exist in such 1D negative-n materials.

References

[1] Weijia Wen, Lei Zhou, Jensen Li, Weikun Ge, C. T. Chan and Ping Sheng, *Phys. Rev. Lett.* **89** (2002) 223901.
[2] Lei Zhou, Weijia Wen, C. T. Chan and Ping Sheng, *Appl. Phys. Lett.* **82** (2003), 1012.
[3] Jensen Li, Lei Zhou, C. T. Chan and Ping Sheng, Phys. *Rev. Lett.* (2003), in press.
[4] Lei Zhou, C. T. Chan and Ping Sheng, (to be published).

- Work done in collaboration with Weijia Wen, Weikun Ge

EXPERIMENTAL DEMONSTRATION OF THE DEPENDENCY OF THE NEGATIVE OPTICAL INDEX OF A LEFT-HANDED MATERIAL VERSUS THE INCIDENCE ANGLE

A. Djermoun[1], A. de Lustrac[2], F.Gadot[2], E. Akmansoy[2]

[1]Groupe d'Electromagentisme Appliqué, 1 Chemin Desvallières, 92410 Ville d'Avray, France
[2]Institut d'Electronique Fondamentale, Bat. 220, Université Paris XI, 91405 Orsay, France

Abstract: *In this study, we present the results of the characterisation of a left-handed material performed in free space between 7 and 16 GHz with an increasing incidence angle. We demonstrate that this material has a negative optical index depending on the incidence. The value of this index is comprised between -0.1 and -0.5 for an incidence angle between 5 and 25°. The bidimensional structure of the material is the origin of this dependency.*

In 1999 J. Pendry proposed the concept of a magnetic material having a negative permeability with a non classical magnetic material (see Fig.1(a))[1]. As depicted in the diagram, this structure is composed of discontinuous "Swiss-roll" disks. Earlier he had demonstrated that a tridimensional thin wire metallic lattice has a negative permittivity in the first forbidden band (Fig.1(b))[2]. A structure associating such materials can present both negative permittivity and permeability in a given frequency band. D.R. Smith has experimentally demonstrated this property in 2000 for a guiding structure[3]. Then the optical index was measured only for one incidence angle [4]. In this study we present the results of the measurements on a material tested in free space with an increasing incidence angle. We demonstrate that this material has a negative optical index depending on the incidence.

(a)   (b)

Figure 1: 1a: Theoretical negative permeability material. 1b: theoretical negative permittivity material. Both are metallic.

We have designed and simulated the material (Fig. 2) with a standard finite element software (HFSS from Ansoft). This simulation allows us to verify the existence of a resonance corresponding to a zone with a negative permeability (Fig. 3). The material is realised with printed epoxy boards, whose thickness and length are respectively 0.25mm and 150mm. The metallic disks have a diameter of 2 mm for the internal disk and 4 mm for the external one (Fig.1a)). The metallic stripes are 1mm wide and are spaced 5mm apart. The stripe boards are stacked alternatively with the disk boards. This metallic stripe lattice has a first forbidden band extending from DC up to 30 GHz. The distance between two boards is 2.5 mm. The incident wave is transverse magnetic (the electric field is parallel to the metallic stripes) and the wave vector is parallel to the boards at normal incidence. In the direction of propagation the thickness of the material is 1.5 cm, corresponding to three disk and stripe layers.

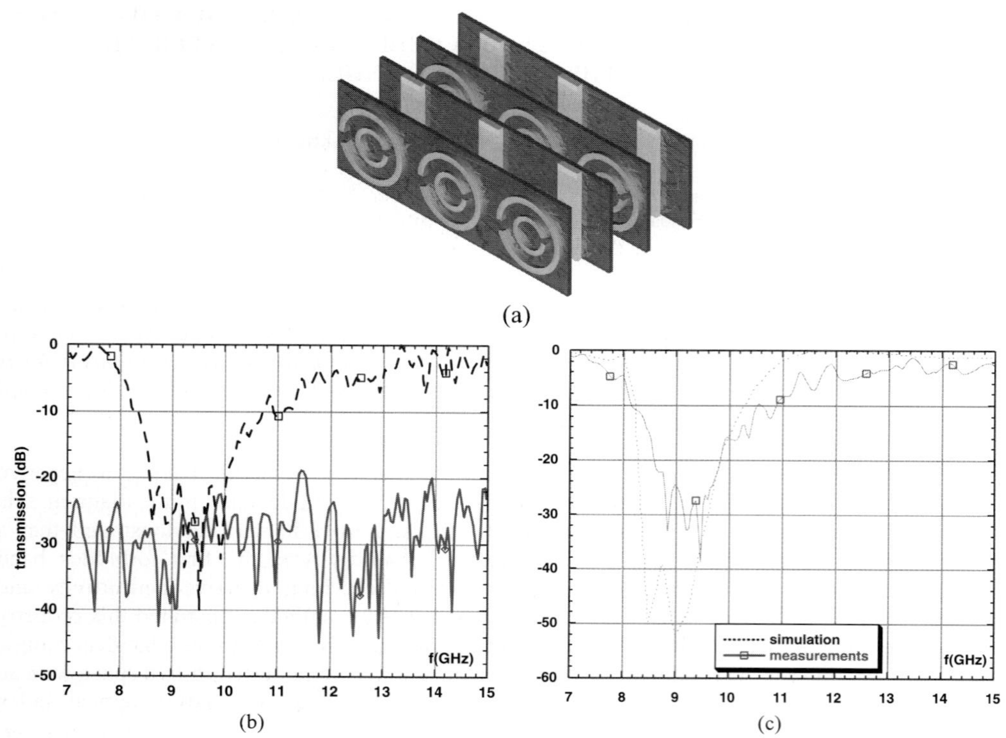

Figure 2: 2a: Theoretical structure of the left handed material: the material is composed of stacked printed boards with alternatively copper disks and stripes.
2b: Measured transmission ratio of the disk lattice only (dotted line) and the stripes only (continuous line).
2c: comparison between the calculated (dotted line) and the measured (continuous line) transmission ratio of the disk lattice.

The transmission ratio is measured with a scalar network analyser Marconi 6400 and two horns between 7 and 16GHz. A Teflon lens in front of each horn, with an aperture of 8cm, focalises the wave on the material. The device under testing is placed between these two horns. First we have measured the transmission ratio of the disk lattice alone, then that one of the metallic stripes alone (Fig. 2(b)). The curve corresponding to the disks shows a transmission close to zero dB, except in the frequency range from 8 to 12 GHz, where a resonance occurs. The resonance frequency is linked to the dimensions of the disks and the lattice. We have adjusted these dimensions to obtain this resonance in the frequency range of our experimental set-up. In fig.2(c) we have compared the measured transmission curve to the calculated one. We observe that the measured resonance is weaker than the calculated one. This weakening is due to the epoxy substrate which has non-negligible losses at these frequencies and to the defect in the lattice board parallelism (the printed boards are too thin and too long). Then we measured the transmission ratio of the whole structure (disks and stripes) and compared it to the calculated one (Fig.3). The calculated resonance corresponding to the frequency domain where the material has a negative permeability extends from 8 to 9.3 GHz, with a splitting of the resonance around 8.6 GHz. The measured resonance was weaker and shifted to the low frequencies with a maximum around 8.3GHz (Fig.3). At the resonance, the measured transmission level is low (-20dB). This weak transmission explains why we only used three layers of disks and stripes in the direction of the wave propagation. We are in the first forbidden band of the metallic stripe lattice. But the weakening of the resonance is also linked to the lowering of the disks resonance. As this frequency corresponds to a resonance between a capacitive circuit (the disks) and an inductive one (the stripes), some dispersion in the fabricated structure may weaken this resonance.

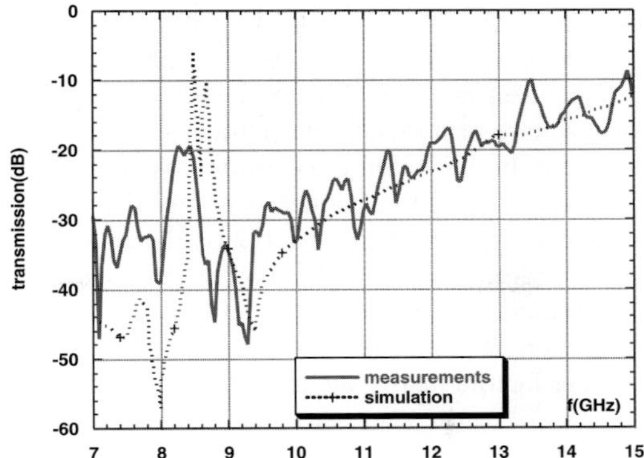

Figure3: Calculated (dot line) and measured (continuous line) transmission ratio of the disks associated with the stripes with a resonance located at 8.6GHz in the simulation and 8.3 GHz in the measurement corresponding to the negative permeability zone.

Finally we have measured the transmission ratio of the material for different incidence angles at the resonance frequency (8.3 GHz). The wave polarisation remains transverse magnetic. Due to the negative optical index of the material, the microwave beam is shifted by a length d from the direction of the incident wave (fig.4). This distance d depends on the thickness e of the material, the incidence angle α and the negative optical index of the material. Then this index can be deduced from this distance d. Here the thickness e of the material is 1.5 cm. The emitting horn is oriented with an angle α from the normal direction to the material (fig.4 (a)). The receiving horn is moved along a line parallel to the edge of the slab. We searched the maximum of the received signal to detect the direction of the transmitted wave and the distance d between the refracted wave and the incident wave (Fig.4b). On this figure the continuous and discontinuous curves correspond respectively to the transmitted wave with and without the left-handed slab.

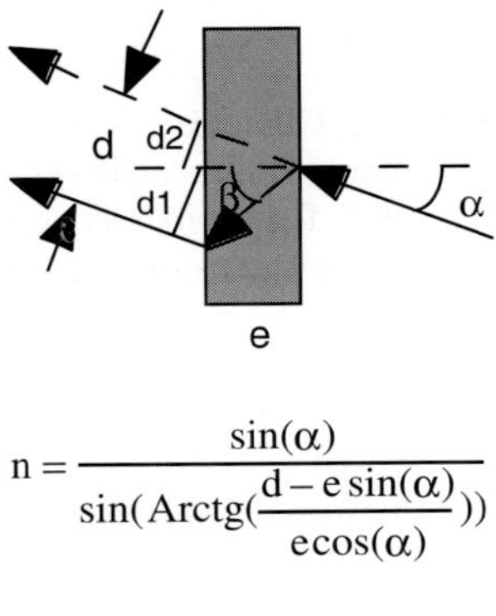

$$n = \frac{\sin(\alpha)}{\sin(Arctg(\frac{d - e\sin(\alpha)}{e\cos(\alpha)}))}$$

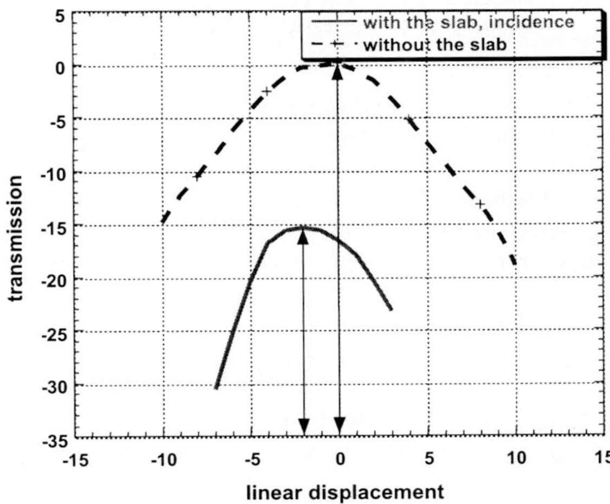

Fig. 4(a): Calculation of the effective optical index n of the material deduced from the measurement of the shift d of the microwave beam transmitted through the material.

Fig. 4(b): Measurement of the linear shift d due to the negative index n of the material for an incidence angle of 11°. The continuous and discontinuous curves correspond respectively to the transmission with and without the left-handed slab.

Deducing the index from the distance d, we obtained the curve of the figure 5. The circles correspond to the measurements whereas the curve is a polynomial fit between these points. The modulus of the negative effective index of the material is lower than unity and increases with the incidence angle α with steps around 6, 13 and 24°. The index clearly depends on the incidence angle. This could be simply explained by the internal symmetry of the structure. For a bidimensional structure, the transmission properties often vary with the incidence angle. This dependency is higher for a square or a rectangular lattice, such as our material, than for a hexagonal or triangular lattice due to the higher number of symmetry axes of these lattices compared to the square one. The precision of the measured values is connected in part to the measurement method (Fig.4). The dimensions of the incident beam are relatively wide (3 cm diameter) compared to the dimensions of the material (width: 22 cm, height: 16cm). As a result, the width of this beam induces an averaging in the measurement of the index, particularly for the lowest incidences. Moreover as the emitting and receiving horn are placed close to the sample (10cm), the incident wave is not really a plane wave. All these factors justify the uncertainties of the measured values.

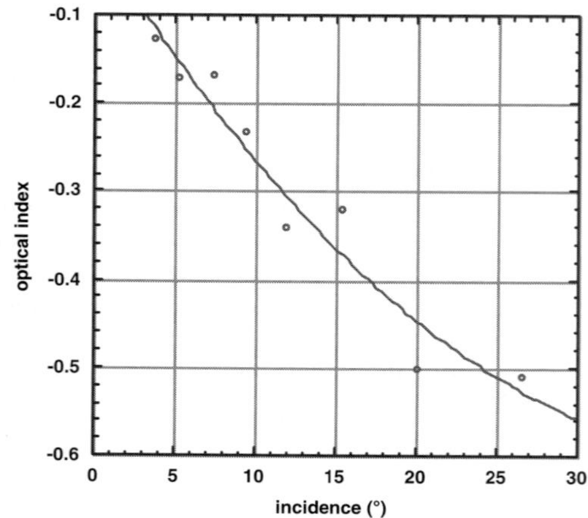

Fig. 5: Measured optical index of the material at 8.3 GHz versus the incidence angle of the wave. The circles correspond to the measured value. The curve is only a polynomial fit between these points.

In conclusion we have built a left- handed material at 8.3 GHz. This material has been simulated with a finite element software and characterised with a scalar network analyser. This material has a negative optical index, which varies between -0,1 and -0,5 for an incidence angle comprised between 5 and 25°. This left-handed material has a rectangular bidimensional geometry. This geometry partly accounts for the variation of the optical index with the incidence angle. With a triangular or hexagonal, bidimensional or tridimensional crystal, this dependency could be lower.

References

[1] J.B. Pendry, A. Holden, D. Robbins, and W. Stewart, IEEE Trans. Microwave Theory and Tech., 47, pp. 2075, 1999.

[2] J.B. Pendry, A. Holden, W. Stewart, and I. Youngs,, Phys. Rev. Lett., 76, pp. 4773, 1996.

[3] D. R. Smith, W. Padilla, D. Vier, S. Nemat-Nasser, and S. Schultz, Phys. Rev. Lett. 84, pp.4184, 2000.

[4] R. A. Shelby, D. R. Smith, S. C. Nemat-Nasser, and S. Schultz, Appl. Phys. Lett., 78, n°4, pp. 490-491, 2001.

Numerical and Experimental Investigation of the Superresolution in a Focusing System Based on a Plate of "Left-Handed" Material

A.N. Lagarkov and V.N. Kissel[1]

Institute for Theoretical and Applied Electromagnetics
Russian Academy of Sciences, Moscow, Russia

The unique properties of so-called "left-handed" materials (LHM) with simultaneously negative real parts of the permittivity $\varepsilon = \varepsilon' - i\varepsilon''$ and of the magnetic permeability $\mu = \mu' - i\mu''$ (including the possibility of focusing the radiation of point sources by a plane-parallel plate) were predicted by Prof. V.G. Veselago back in the late 1960s [1]. After that, various options were suggested for the practical realization of composites with negative values of ε' and μ' [2-5]. Special interest was aroused by the outstanding idea of Prof. J.B. Pendry [6] about the possibility of developing a "superlens" whose resolution is not restricted by the well-known diffraction limit. Our ICMAT 2003 presentation [7] touches upon the theoretical clarification of the conditions in which use of (dissipative) LHM-plate results in a higher resolution of the focusing system compared to conventional ones. In this paper we present the results of both computer modeling and experiment, which demonstrate the effect of superresolution.

According to Pendry [6], in the case of $\varepsilon = \mu = -1$, a plane-parallel LHM-plate enables one to obtain an "ideal" image of a point source owing to the amplification of the evanescent components of the spectrum of spatial frequencies of electromagnetic radiation. Theoretical analysis of the focusing properties of the system based on the realistic dissipative LHM-plate [7] shows that given a plate thickness comparable to the wavelength, extremely high requirements are placed on the quality of the "left-handed" material to make superresolution noticeable. On the other hand, the use of thin LHM-plates brings about a drastic reduction of loss requirements and leads one to hope for the possibility of practical realization of that effect.

In particular, even in the case of relatively high losses in the "left handed" material, when $\varepsilon'' = \mu'' = 0.1$, one can obtain a sharp separable image of sources separated by a distance of one tenth (!) of wavelength. This is confirmed by the results of calculation by the method of volume integral equations (Fig. 1) obtained for the plate thickness $2k_0 a = 0.2$, width $2k_0 b = 2.4$, and distance between the sources $k_0 d = 0.6$ (i.e. $d \approx \lambda/10$). Note the peaks in the figures that correspond to the maximal values of the field amplitude on the front and rear faces of the plate due to the accumulation of the energy of evanescent modes (the peaks on the front face disappear at $\varepsilon'' = \mu'' = 0$, $\varepsilon = \mu = -1$).

An experiment was performed to check the possibility of attaining the superresolution in practice. The basic components of the experimental facility are shown in Fig. 2 (the geometric dimensions are given in mm). The facility includes, vertically arranged, two transmitting antennas 1 and 2 and one receiving antenna 3 (parallel half-wave vibrators), as well as an LHM-plate 4. The scheme of their arrangement copies the geometry of the treated problem. The microwave image of the transmitting antennas is registered by measuring the level of the signal received by the antenna 3 during its movement in the horizontal direction parallel to the plate.

The LHM-plate is made of a polystyrene foam-based composite material. The artificial magnetic properties of the composite are provided by resonance inclusions in the form of spirals, similar to the media suggested and investigated in [3, 4]. For the composite not to show the effect of chirality (in this case, the emergence of the horizontal component of electric field because of the electric moment of the spirals), equal numbers of left- and right-hand spirals are used, staggered as in [4] (Fig. 2b). Note that bifilar spirals may be used for the same purposes [3]; in the particular case of one turn and zero pitch, these spirals reduce to split-ring resonators. The choice of

---

[1] Corresponding author. E-mail: kis_v@mail.ru

"standard" spirals was governed by the simplicity of their individual adjustment to one and the same frequency (which is important for the success of the experiment) by way of shortening the wire or slightly changing the distance between turns. Each spiral is 5 mm in diameter, contains approximately two turns of insulated copper wire 0.74 mm in diameter, and is tuned to the resonant frequency of 1.6 GHz due to a little pitch chosen.

Different ways exist to ensure the effective dielectric properties of a composite $\varepsilon' < 0$. A highly homogeneous medium may be produced using elements of one type, namely, spirals arranged in a special manner [4]. However, for the purposes of our experiment, it is quite sufficient to use the simplest system of parallel conductors arranged vertically between polystyrene foam holders (Fig. 2b); the resultant structure turns out to be close to that described by Smith et al. [2]. We used conductors made of copper wire 0.74 mm in diameter, tuned to the resonance frequency of 1.6 GHz (the length of each conductor, 89 mm).

Since ε' and μ' can reach negative values simultaneously in a narrow frequency band, the measurements were carried out in the bandwidth sufficiently wide to include that frequency region. Figure 3 gives the levels of recorded signal (in dB) as a function of frequency (in GHz) and location of the receiving antenna (coordinates y, mm). The following options were examined: (a) no plate between the antennas (Fig. 3a), (b) the LHM-plate described above is located between the transmitting antennas and the receiving antenna (Fig. 3b), and (c) a plate of quartz glass 6 mm thick is placed between the transmitting antennas and the receiving antenna ("standard" dielectric). In the first and third cases, two sources cannot be resolved by readings of the receiving antenna. A signal maximum is observed at point $y = 0$ equidistant from both sources. The frequency at which the signal is maximal is determined by resonant properties of the receiving and transmitting antennas. The presence of a quartz plate somewhat reduces this frequency; however, the pattern does not change qualitatively. Entirely different results are observed when LHM-plate is used. Three characteristic frequency ranges may be identified in the diagram (Fig. 3b). At frequencies below approximately 1.62 GHz, the superresolution is absent again, because, in this frequency range (below the resonance frequency of the composite elements), the effective ε' and μ' are positive [2-4]. Between the frequencies of 1.62 and 1.65 GHz, the signal level in the receiving antenna drops abruptly, which is attributed to high values of the effective ε'' and μ''. Finally, at frequencies of approximately 1.65 to 1.8 GHz, two signal maxima are registered whose coordinates y exactly correspond to the position of the sources. So we observe superresolution of two sources separated by the distance of $\lambda / 6$. For different distances between the sources, measurements likewise reveal this correspondence. Characteristic sections of the diagram at fixed frequencies are given in Fig. 3c. Curves 1, 2, and 3 are given for options (a), (b), and (c), respectively, at a frequency f = 1.71 GHz on which the signal in the receiving antenna at $y = \pm 15$ mm is maximal and several dB higher than the level recorded in the absence of a plate. In the vicinity of point $y = 0$, curve 3 exhibits a dip up to 4.5 dB deep (with respect to the maximum). One can select such a frequency on which the dip is more pronounced and, consequently, better conditions are provided for separate observation of the sources (see curve 4 in Fig. 3c, frequency of 1.664 GHz). A minor asymmetry in the shape of the curves is associated with the natural imperfection of the experimental sample of LHM-plate.

Worthy of note is that neither set of spirals nor wire grid used separately (to result in $\mu' < 0$ or $\varepsilon' < 0$ correspondingly) did not reveal the possibility to achieve superresolution at the given thickness of the LHM-plate sample (both in computer modeling and experiment). On frequencies where the effect of superresolution shows up, the manufactured plate sample indeed exhibits the properties of "left-handed" material (in particular, with respect to the negative phase velocity of the wave in such a medium). To confirm this, a plate, first of quartz glass and then of LHM-composite, was introduced into the spacing between the field source and receiving antenna. The phase advance $\Delta\varphi$ of the received signal was registered in the process of plate travel. The experimental scheme is shown in Fig. 4a, and the curves of phase advance as a function of travel t for a frequency f = 1.71 GHz are given in Fig. 4b. One can see that the emergence of a plate of "standard" material (quartz) between the source and receiving antenna leads to a positive phase increment (curve 1). The presence of the LHM-plate causes a decrease in the phase advance (curve 2), which confirms the realization of negative effective ε' and μ'. In so doing, the wave-like behavior of curve 2 at $t > 15$ mm demonstrates the degree of inhomogeneity of the sample (the period of wave-like variations of curve 2 coincides with the period of the LHM-structure).

Note that a significant value of negative phase advance of over 100 degrees was reached when using the manufactured LHM-plate. This is indicative of high values of effective $|\varepsilon|$ and $|\mu|$ of the composite (an approximate estimate is $\sqrt{\varepsilon\mu} \approx 8.5$) and, as was demonstrated by the calculation results, is important from the standpoint of ensuring a high resolution when the LH-material is not isotropic. The composite employed in the experiment has the xx-component of the tensor of effective magnetic permeability $\mu'_{xx} \approx +1$, because the resonators are hardly excited by the x-component of magnetic field. Computer simulation revealed that, at $\mu'_{xx} = +1$ and $\varepsilon'_{xx} = \varepsilon'_{yy} = \mu'_{yy} = -1$, the image quality of two filamentary sources greatly deteriorates (compared to the results given in Fig. 1b): the field amplitudes in the focusing zones are low, and the dip between the images of the sources is almost unnoticeable. Nevertheless, the image quality may be sharply improved by increasing the values of parameters, for example, to $\varepsilon'_{xx} = \varepsilon'_{yy} = -10, \mu'_{yy} = -6$ (other relationships between ε and μ may be selected as well). Therefore, in designing technical devices involving the use of LH-materials, there is no need to develop an isotropic medium and exactly satisfy the condition $\varepsilon = \mu = -1$. Acceptable results may be attained using anisotropic samples which are simpler to manufacture. Note that high values of $|\varepsilon|$ and $|\mu|$ in the case of an isotropic LH-material disturb the focusing properties of the system (in particular, with $\varepsilon = -10 - i0.1, \mu = -6 - i0.1$, source images are no longer separately observed); however, small deviations from $\varepsilon = \mu = -1$ (for example, $\varepsilon = -1.5, \mu = -0.7$) may be permitted depending on the requirements placed on the image quality.

Therefore, theoretical, numerical and experimental data demonstrate that one can use thin-layer "left-handed" materials to develop a system with an improved (compared to conventional devices) resolution, which is not restricted by the diffraction limit.

References

1. V.G. Veselago, "The electrodynamics of substances with simultaneously negative values of ε and μ", Sov. Phys. Usp. **10**, 509 (1968)
2. D.R. Smith, W.J. Padilla, D.C. Vier, S.C. Nemat-Nasser and S. Schultz, "Composite medium with simultaneously negative permeability and permittivity", Phys. Rev. Lett. **84**, 4184 (2000).
3. A.N. Lagarkov, V.N. Semenenko, V.A. Chistyaev, D.E. Ryabov, S.A. Tretyakov and C.R. Simovski, "Resonance properties of bi-helix media at microwaves", Electromagnetics **17**, 213 (1997).
4. A.N. Lagarkov, V.N. Semenenko, V.N. Kisel and V.A. Chistyaev, "Development and simulation of microwave artificial magnetic composites utilizing nonmagnetic inclusions", J. Magn. Magn. Mater. **258-259**, 161 (2003).
5. G.V. Eleftheriades, A.K. Iyer and P.C. Kremer, "Planar negative refractive index media using periodically L-C loaded transmission lines", IEEE Trans. Microwave Theory Tech. **50**, 2702 (2002).
6. J.B. Pendry, "Negative refraction makes a perfect lens", Phys.Rev.Lett. **85**, 3966 (2000).
7. A.N. Lagarkov, V.N. Kissel, A study into the possibility of field focusing using "left-handed" materials – to be presented at the International Conference on Materials for Advanced Technologies ICMAT 2003, December 7-12, 2003, Singapore.

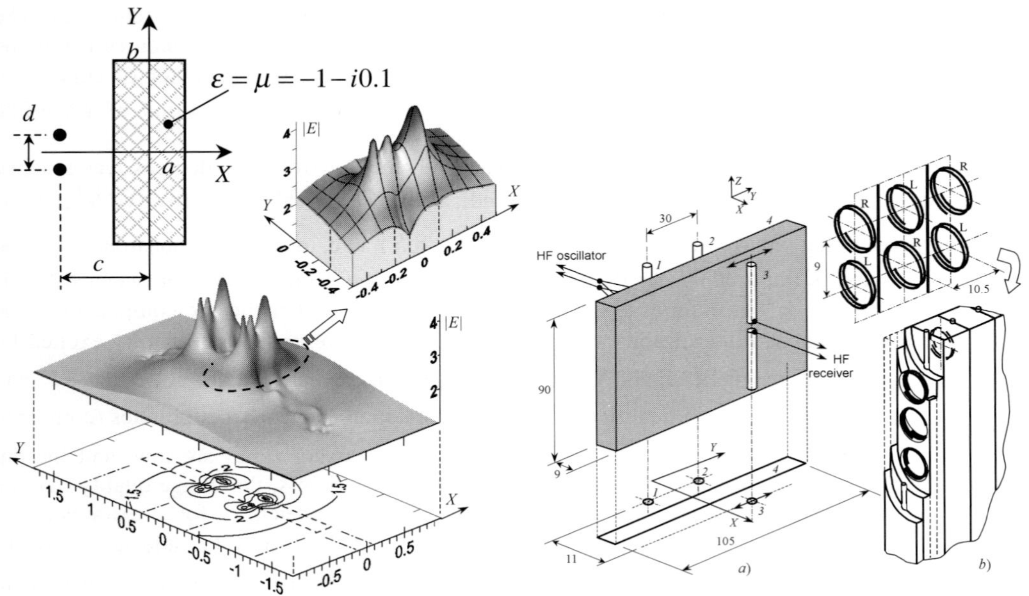

Fig. 1

Fig. 2

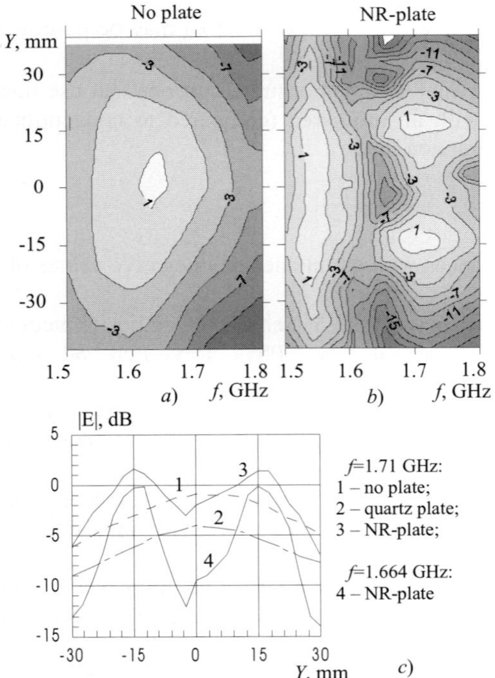

Fig. 3

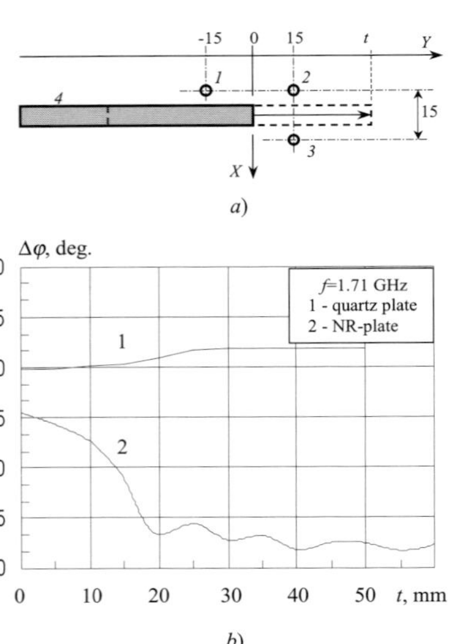

Fig. 4

Session F9

Periodic Structures

Chair: B.A. Munk

ON ABSORBERS

Ben A. Munk, Jonothan B. Pryor and Gan Yeow Beng *

The Ohio State University ElectroScience Laboratory, 1320 Kinnear Rd., Columbus, Ohio 43221, USA
* Temasek Laboratories, National University of Singapore, 10 Kent Ridge Crescent, S(119260), Singapore

INTRODUCTION

When a target is exposed to an incident signal, it will be partly scattered in all directions and partly absorbed if the target is lossy. We are most often interested in the backscattered signal (i.e. the monostatic RCS) although the bistatic signal sometimes is of interest as well.

In order to reduce the RCS level, we may apply one or both of the following techniques:

1. Shaping. This may result in the reflected signal being diverted into directions different from the backscatter and thereby producing a lower monostatic RCS.

2. Absorption. Use of lossy material can lead to a lower scattering level in some directions but not necessarily in all. It is the designer's responsibility to use the absorber such that a lower RCS is obtained in the general backscatter direction. The forward scattering will in general not be affected significantly.

Basically there are at least two kinds of absorber. The first type is intended to reduce the backscatter as well as the bistatic signals. The second type is designed to attenuate surface waves that may be present on the target. These waves will radiate and can therefore lead to an increase in the RCS level as discussed in reference [1]. The reduction of these waves is usually obtained by using absorber along the surface or by terminating these in resistive loads.

In this paper, we shall investigate only the first type, namely the specular absorptive. It comes in at least three different versions:

1. The ideal absorber (that is somewhat illusory).

2. The Jaumann absorber. It can be considered a generalization of the Salisbury screen.

3. The Circuit Analog Absorber (CA). Here, the resistive sheets are provided with a geometrical pattern resulting in a reactive component in addition to the resistive.

THE IDEAL ABSORBER

Let us assume that the ideal absorber is made of a material with permeability μ_1 and permittivity ε_1. Then, a necessary requirement for the ideal absorber is that its intrinsic impedance Z_1 is equal to that of free space Z_0, i.e.

$$Z_1 = \sqrt{\frac{\mu_1}{\varepsilon_1}} = Z_0 = \sqrt{\frac{\mu_0}{\varepsilon_0}} \tag{1}$$

If the absorber is "infinitely" thick, condition (1) is sufficient. However, since this is hardly realistic, we must also require that the signal is sufficiently attenuated before it reaches the backwall of the absorber, where we usually place a ground plane.

In that event, we must use lossy material, i.e., μ_1 and ε_1 must be complex. Thus, in addition to condition (1) we must require that the imaginary components of μ_1 and ε_1 produces sufficient attenuation of the incident signal as it makes its way from the front to the ground plane and back. How much attenuation is required? That simply depends on what level of reflectivity can be tolerated. Note, the necessary condition is still given by (1) where μ_1 and ε_1 are complex.

Thus, the concept for the ideal absorber is very simple. However, actually producing an ideal absorber has proven to be very difficult. In fact, it has been the subject of sometime intense research for at least half a century. Although quite remarkable progress has been demonstrated at the lower frequencies, it is still possible to gain fame and glory by making a contribution at the higher frequencies.

In the meantime, we shall next discuss types of absorber that realistically can be made today at practically all frequencies.

THE JAUMANN ABSORBER

The simplest Jaumann absorber is the Salisbury screen depicted in Fig. 1. It is comprised of a resistive sheet mounted one-quarter wavelength in front of a ground plane (all absorber should in general have a ground plane in the back to prevent any ambiguity as caused by any objects located behind the absorber sheets). This type of absorber was named after its inventor, W.W. Salisbury of the MIT Radiation Laboratory, who was issued a patent in 1952 [2].

The workings of the Salisbury screen are easily understood by inspection of the equivalent circuit shown in Fig. 1 middle. At the frequency where the spacing from the resistive screen to the ground plane is $\lambda/4$, the input admittance of the equivalent short-circuited transmission line denoted (1) is zero. We merely see the resistive sheet alone, which has an admittance close to that of free space Y_0. In other words, we obtain a total input admittance located close to the center of the Smith chart as shown in Fig. 1 bottom. However, at lower frequencies, the input admittance of the equivalent transmission line (1) becomes inductive, causing the total admittance to be located in the lower half of the Smith chart. A typical reflection curve is readily obtained, as shown in Fig. 1 bottom-right. Use of a dielectric slab between the screen and the ground plane leads to a higher input admittance in the equivalent transmission line (1) (except for $\lambda/4$). Thus, this is seen to reduce the bandwidth of the Salisbury screen. Ideally, we could use ferrite or some other material with a lower intrinsic admittance than free space Y_0. However, that approach is usually not very popular because of the weight penalty and because better solutions are obtained by other means (as will be discussed later). Use of dielectric and/or a high μ material will of course lead to designs of reduced thickness.

The general Jaumann absorber is made of two or more homogeneous resistive sheets mounted in front of a ground plane as shown in Fig. 2 top. These screens are separated by dielectric slabs of about $\lambda/4$ thickness measured in the respective slabs at the center frequency.

The workings of a Jaumann absorber with air separations is readily understood by inspection of the Smith charts and the equivalent circuit also shown in Fig. 2. The admittance at position (1) looking

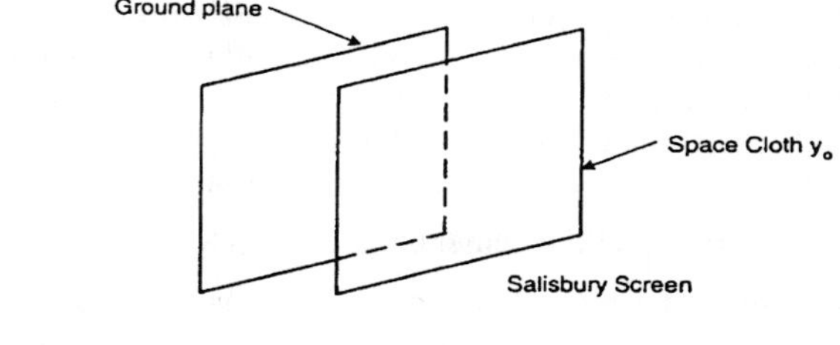

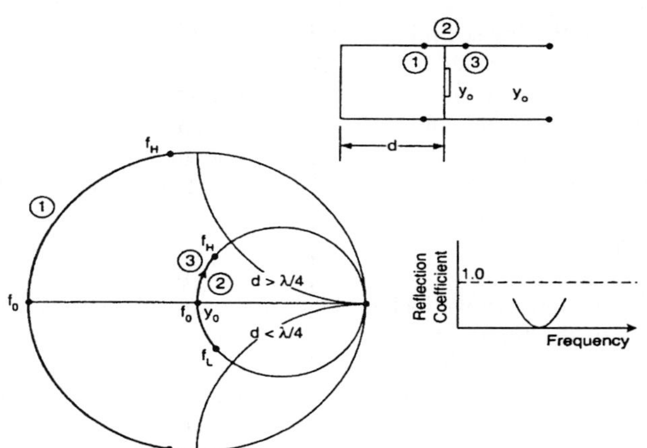

Fig. 1. *Top*: Simple Salisbury screen comprised of a resistive sheet in front of a ground plane. *Middle*: Equivalent circuit. *Bottom*: Smith chart showing the derivation of the input admittance as a function of frequency (*left*). Reflection coefficient as a function of frequency (*right*).

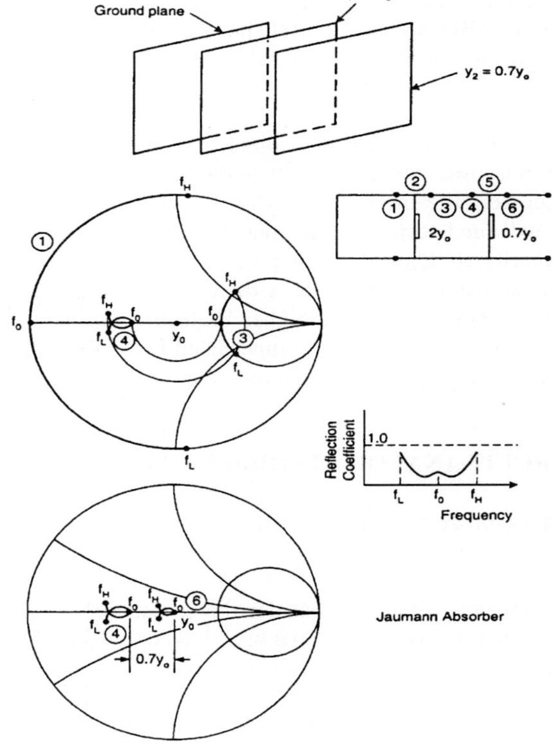

Fig. 2. *Top*: Typical absorber comprised of two resistive sheets in front of a ground plane. *Middle*: Smith chart (*left*) showing the beginning of how to obtain the input admittance of the absorber and equivalent circuit (*right*). *Bottom*: Smith chart (*left*) showing the final calculation of the input admittance and typical reflection coefficient as a function of frequency (*right*).

toward the ground plane in the equivalent circuit is depicted by curve (1) at the rim of the Smith chart in the middle. To this pure reactance, we now add the sheet admittance $2Y_0$ as denoted by curve (2). The sum of curves (1) and (2) yields curve (3) where we note that the highest frequency f_H is located in the upper half of the Smith chart while the lowest frequency f_L is located in the lower half. We next rotate curve (3) into curve (4) corresponding to the separation between the two resistive sheets. If we choose this separation to be $\lambda/4$ at the center frequency f_0, we will rotate less than $180°$ at the lower frequency f_L and more at the higher f_H. The result is seen to be that the frequencies f_L, f_0, and f_H are more clustered together in curve (4) than in curve (3). We finally add the last sheet admittance $0.7 Y_0$ as shown in the lower Smith chart in Fig. 2 and obtain curve (6). This curve is seen not only to be located at the center of the Smith chart but also to have the frequencies further clustered together due to the nature of the Smith chart. Recalling that the reflection coefficient is simply the distance from the center of the Smith chart to the respective frequency points of curve (6), we readily obtain the typical reflection curve shown in Fig. 2 bottom-right. We observe that contraction of the admittance curves (4) and (6) is obtained only when the first sheet admittance $Y_1 > Y_0$. If $Y_1 < Y_0$, no contraction is observed; in fact we cannot even land in the center of the Smith chart unless Y_2 is negative.

The simple example of a Jaumann absorber shown above had air between the resistive sheets. This was chosen to facilitate the calculations in the Smith chart. Under most practical circumstances dielectric slabs will usually be required at least for mechanical reasons. However, use of dielectric cannot only reduce the total thickness of a Jaumann absorber, it can actually extend the bandwidth. This can easily be observed by pursuing the calculations in a Smith chart. An example is shown in Fig. 3 where the dielectric constants of 3 and 2 were given as shown in the schematic. Straightforward calculations in the Smith chart easily yield curve 4 that is seen to be centered around Y_0 located at $0.7 Y_2$. We easily obtain a reflection level below 20 dB in the frequency range 5 to 15 GHz and for a total thickness of about 1.2 cm. This would be a good place to start an optimization process for the resistive sheets and the optimum values of the intrinsic admittances Y_1 and Y_2.

Note that we have applied a dielectric slab on the outside of the outermost resistive sheets. It serves two purposes. First, it improves the bandwidth of the absorber, and second, it shields the resistive sheet from the environment (this is particularly important in the cases where we use a CA sheet as we will see later). Note also the 30 mil dielectric layer on the very outside of the absorber. It serves primarily as a mechanical protection for the lower dielectric constant material underneath. In general, the ruggedness of a dielectric corresponds with its dielectric constant, that is, a more rugged dielectric has a higher dielectric constant. It is of course part of the outer matching transformer and is easily compensated for in this case by simply making the thickness of the outer transformer slightly less than one-quarter wavelength.

CIRCUIT ANALOG ABSORBER (CA)

The absorber considered above, namely the Jaumann absorber, is characterized by using purely resistive sheets.

We are now going to consider the so-called CA absorbers. These are characteristically made of sheets that not only contain a resistive component but are reactive as well. This is accomplished by using periodic surfaces made of lossy material, as shown in Fig. 4 top. We have in [3] developed the equivalent circuit for an array of lossless elements loaded with Z_L. Based on this circuit, it is fairly obvious that the equivalent circuit for a CA sheet in either principal plane will consist of an RLC series combination as indicated in the equivalent circuit shown in Fig. 4 middle. The resistive component is a result of the lossy element material, while the inductance is associated with the

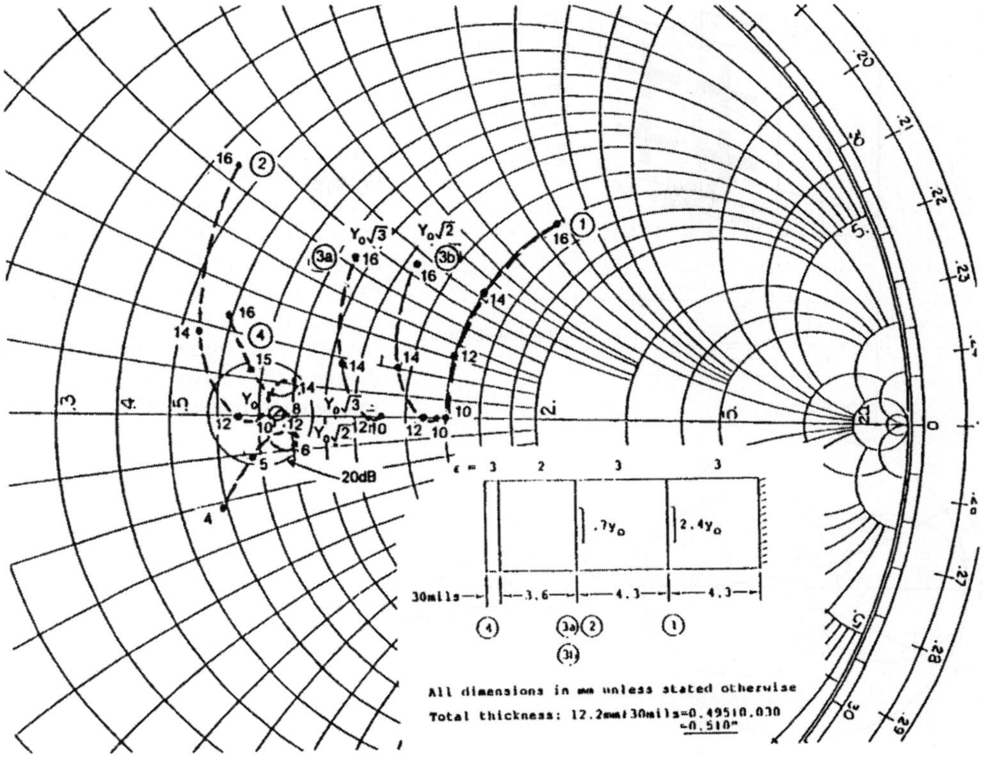

Fig. 3. Smith chart showing calculations of a Jaumann absorber comprised of two resistive sheets sandwiched between three dielectric slabs and an external skin.

straight part of the elements and the capacitance with the gaps between the elements, as shown at the top of Fig. 4. We emphasize that this equivalent circuit is only an approximation that merely serves to explain the workings and the name of the CA absorber. The more precise calculation is obtained from special computer programs. In fact, it is not even recommended to obtain the exact equivalent circuit but merely to plot the CA sheet admittance directly on a Smith chart and simply work from there.

Based on these facts, the basic idea of a CA absorber becomes quite clear. At the center frequency f_0 the distance between the ground plane and the CA sheet is approximately $\lambda/4$ while the RLC-series circuit resonates. In other words, it is similar to the simple Salisbury screen discussed earlier. However, at the lower frequency f_L, the input admittance Y_g toward the ground plane becomes inductive as indicated by curve (1) of Fig. 4 at the rim of the Smith chart, while the RLC sheet Y_a becomes capacitive as indicated by curve (2). The total admittance of the complete CA absorber is simply the sum of Y_a and Y_g, as indicated by curve (3).

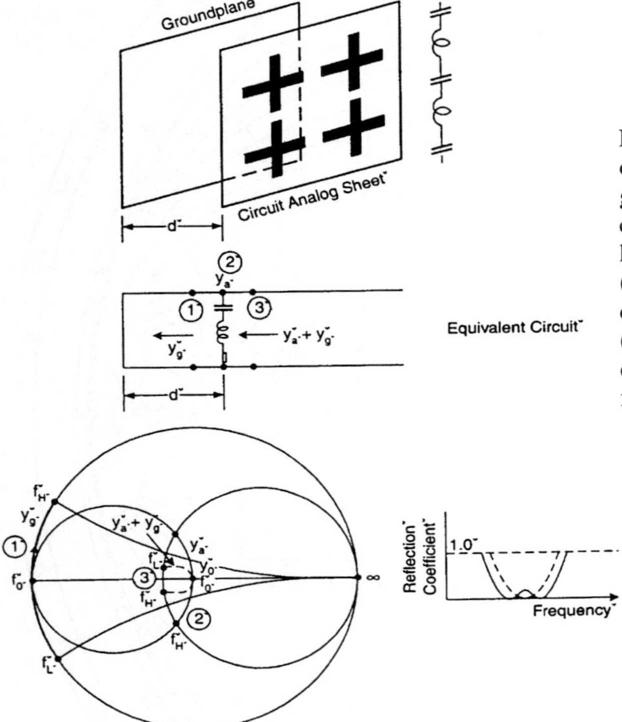

Fig. 4. *Top*: CA absorber comprised of a single CA sheet in front of a ground plane. *Middle*: Equivalent circuit. *Bottom*: Smith chart depicting how to calculate the input admittance (*left*) and typical reflection coefficient as a function of frequency (*right*). The frequencies f_L and f_H denote the lowest and highest frequencies, respectively.

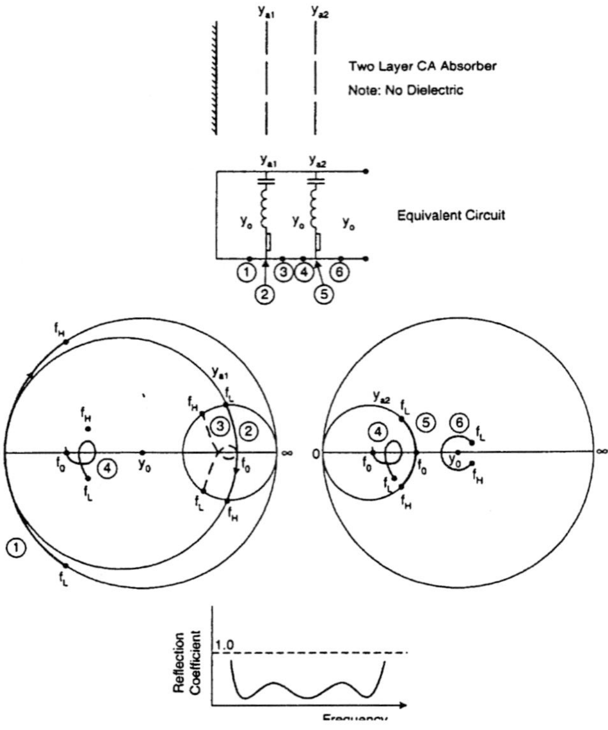

Fig. 5. *Top*: CA absorber comprised of two CA sheets in front of a ground plane. *Middle*: Equivalent circuit (*above*) and two Smith charts showing the calculation of the input admittance (*below*). *Bottom*: Typical reflection curve as a function of frequency. The intricate details of curve (6) are not shown, only an outer bound.

We observe that the reactive parts of Y_a and Y_g to a large degree cancel each other resulting in a location for f_L very close to the center of the Smith chart; it leads to a small reflection. The same is true for the higher-frequency f_H.

It is easily noted that choosing the real part of the CA sheet admittance equal to Y_0 is not necessarily the best choice when designing for the largest bandwidth. In fact, if chosen slightly higher than Y_0, it can result in a somewhat higher bandwidth as indicated by the full line reflection curve in Fig. 4 bottom-right.

Just like the Jaumann absorber was made of more resistive sheets, a CA absorber can be made from an arbitrary number of CA sheets. A simple example comprised of two CA sheets is shown in Fig. 5. The best understanding is obtained by simply calculating the various admittances in the Smith chart starting from the ground plane and continuing to the front. As you progress, you will see where the CA sheets should be modified to obtain the largest bandwidth at the front of the absorber. You next aim to change the dimensions of the CA sheets to obtain these ideal values of the CA sheets.

In short, you first sketch your design in a Smith chart. You next use a computer program simply to reduce the workload and improve accuracy. It is strongly recommended, however, that each additional layer in a Smith chart be plotted in order to keep track of the computation activity.

CONCLUSION

We have discussed two absorbers of the specular type, namely the Jaumann and the Circuit Analog. They can in principle be made to operate at any frequency in contrast to the ``ideal" absorber that favors the lower frequencies. However, the Jaumann and Circuit Analog will not attenuate surface waves at all angles of incidence and all polarizations while the ideal absorber will. No question about it, absorbers are still a ``hot" research subject.

REFERENCES

[1] B.A. Munk, Finite Antenna Arrays and FSS, Chapter 4, Wiley, NY, 2003.

[2] W.W. Salisbury, United States Patent 2,599,944, Absorbent body of electromagnetic waves, June 10, 1952.

[3] B.A. Munk, Frequency Selective Surfaces, Theory and Design, Section 4.11.1, Wiley, NY, 2000.

The Electromagnetic response of two-dimensional Gratings
of dielectric and magnetic fibers : Infinite and finite cases

H. Roussel and W. Tabbara
Département de Recherche en Électromagnétisme/ L. 2. S., Supélec
Plateau de Moulon, 91 190 Gif-sur-Yvette, France

Abstract :
The last few years have seen a growing interest in the analysis of heterogeneous materials. These investigations lead to many applications as for example the development of absorbing materials or frequency selective surfaces. To the diversity of the applications corresponds a variety of approaches used in the modeling of the fields scattered by heterogeneous materials. We have mainly investigated the case of 2D gratings of dielectric and magnetic fibers embedded in a multilayered medium placed or not over a perfectly conducting plane.

1. Introduction

Grating like structures have received continuous attention since many years. They are used in various fields of physics and engineering, and have been modeled and analyzed by means of exact and approximate approaches. Among the more recent applications of grating are the photonic band-gap materials (PBG) and absorbing structures used in anechoic chambers, the analysis of which will be considered in the present paper.

The PBG structures are of interest due to the applications in several scientific and technical areas such as narrow filters, optical switches, cavities, design of more efficient laser [1-4]. The position and width of the forbidden gap of the PBG depend on many parameters such as the geometry of the structure, and the dielectric properties of the inclusions. Previous modeling is then necessary to predict the response of such complex structures. In this paper we present a model based on the volume integral representation of the electromagnetic field, from which we determine the reflectivity and the transmittance versus frequency for different configurations and analyse the influence of each parameter driving the behavior of the PBG. These first results are given for infinite gratings. In a way to analyse waveguides we consider also finite gratings to obtain a map of the electric field inside the grating for various frequencies and configurations.

Absorber used in anechoic chambers to simulate free-space environment at microwave frequencies are also analyzed here. Optimal properties of these absorbers are obtained by adjusting their shape or their dielectric/magnetic properties. Recently, the interest presented by the analysis of low frequency absorbers has increased with the developpement of measurment in electromagnetic compatibility [5-6]. The model proposed here is also well adapted to adjust the dielectric/magnetic and geometrical properties of the gratings in a way to optimize the performances of these absorbers.

2. Modeling of the electromagnetic field

2.1. Geometry of the 2D infinite grating

We consider a 2D structure, infinite along the axis z, made of an infinite dielectric slab of thickness h, backed or not by a perfectly conducting plane and placed in the air, characterized by its relative permittivity ε_2 and its relative permeability μ_2. This geometry is called the reference structure. It is now loaded by one or more layers of periodic arrays of fibers having the same period Δx (see Fig. 1.a). The fiber's cross-section may be different in each layer. The electromagnetic parameters of the ith fiber in the reference cell are ε_i and μ_i, and s_i represents the cross section of this same fiber.

For the analysis of the electric field inside a waveguide, we consider also the case of gratings including a defect of periodicity. Hence by breaking the periodicty of the structure we need to introduce a new model for finite gratings. On figure 1.b) we removed the fourth column of a stacked periodic grating to analyse the efficiency of the waveguide inside the bandgap .

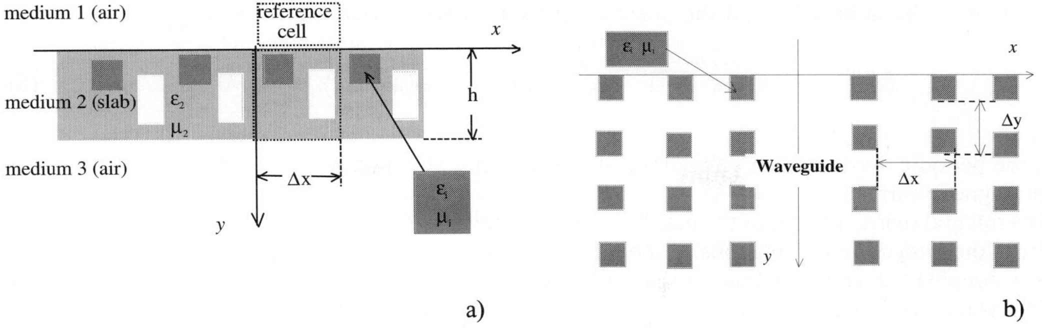

Figure 1: Geometry of the stacked infinite gratings a) and the finite one b)

2.2. Domain integral representation of the electric field

The incident field is a plane wave propagating in the medium 1, in a plane perpendicular to the fiber's axis at oblique incidence, θ, with respect to the normal to the slab upper interface. Then we have to solve the problem for both TM (electric field parallel to the fibers' axis) and TE (magnetic field parallel to the fibers' axis) polarizations. In the following, the equations will be established for the TM polarization only and the time dependence is exp(-jωt).

The incident plane wave in the TM case is given by:

$$E_z^i(x,y) = E_1 \exp\left[j\left(k_{1x}x + k_{1y}y\right)\right] \tag{1}$$

and for the reference structure, the total electromagnetic field in each medium i (i =1,2,3) is denoted by $\left(\mathbf{E}^{slab}, \mathbf{H}^{slab}\right)$, is well known and given by:

$$E_z^{slab}(x,y) = E_1\exp(jk_{1x}x)\left[B_i\exp(jk_{iy}y) + A_i\exp(-jk_{iy}y)\right] \tag{2}$$

with
$$\begin{cases} k_i = \omega\sqrt{\varepsilon_i\mu_i\varepsilon_0\mu_0} \\ k_{1x} = k_{2x} = k_1\sin\theta \\ k_{iy} = \sqrt{k_i^2 - k_{ix}^2} \end{cases}$$

Where A_1 is the reflection coefficient of the slab, $B_1 = 1$, $A_3 = 0$ and B_3 is the transmission coefficient of the slab.

The equations for the magnetic field can be easily derived from (2).

Let $\mathbf{E}(x,y)$ be the total electric field when the fibers are placed in the slab. The scattered field is then given by:

$$\mathbf{E}^d(x,y) = \mathbf{E}(x,y) - \mathbf{E}^{slab}(x,y) \tag{3}$$

By making use of the theory of distribution and the Floquet's theorem, we obtain the integral representation of the electromagnetic field in terms of the polarization and magnetization currents inside the fibers [7]. The total electric field is given by:

$$\mathbf{E}(x,y) = \mathbf{E}^{slab}(x,y)$$

$$+\left[\text{grad div}_{x,y} + k_i^2\right]\iint_{S_0} \Delta\varepsilon_i(x',y')\overline{\overline{G}}_p^{el}(x-x',y,y')\mathbf{E}(x',y')dx'\,dy' \tag{4}$$

$$+j\omega\mu_i\mu_0\,\text{rot}_{x,y}\iint_{S_0}\Delta\mu_i(x',y')\overline{\overline{G}}_p^{ma}(x-x',y,y')\mathbf{H}(x',y')dx'\,dy'$$

where the integration domain is the fibers' cross-sections in the reference cell $S_0 = \sum_i s_i$, and

$$\overline{\overline{G}}_p^{el,ma}(x-x',y,y') = \sum_{n=-\infty}^{n=+\infty} \overline{\overline{G}}^{el,ma}(x-x'-n\Delta x,y,y')\exp(jk_{1x}n\Delta x) \qquad (5)$$

is the periodic dyadic Green's functions of the reference structure related to an elemantary electric or magnetic current.

The integral representation of the magnetic field is similar to the above one.

From the integral representations of the electric and magnetic fields, we first establish a system of two coupled integral equations by considering (x,y) in the domain S_0. The unknowns are the polarization and magnetization currents in the fibers. We apply a method of moment to solve these equations, where the expansion functions are rectangular pulse functions and the test functions are Dirac's delta distribution. The domain S_0 is divided into square subcells of side c equal to $\lambda_{fiber}/15$ in the TM case and $\lambda_{fiber}/30$ in the TE case to ensure the convergence of the results. If N is the total number of sub-cells, we obtain a system of linear equations of order 3N. From the knowledge of the currents, one can compute the total electromagnetic field at all points inside or outside the slab.

When we consider finite structures, the integration domain increases becoming the all fibers' cross-sections of the structure. Since we have analysed finite gratings in the air the Green's function used for this application is the well known 2D free space Green's function given by:

$$G(x-x',y-y') = \frac{j}{4} H_0^{(1)}\left(k\sqrt{(x-x')^2+(y-y')^2}\right) \qquad (6)$$

2.3. Reflection and transmission coefficients for infinite gratings

As we consider the case of a stack of infinite gratings, the diffracted field $\mathbf{E}^d$ is an infinite sum of terms representing a discrete spectrum of modes. Here we only calculate the diffracted field of propagating modes $\left(\mathbf{E}^d\right)_n$.

The reflection coefficient of mode n, R_n^{TM} , for a point (x,y) located in medium 1 is defined by:

$$R_n^{TM} = \frac{\left(E_z^d(x,y)\right)_n + E_z^r(x,y)}{E_z^i(x,y)} \qquad (7)$$

where $E_z^i(x,y)$ is given by (1), and (2) provides the reflected field of the unloaded slab:

$$E_z^r(x,y) = A_1 E_1 \exp j\left(k_{1x}x - k_{1y}y\right) \qquad (8)$$

We define the transmission coefficient T_n^{TM} of mode n at a point (x,y) located in medium 3 as:

$$T_n^{TM} = \frac{\left(E_z^d(x,y)\right)_n + E_z^t(x,y)}{E_z^i(x,y)} \qquad (9)$$

where the field transmitted by the slab $E_z^t(x,y)$ is obtained from (2):

$$E_z^t(x,y) = B_3 E_1 \exp j\left(k_{3x}x + k_{3y}y\right) \qquad (10)$$

3. Results

3.1 Infinite gratings

In a first step we have considered a dielectric photonic band gap structure made of 5 rows of square cylinders placed according to a square lattice. The size of each cylinder is 5mm x 5mm and the periodicity of the grating is 15 mm. The relative permittivity of the fibers is 10. On figure 2 we show the variation of the transmittance (in dB) *versus* the frequency of the incident wave (4 GHz<f<18 GHz). We observe two photonic band-gaps centered on 7.5 GHz and 12 GHz.

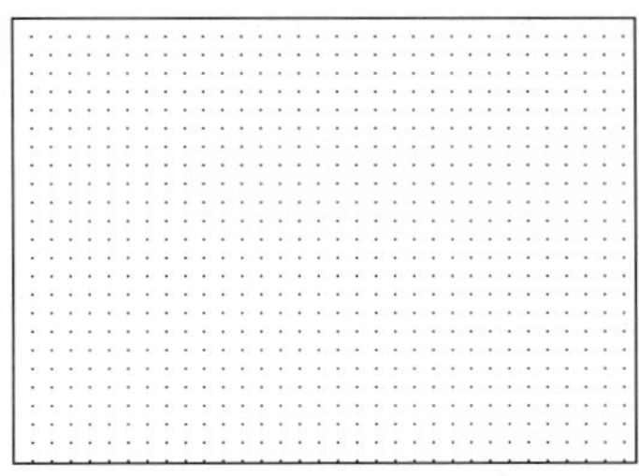

Figure 2 : Transmittance of a 2D dielectric photonic band-gap versus frequency for TM polarization at $\theta = 0°$.

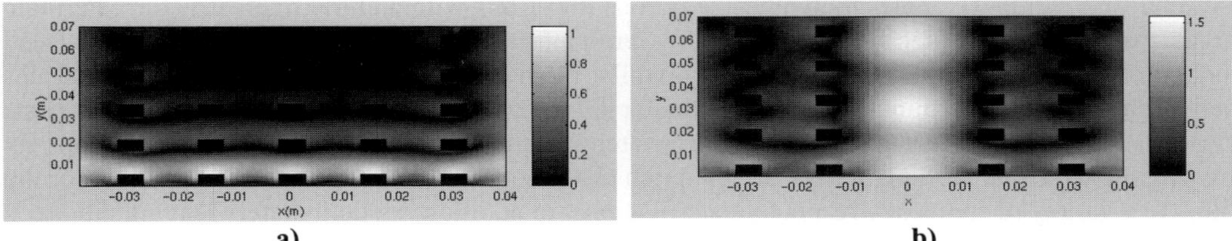

a)   b)

Figure 3 : Modulus of the Electric field inside the gratings at 7.5 GHz , with a) and without the third column b).

3.2 Finite gratings

In a second case we have considered a finite grating by keeping the same geometry as in the previous case and limiting the structure to 5 columns. Figure 3 shows the modulus of the electric field between the cylinders (the fibers are in black) at 7.5 GHz when we keep the five columns (Fig. 3.a) and without the third one (Fig3.b). We can notice that at this frequency the periodic structure behaves like a waveguide.

4. Conclusion

In this paper we have presented a model, based on an integral representation of the electromagnetic field, which gives the transmittance and the reflectivity of dielectric or metallic stacked gratings illuminated by a TM or TE plane wave. We showed that this approach is well adapted to the analysis of dielectric or metallic PBG. We will present also results on the analysis of wedge-shaped absorbers.

REFERENCES

1. E.R.Brown, C.D.Parker, " Radiation properties of a planar antenna on a photonic-crystal substrate ", *Journal of Optic Society of America B*, Vol. 10, n°2, 404-407, 1993.

2. H.Y.D.Yang, N.G.Alexopoulos, E.Yablonovich, » Photonic Band-gap materials for High-Gain Printed Circuit Antennas », *IEEE Trans. Antennas and Propagation*, Vol 45, n° 1, p.185-187, 1997.

3. M.M.Sigalas, C.T.Chan, K.M.Ho, C.M.Soukoulis, Physical review B, vol.52, n° 16, 1995.

4. F. Terracher, D. Berginc, »broadband dielectric microwave absorber with periodic metallizations », *JEWA*, Vol. 13, pp 1725-1741, 1999.

5. R. R. DeLyser, C. L. Holloway, R. T. Johnk, A. R. Ondrejka and Motohisa Kanda, « Figure of Merit for Low Frequency Anechoic Chambers Based on Absorber Reflection Coefficients », *IEEE Transactions on Electromagnetic Compatibility*, Vol. 38, pp. 576-583, 1996.

6. C. L. Holloway, R. R. DeLyser, R. F. German, P. McKenna and Motohisa Kanda, « Comparison of Electromagnetic Absorber Used in Anechoic and Semi-Anechoic Chambers for Emissions and Immunity Testing of Digital Devices », *IEEE Transactions on Electromagnetic Compatibility*, Vol. 3, pp. 33-46, 1997.

7. H. Roussel, F. Jouvie and W. Tabbara, « Reflection from buried dielectric and magnetic gratings of fibers: a domain integral equation approach », *Journal of Electromagnetic Waves and Applications*, Vol. 8, pp. 1645-1668, 1994.

8. C. L. Holloway and E. F. Kuester, »A low-Frequency Model for Wedge or Pyramid Absorber Arrays-II: Computed and Measured Results », *IEEE Transactions on Electromagnetic Compatibility*, Vol. 36, pp. 307-313, 1994.

9. H. Roussel and W. Tabbara, »Domain integral representation vs. homogenization methods for the analysis of photonic band gaps and absorbers », Eur. Phys. J. AP., Vol 6, pp 33-39, 1999.

Fabrication of Periodically Structured Thin Films with Electromagnetic Band Gaps

Martin O. Jensen and Michael J. Brett
Department of Electrical and Computer Engineering, University of Alberta,
Edmonton, Alberta, Canada T6G 2V4

ABSTRACT

We present a method for fabricating thin films with three-dimensional electromagnetic band gaps at optical frequencies (i.e., photonic band gaps). The films consist of periodic tetragonal arrays of nanometre scale square helices, and are produced using a thin film deposition method known as Glancing Angle Deposition (GLAD). This technique relies on advanced substrate motion and extreme flux incidence angles to synthesize porous thin films with a chiral microstructure. Periodic organization of the material is achieved by pre-patterning the substrate surface with a seed topography using electron beam lithography, whilst the helical microstructure of the film provides periodicity in the third dimension, perpendicular to the substrate. The square spiral GLAD films closely match a near-ideal photonic band gap architecture, as proposed by Toader and John. By varying the substrate seeding period and the pitch of the square helices, we can potentially tailor the films to a variety of frequency response specifications. In addition, by deliberately leaving out seeds in the substrate pattern we can engineer line defects in the film. These properties will be crucial for using the GLAD films in electromagnetic components.

INTRODUCTION

Glancing Angle Deposition is a physical vapour deposition (PVD) technique, in which the substrate is held at a highly oblique angle relative to the direction of the flux incident from the vapour source [1-4]. This produces thin films with high degrees of porosity (> 50 %), giving GLAD films significantly different properties than conventional thin films deposited at near-normal flux incidence angles. Rather than being continuous, dense films, GLAD films consist of individual nanometre-scale structures, whose size and spacing depend on deposition parameters such as the flux incidence angle, temperature, and material. The shape of the structures can be precisely controlled by advanced three-dimensional motion of the substrate, producing vertical and slanted posts, chevron structures, and circular and polygonal helices.

The unique topography of GLAD films is the result of pronounced self-shadowing among the growing structures at high flux incidence angles (see figure 1b below), as well as limited adatom mobility and diffusion [5-6]. When using smooth substrates, self-shadowing is initially enforced by clusters of atoms nucleating on the substrate, creating a stochastic array of structures. However, if the substrate is pre-patterned with an array of high aspect ratio seeds, film growth conforms to this pattern to yield a periodic film [7].

Numerous materials related applications are being explored for GLAD thin films, including mechanical resonators, humidity sensors, and thermal barriers. For electromagnetic applications, GLAD circular helices have previously been reported to be optically active, and to induce a chiral nematic phase in liquid crystals added to the film, enabling electro-optic switching of the liquid crystal component [8-9]. In this paper we present work conducted in pursuit of developing square spiral GLAD helices for photonic bandgap (PBG) crystals. PBG materials, which are not found in nature, are characterized by inhibiting the propagation of bands of electromagnetic (EM) modes. If defects are intentionally introduced into the crystal, defect states lying within the stop band may arise, so that EM radiation of certain frequencies can propagate along the defects with minimal losses and superior confinement [10-12]. In fabricating photonic bandgap crystals, the central requirement is to have a high dielectric contrast varying with nanometre scale periodicity. GLAD helical films successfully address this requirement: Our films inherently have a high degree of porosity for providing dielectric contrast, and periodicity arises from using pre-patterned substrates

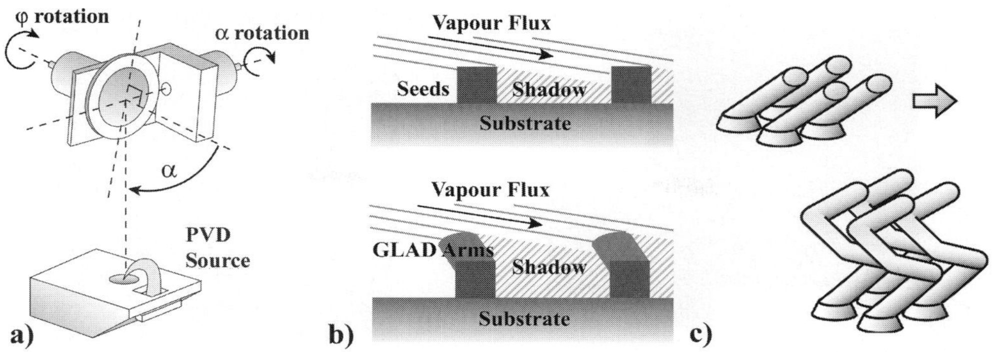

Figure 1. a) Schematic drawing of Glancing Angle Deposition, indicating the flux incidence angle α and the substrate rotation φ. b) Seeded substrates give periodic GLAD growth due to enforced shadowing. c) Square helices are fabricated by keeping α constant and incrementing φ in 90° steps.

and from the regular turns of the interwoven square helices (the latter ensuring periodicity perpendicular to the substrate). The result is a three-dimensional EM material.

Toader and John recently identified GLAD square helices deposited on a tetragonal array of substrate seeds as a promising approach to the fabrication of robust PBG crystals. Using the plane waves expansion method, their theoretical calculations predict tetragonal square spiral GLAD films to have a large bandgap of 15 % of the centre frequency [13-15]. For inverse square spiral structures, i.e., helical voids perforating a slab of dielectric, the estimated band gap is even higher at 24 % of the centre frequency.

EXPERIMENTS

In the present study we used electron beam lithography for substrate pattern generation. As e-beam resist we used a specially formulated cyclopentanone solution of the epoxy novolak SU-8 (provided by MicroChem Corp.). SU-8 is normally used as a very thick resist for micro-electro-mechanical systems (MEMS), but in a diluted formulation it yields a thickness after spin coating of only 150 nm, and in this case we have found it to be an excellent negative e-beam resist [16]. The lithography process was performed in a LEO 440 scanning electron microscope (SEM), equipped with the Nanometer Pattern Generation System from JC Nabity. The fabricated substrate patterns all had a tetragonal base geometry, with lattice periods ranging from 333 nm to 1000 nm, and seed diameters between 100 nm and 700 nm. The patterned areas were as large as 900x900 μm^2.

GLAD films were deposited in an electron beam evaporation system fitted with one stepper motor controlling the flux incidence angle α, and another motor controlling the substrate rotation φ. A schematic of the set-up is shown in figure 1a, and the growth mechanisms are illustrated in figure 1b. Together, α and φ determined the porosity of the film, the pitch of the helices (defined as the vertical distance between each turn of a helix), the helical radius or side length, and the helix handedness. To make a square helix film, the flux incidence angle α was kept constant, while φ was incremented 90° at regular intervals (see figure 1c). A feedback crystal thickness monitor compensated for variations in the deposition rate.

All GLAD square spiral films described in this study were made in silicon, since it has a reasonably high refractive index (yielding films of high dielectric contrast), is straightforward to deposit, and produces helices with a well-defined microstructure and smooth surfaces. The flux incidence angle α was kept at 84°, and the height of each helix arm was set to 325 nm, resulting in a pitch of 1300 nm. The films were made with right-handed helices. Typical deposition parameters were a system pressure on the order of 10^{-4} Pa, and deposition rates of 3-10 Å/s.

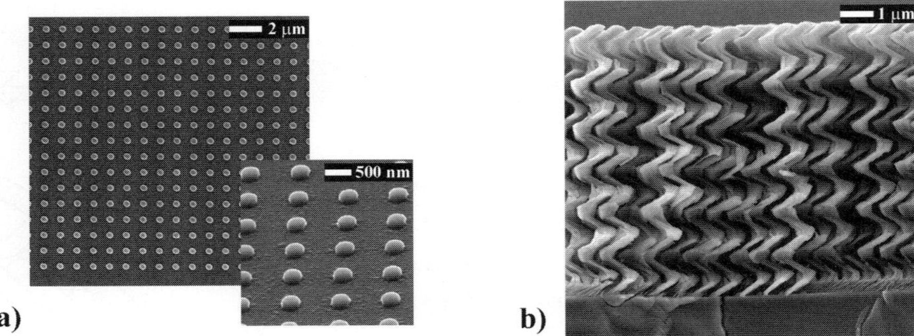

Figure 2. a) Substrate pattern with 400 nm diameter seeds in a tetragonal lattice with 1000 nm period, made by e-beam lithography. The inset shows an oblique view of the seeds. b) Side view of GLAD square spiral film on substrate pattern with 250 nm diameter seeds in a 600 nm period lattice. Note how the GLAD helices are forced to grow off the seeds.

RESULTS

Figure 2a shows SEM micrographs of a typical substrate seed topography before thin film deposition. The period of the tetragonal lattice is 1000 nm, and each seed has a diameter of approximately 400 nm. The seeds are uniform and well defined, and the high contrast between seeds and substrate is indicative of full resist development, providing maximum seed aspect ratios (crucial to efficient shadowing during subsequent GLAD).

A side view of a square spiral GLAD film is shown in figure 2b, in this case with a seed lattice period of 600 nm. Some seeds are visible on the cleaved substrate (although most are obscured by debris from the PVD process), and they are found to be approximately 130 nm tall. It is evident that the square helices initiate off the seeds with only one helix per seed, and that the seed layer periodicity translates all the way through several complete, uniform helix turns to the top of the 7.5 μm thick film. Along with the high porosity and the interweaving of individual helices, also visible in figure 2b, the three-dimensional periodic structure is what makes the GLAD square helix films a potential electromagnetic material.

Figure 3a shows a substrate seed design engineered with intentional defects, as the 1000 nm lattice now contains line defects one and two rows wide, corresponding to 2 and 3 μm gaps. A GLAD square spiral film deposited on this seed topography is shown in figures 3b and 3c. Clearly the GLAD film conforms to the line defects without filling in the gaps, but an equally important point is that the square spirals bordering the defects are largely unperturbed. Only the first row of

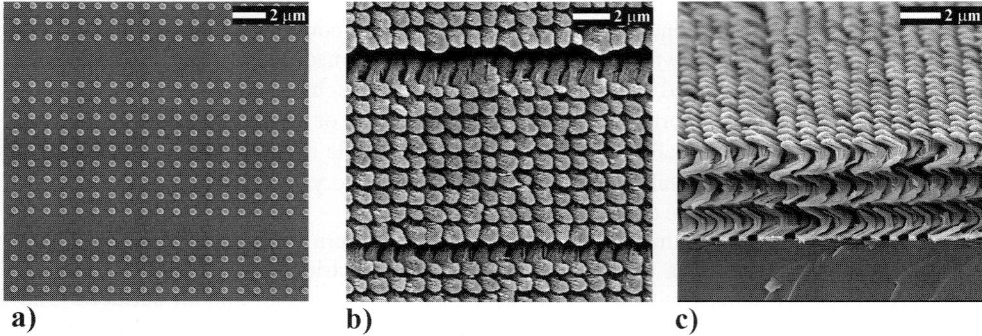

Figure 3. a) Substrate pattern similar to fig. 2a, but now with 1 and 2 row wide line defects. b) Top view of GLAD square spiral film deposited on the same defect engineered seed pattern. c) Oblique view of GLAD square spiral film, incorporating a one-row line defect.

helices is slightly affected by the line defects, and the impact is on the microstructure of these helices rather than on the periodicity of the film. This robustness will be vital to maintaining both the electromagnetic band gap properties of the film, and the guiding of EM modes along defects. However, if the defect lines are three or four rows wide (i.e., 4-5 μm), the GLAD process results in the growth of aperiodic square spirals in the defect gaps. Thus, successful GLAD growth over intentional defects is limited to lateral dimensions of 3-4 μm.

CONCLUSION

With this paper we have presented our preliminary work on the fabrication of periodically structured thin films designed to have three-dimensional electromagnetic band gaps at optical frequencies. The advantages of our approach, based on the Glancing Angle Deposition method, are the ease of fabrication, its applicability to a multitude of substrate and film materials, and the capability of tailoring the frequency response by changing the seed layer geometry or the film microstructure. In a proof of principle experiment we have shown that GLAD films can incorporate moderately sized intentional line defects, which may potentially be used for guiding radiation in electromagnetic components.

We are currently focussing our efforts on characterizing the electromagnetic properties of the square spiral films, including the size and shape of band gaps, and their centre frequencies. We are also working on gaining more insight into the materials related issues governing the GLAD process, such as how crystallography and diffusion during growth may affect the fine structure of the square helices. Fortunately, the frequency response of the square spiral architecture is predicted to be very robust to non-idealities in the microstructure of the helices, provided that the long-range periodicity is maintained.

Eventually we would like to master the production of GLAD square spiral films to the point where we can synthesize films to meet a range of electromagnetic specifications. This would open up numerous exciting applications, as well as prove the importance of materials research for advanced technologies.

The authors would like to acknowledge Scott R. Kennedy for his advice and initial work in the area, and George Braybrook for his SEM imaging. We also wish to thank NSERC, iCORE, the Alberta Ingenuity Fund, and Micralyne, Inc for their generous support of our research.

REFERENCES
1. K. Robbie, M.J. Brett, A. Lakhtakia, Nature **384**, 616 (1996).
2. K. Robbie and M.J. Brett, J. Vac. Sci. and Tech. **A15**, 1460 (1997).
3. R. Azzam, Appl. Phys. Lett. **61**, 3118 (1992).
4. K.J. Robbie and M.J. Brett, U.S. Patent No. 5 866 204 (2 February 1999).
5. D. Vick, B. Dick, S. Kennedy, T. Smy, and M.J. Brett, Mater. Res. Soc. Proc. **648**, Boston, MA 2000.
6. K. Robbie, J.C. Sit, M.J. Brett, J. Vac. Sci. Technol. B **16** (3), 1115 (1998).
7. M. Malac, R.F. Egerton, M.J. Brett, B. Dick, J. Vac. Sci. Technol. B **17** (6), 2671 (1999).
8. K. Robbie, D.J. Broer, M.J. Brett, Nature **399**, 764 (1999).
9. J.C. Sit, D.J. Broer, M.J. Brett, Liquid Crystals **27**, 387 (2000).
10. S. John, Phys. Rev. Lett. **58**, 2486 (1987).
11. E. Yablonovitch, Phys. Rev. Lett. **58**, 2059 (1987).
12. J.D. Joannopoulos, P.R. Villeneuve, S. Fan, Nature **386**,143 (1997).
13. O. Toader and S. John, Science **292**, 1133-1135 (11 May 2001).
14. S. Kennedy, M.J. Brett, O. Toader, S. John, Nanoletters **2** (1), 59 (2001).
15. O. Toader and S. John, Phys. Rev. E **66**, 016610 (2002).
16. M. Aktary, K.L. Westra, M.O. Jensen, M.J. Brett, M.R. Freeman, submitted to J. Vac. Sci. Technol. B.

LASER RAPID PROTOTYPING OF 3-D ALUMINA PHOTONIC BANDGAP STRUCTURES. APPLICATION TO THE DESIGN OF DIRECTIVE ANTENNAS.

T. JAFFRÉ*[1], L. LEGER*[1], B. JECKO*[1], J. CLAUS*[2], C. CHAPUT*[2]
*[1] IRCOM – CREAPE / UMR CNRS n°6615 – Electromagnetism team
Faculté des sciences, 123 Albert Thomas 87 060 Limoges Cedex – France
Tanguy.Jaffre@unilim.fr
*[2] CTTC – Centre de Transfert des Technologies Céramiques
ESTER Technopole – BP 6915 87069 LIMOGES Cedex – France
Claus@ceramic-center.com

I.   DEFINITIONS AND BACKGROUND

This paper highlights some applications areas of **P**hotonic **B**and**G**ap (PBG) or **E**lectromagnetic **B**and**G**ap (EBG) technology at (sub)millimetre wave frequencies. PBG structures are periodic dielectric or metallic materials which give them the property of permitting or preventing the propagation of electromagnetic waves at a known frequency. They are usually made of dielectric or metallic materials with 1, 2 and 3-D periodicity. PBG structures have numerous promising applications, including special antennas, broadband absorbers, frequency selective surfaces (FSS), single structure-multi-channel filter...

Indeed, PBG materials working at centimetre or millimetre frequencies have generated a growing interest for antenna's applications to design a new kind of high gain antenna with a low cumbersome. These materials are used for their frequency properties and their ability to enhance the behaviour of a single patch antenna in order to have a higher gain. The frequency bandgap depends on the permittivity of the dielectric materials used and its Q factor, their dimensions, their periodicity and the incidence angle of electromagnetic waves.

In 1987, two researchers, Eli Yablonovitch [1] of the University of Los Angeles and Sajeev John of the University of Toronto, had the idea to trap photons inside a crystalline structure. They had widened the concept of BRAGG's mirrors to the microwave frequencies and that for different propagation's direction.

II.   MANUFACTURE OF 3-D ALUMINA PBG STRUCTURES

Digital manufacturing technologies, developed and used for rapid prototyping applications exhibit outstanding features in the field of ceramic material shaping and mainly. They are able to produce prototype parts or small runs without the use of any tooling like in classical ceramic processes (Figure 1). CTTC have developed ceramic photocurable paste that could be use in a layered process similar to stereolithography. At the end, fully dense ceramic parts are obtained that exhibit similar properties than those obtained by classical processes.

Rapid prototyping permits to valid different stages of the development of a new product before manufacturing. This technology reduces the time to market and helps innovation. Knowledge in rapid prototyping permits to manage the different processes (and their evolution) and can be integrated within an expert system.

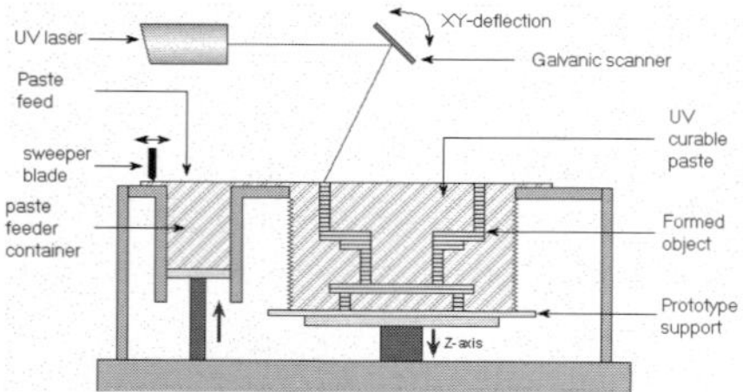

Figure 1 : Rapid prototyping process.

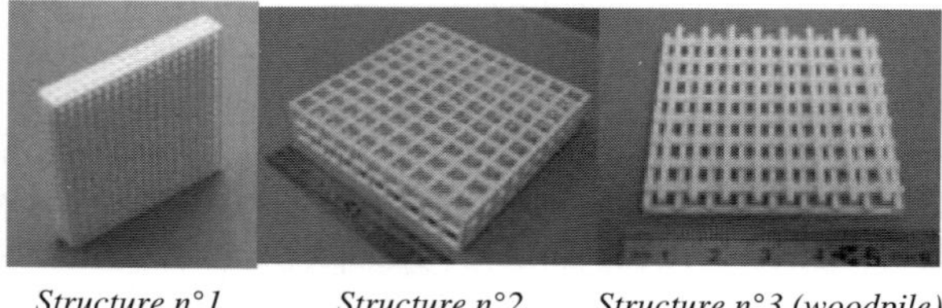

Figure 2 : 3-D PBG structures realized

Table I : Dimensions of PBG structures.

	Length/width/height	Dielectric rod size	Air gap size	Filling Factor	BandGap (GHz)
Structure 1	50.5 / 50.5 / 6.2 mm	1.1 mm	1.5 mm	0.42	20-34
Structure 2	95.2 / 95.2 / 16 mm	1,6 mm	5.6 mm	0.22	11-21
Structure 3	56 / 56 / 3.2 mm	1.6 mm	4.8 mm	0.25	18-34

Rapid prototyping technology enables manufacturing ceramic pieces, whose properties are similar to properties of classic processes of manufacturing, like pressing or injection. The stereo lithography process produces exact replicas of computer solid models; such an object can be made within hours, by solidifying a photo-curable paste, layer by layer, with a UV laser. This paste is a mixture of a photo-curable resin and a ceramic powder. The binder is eliminated by heating, and then the object is sintered in a classic way so as to get the final piece.[2]

Three 3-D alumina-based photonic bandgap structures with a bandgap in the millimetric range were designed and then fabricated by using the rapid prototyping technique (Figure 2). The PBG structure geometries that were modelled use stacked rectangular rods, parallel to each other in each layer, and perpendicular to the direction of the immediate neighbouring layers. In some structures, for every second layer, there is a gap in the position of the rods by a half lattice constant as the woodpile structure. Alumina (Al_2O_3) was used as the high permittivity dielectric material (Er=9.3 ; $tan\delta=8.10^{-4}$) and air as the low permittivity one.

III. SIMULATION AND CHARACTERIZATION OF 3-D ALUMINA PBG STRUCTURES

III-1. Free space microwave measurement system :

The free space measurement system is a highly versatile and mobile piece of equipment used for characterization of dielectric materials. Our objectives are to check transmission factor of PBG structures with simulation to ratify the measurements. The free space measurement permits a non-destructive and in situ testing method over a wide frequency range. This system use two horn antennas, cables, Plexiglas holder, a rail and a network analyser to do the measurements. After several measurements, lenses have been used to focus electromagnetic waves on the photonic band gap structures.[3]

Three dimensional photonic bandgap structures using alumina as the high permittivity material were characterized by the free space microwave measurement system. The finite element and electromagnetic modelling of the fabricated structure resulted in a 3-D photonic bandgap in the 12-40 GHz range, which agreed with experimental measurements (Figure 3). These structures are used to build 3-D alumina PBG resonator antennas (except for the woodpile structure) describe in the next part. We can notice the woodpile structure is a perfect stop band filter. The slope associated in edge of bandgap indicates excellent properties of filtering. These properties are strengthened by a high transmission to the frequencies below the bandgap and a maximal attenuation of 40dB in the bandgap for only few periodic layers.

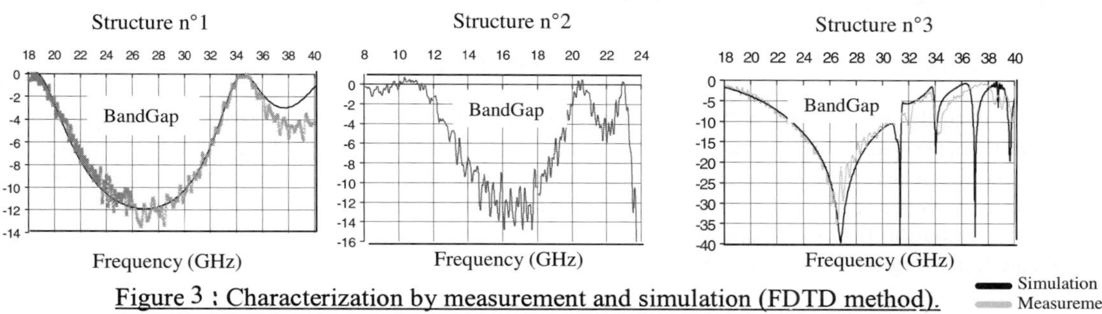

Figure 3 : Characterization by measurement and simulation (FDTD method).

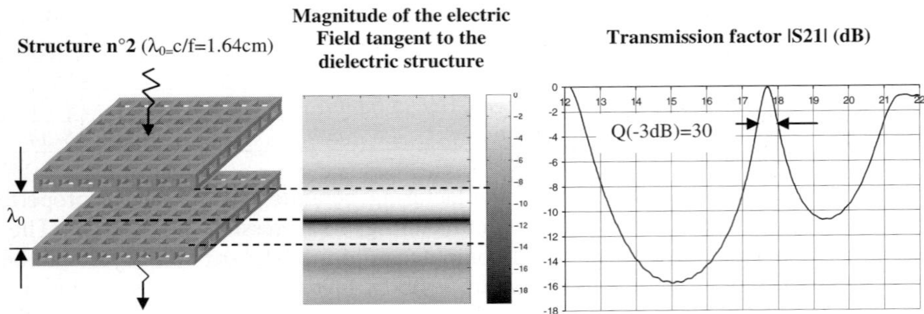

Figure 4 : Electric field distribution at 17.75 GHz frequency and its transmission factor.

PBG materials working at the centimetre or millimetre frequencies are easier to obtain and they have generated a growing interest for applications in the antennas domains. The structure has been designed with a computational code based on the well known finite-difference time-domain (FDTD) method. A defect mode was introduced in the PBG structure. Due to this defect, a narrow transmission band appears inside the bandgap. The defect dimension must be of λ_0 (air) to create an allowable band in the middle of our bandgap. The directive antenna described in this section uses the defect mode of this photonic material to achieve electromagnetic radiation. Then, 3-D PBG material is used as a cover to increase the gain of a usual patch antenna.[4]

This defect structure presents a physical symmetry. The following figure displays the magnitude part of the dielectric field tangent to the dielectric layer. The magnitude result presents a perfect symmetry plane on its middle with a null value (Figure 4). All of these conditions allow us to insert a ground plane at this location.

Photonic crystals can provide significant advantages for suppressing and directing radiation when used in antennas. Thus far, most research has focused on one and two dimensional photonic crystals because they are easier to build and to study. Three dimensional photonic crystals have the potential to provide greater control of the radiation properties of antennas due to their complete bandgap (Three dimensional photonic crystals provide three dimensional gaps, that is to say any wave from any direction will be completely reflected.) but dielectric permittivity of the Alumina used for our structure is not high enough (Er=9.3) to create a complete bandgap.

A thin, high gain antenna can be created from a three dimensional photonic crystal and a metallic ground plane fitted with a patch antenna. The patch antenna seems a popular choice at these frequencies as it is robust, conformable if required and inexpensive to manufacture.

To obtain a radiation pattern with one lobe in the axis and very low side lobes it is necessary to tune the equivalent aperture of the antenna function of the surface with the quality factor of the 3-D PBG material with defect. In fact, the more the quality factor of the structure is high the more the surface must be large. Then the directivity of the antenna will increase with this two parameters. The quality factor of the structure increases with the relative permittivity of the material or/and with the number of periodicities in height (dielectric layers).

III-3 : Experimental results of a 3-D resonator antenna prototype

We describe measured results for a prototype antenna to support the concept. The antenna provides several advantages over existing technologies. It reduces the antenna height and this reduction also leads to a considerable decrease in the volume and weight of the antenna. In addition, the antenna has the advantage of being highly efficient, since the dielectric losses are very low. The measured radiation patterns, gain, directivity and efficiencies of the antenna prototype will also be presented, proving the innovative concept. The prototype has been designed to obtain a directivity

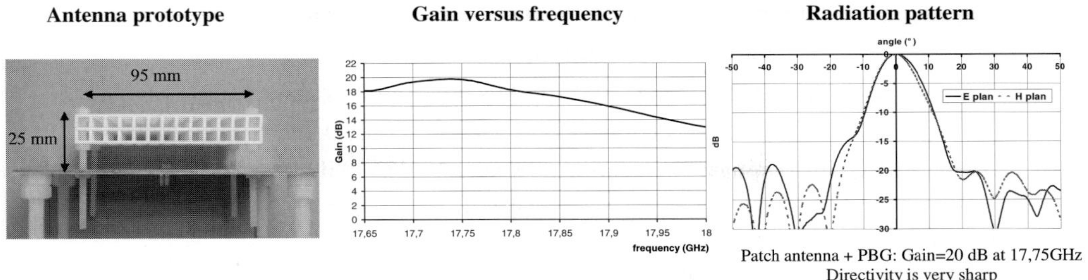

Figure 5 : 3-D PBG resonator antenna prototype : its gain and its radiation pattern.

about 20dB at 17,7GHz (Ku band). The PBG material is made of alumina with a relative permittivity equal to 9,3 and is based on the design of the structure n°2. The structure has a periodicity and dimensions for the rods that give a bangap arround 18GHz. The double periodicity in height gives a quality factor about 56 wich means a directivity of 21dB. The 3-D PBG structure is excited by a patch and a coaxial probe.[5]

The measurement results are presented (Figure 5). We can see a good functionning of the antenna because the gain is about 20dB as expected and the radiation pattern has one directive lobe in the axis. The space between the patch antenna and our PBG structure of 8.2 mm allow us to tune the global antenna working frequency. The patch is too high and we can see the influence of the probe in the left of the principal lobe on the radiation pattern for the E plane. This problem doesn't appear if the patch is thin enough. Unfortunately, we didn't have the time to correct it for the moment but better experimental results will be shown at the symposium.

IV. CONCLUSION AND PERSPECTIVES

We saw how to design 3-D structure with rapid prototyping. This new manufacturing technology opens a large investigation field for the PBG ceramical structures. New geometries can be now studied and measured and also new materials will be manufactured in the same way. In the future, Zircon will be used to realize our 3-D PBG structures with a dielectric constant of 30. This will be excellent to obtain complete bandgap and high gain antennas. We notice that the uses of 3-D PBG structures in the antenna domain is interesting because it reduces the mass of the system, increases the efficiency and is simple to manufacture with rapid prototyping technology. What's more, 3-D structures present the advantage to reduce the amount of material so to reduce the losses compared to a 1-D or 2-D PBG structure.

Références :
[1] **E. YABLONOVITCH**, "Inhibited spontaneous emission in solid state physics and electronics", Phys. Rev. Lett., Vol. 58, 1987, pp 2059-2062.
[2] **M. C. WANKE, O. LEHMANN, K. MÜLLER, Q. WEN, M. STUKE** « Laser rapid prototyping of photonic bandgap microstructures » Science, Vol. 275, 28 February 1997.
[3] **Y. CHEN, D. BARTZOS, Y. LU, E. NIVER, M. E. PILLEUX, M. ALLAHVERDI, S. C. DANFORTH, A. SAFARI** « Simulation, fabrication, and characterization of 3-D alumina photonic bandgap structures » Microwave and Optical Technology letters, Vol. 30, No 5, September 5 2001.
[4] **M. THEVENOT, C. CHEYPE, A. REINEIX, B. JECKO** « Directive Photonic-Bandgap Antennas » IEEE Transactions on Microwave Theory and Techniques, Vol. 47, No 11, November 1999.
[5] **G. S. SMITH, M. P. KESLER, J. G. MALONEY** « Dipole antennas used with all-dielectric, woodpile photonic-bandgap reflectors: gain, field patterns , and input inpedance » Microwave and Optical Technology letters, Vol. 21, No 3, May 5 1999.

Comparative study of ferrite and garnet material for design of tunable microstrip patch antenna for high frequency applications

Nitendar Kumar,*   Takeshi Fukusako[1] and Nagahisa Mita[1]

*   Solid State Physics Laboratory, Lucknow Road, Delhi-110054, India.

[1]Dept of Elect & Computer Engineering. Kumamoto University, 860-8555, JAPAN.

nitendar@rediffmail.com

Abstract : The antenna systems with steerable capabilities are of significant importance for many current communication technologies such as satellite tracking for mobile communication, airport traffic control, radar systems etc. The unique magnetic behavior of ferrites and garnets at microwave frequencies can offer tenability to such systems. Garnets are usually considered as the best material for microwave device applications due to their excellent temperature stability and low magnetic and dielectric losses. On the other hand, lithium ferrites have shown the advantage of having high permittivity and high Curie temperatures. Keeping in view, the importance of garnet and some merits of ferrite materials an attempt has been made to make a comparative study in between these two materials, for tunable antenna applications. The paper describes the investigations carried out on the patch antenna, fabricated on LiTi ferrite and YIG material substrates. The resonant frequency of the antennas for both the cases was found to be changed significantly for about 300 MHz, with the external applied field of 2.5 mT. The transmitting power parameters (S_{21}) were found to be in good agreement with return loss characteristic (S_{11}). The gain offered by the ferrite was found superior to that observed for garnet based antenna. These finding can be of immense use for fabrication of device of specific applications.

1. Introduction: Polycrystalline ferrite and garnet materials are in general known as magnetic semiconductor material having very high resistivity (ρ) and low magnetic and dielectric losses $(\tan\delta)$ at microwave frequencies. These materials are at present commercially available and find number of applications in the area of microwave devices [1-3]. Ferrite and garnet materials have been studied for different type of antenna applications by the researchers due to their tunable magnetic properties with the external applied field [4-7]. Though the garnets are the best available material for microwave device applications, but at the same time they have some limitations in saturation magnetization $(4\pi Ms)$ and permittivity (ϵ). The maximum value of $4\pi Ms$ that can be obtained for garnets is ~1800 Gauss and at the same time they offer low value of ϵ (~13) as compared to lithium ferrites. Low $4\pi Ms$ value restricts their use for higher frequency bands, while the low ϵ in turn increase the volume of material required for devices. However, lithium ferrites offer $4\pi Ms$ values from 500 to 5000 Gauss, high Curie temperatures (Tc) and moderate temperature stability, which enabled them to cover vide range of microwave frequencies. Variety of metal ions substitution have already been studied for lithium ferrite compositions in order to study the effect of substitutions on the magnetic, structural, electrical etc properties of these ferrites [8]. Preparation and properties of lithium ferrites have been discussed in detail and compared with other spinal microwave ferrites and rare earth garnet materials by Argentina et al. [9]

1. Presently working at Multimedia Laboratory, Dept of Elect & Comp. Engg. K.U, JAPAN.

2. Experimental:

2.1 *Material and method.* The garnet and ferrite materials, having the $4\pi Ms \sim 800$ Gauss were chosen for the present investigation. The garnet material was arranged from M/S FDK Corporation, Japan and the LiTi ferrite is our own designed and development. The compositional formula for the lithium ferrite system selected for the investigation is,

$$Li_{0.825}Zn_{0.05}Mn_{0.05}Ti_{0.7}Fe_{1.375}O_4$$

The material was prepared by conventional ceramic technique. The Mn^{3+} has been included in the formula due to its beneficial effects on resistivity and megnetostrictive properties [10]. Bi_2O_3 (0.25 wt %), was added in the initial mixing for lowering the sintering temperature [11]. The final sintering of the sample was done at 1050°C for 4 hrs. The resulting material was subjected to XRD analysis to ensure single phase spinel formations. The room temperature $4\pi Ms$ and its variation up to Tc were determined by PAR 155 VSM. The tanδ and spin wave linewidth (ΔH) measurements were carried out at X-band frequency using by perturbation method.

2.2. *Substrate.* Substrates in the size 20X20X1 mm were made from the both the materials by using appropriate techniques. To remove stress and strain developed during the course polishing process, the substrates were annealed at a temperature of 500 °C for 30 minutes.

2.3 *Device.* The antenna fabricated on the magnetic substrate is shown in fig. 1. A 10 mm square patch antenna made of Copper foil was placed at the centre of polished substrate. The opposite side of the substrate was connected to the conducting ground plate. The feeding microstrip line was fabricated on a FR-4 substrate ($\epsilon = 4.2$ at 2.4 GHz) of 1.6 mm in thickness. External variable magnetic bias field up to the 3.0 mT (max) was applied axially from the opposite side of feeding terminal. The return loss (S_{11}), transmitting power (S_{21}) parameter, antenna gain and radiation pattern were measured using Network analyzer (Aligent 8722ES).

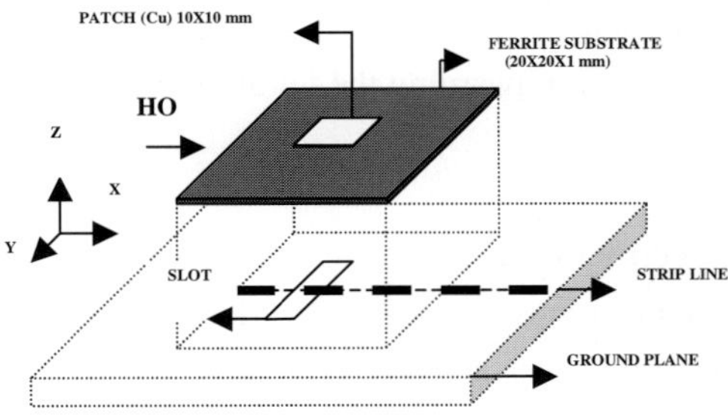

FIG.1 FERRITE PATCH ANTENNA

3. Result and discussion: The XRD examination of the material ensured the single phase spinel formation. The bulk density of the sample was found to be > 96% of X-ray density. These examinations indicate the right quality of the material for all experimental purpose. The magnetic and electric properties of technical importance such as room temperature $4\pi Ms$, Tc, ϵ, ρ, tanδ, ΔH etc were measured and are shown in table-I.

Table. I. Materials characteristic properties

MATERIAL	$4\pi Ms$ (Gauss)	Tc ($^{\circ}$C)	ρ (Ω Cm)	ε	ΔH (Oe)	tan δ
LiTi–FERRITE	800	250	~ x 10^{10}	~ 17	350	$\leq$ 0.0005
YAl–GARNET	800	200	~ x 10^{12}	~ 13	55	$\leq$ 0.0005

The observed behaviors, for the ferrite composition are on the expected lines and can be explained on the basis of cation involved in the system and their distribution over the crystal structure.

The ferrite and YIG substrates were evaluated in a patch antenna configuration in the frequency range 3 to 5 GHz. The important antenna parameters S_{11}, S_{21} and radiation patterns were measured for both the materials and are respectively shown in fig.2 to fig.4. The resonant frequency (f_r) of the antenna was found to be significantly tuned over a f_r – to - f_r + 300 MHz range for an applied field of 2.5 mT. However the effective tuning with LiTi ferrite substrate was found to be 400 MHz, while in the case of garnet, it comes out to be 350 MHz. Similar observations for S_{21} parameters with the applied field were noticed and found to be in good agreement with the S_{11} characteristics. The antenna gains w.r.t resonant frequencies were computed, for LiTi ferrite the gain obtained was in the range of 4 to 5 dBi, while it was ~ 3.5 dBi for garnet based antenna. No significant changes were observed for the radiation patterns of antenna with different magnetic bias field. The observed results are accounts for the anisotropic properties of the ferrite or garnet materials.

Conclusions: The substituted lithium ferrite material developed for the patch antenna applications meets the requirement of a device for specified frequency band. Tunability with low bias is a significant development and will help in miniaturization of a device. The lithium ferrite materials can also used for other higher frequency bands by modifying the composition formula. The findings can be exploited for designing antenna of specific applications.

Acknowledgement: One of the author, Nitendar Kumar wishes to thanks Japan Society for the Promotion of Science (JSPS), Japan for awarding the fellowship.

References:

[1] J. Smit and H. P. J. Wijn., Ferrites (Philips Technical Library, Eindhoven, 1959)

[2] R.F.Soohoo., Theory and applications of ferrites, Prentice-Hall, Englewood, Cliffs, N. J.,1962.

[3] B.Lax and K.J.Button., Microwave ferrites and ferrimagnetics, McGraw-Hill, New York, 1962.

[4] Das N and Chawdhury, S.K., Electronics Lett., 16(1980)817.

[5] A. Handereson, J.R. James and A. Fray., Electronics Lett., 24, no1(1988)45.

[6] D. M. Pozar and V. Sanchez, Electronics Letters, 24(1988)729.

[7] Mishra, R.K, Pattaniak, S.S and Das, N., IEEE Trans. Antenna Prop., 41 no.2(1993)230.

[8] N. Kumar, Thesis, Jamia Millia Islamia, New Delhi, India, 1999.

[9] G.M. Argentina and P.B.Baba, IEEE Trans. MTT 22(1974)652.

[10] P.B.Baba,G.M.Argentina,W.E.Courtney,G.F.Dionne,D.H.Temme,IEEETrans.Mag.8(1972)83.

[11] P.Kishan,D.R.Sagar,S.N.Chatterjee,L.K.Nagpaul,N.Kumar,K.K.Laroia,Adv Cer.15(1985)507.

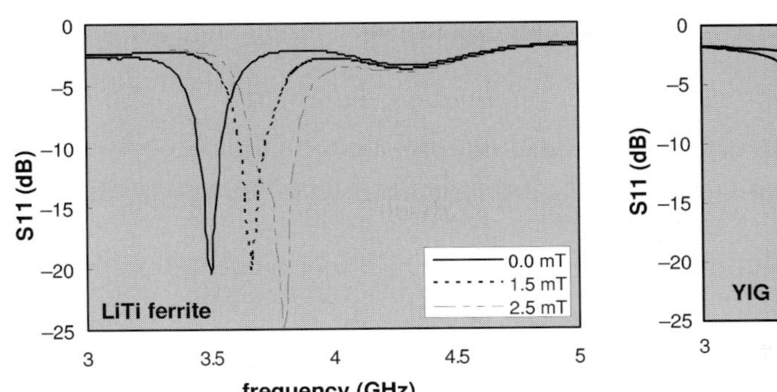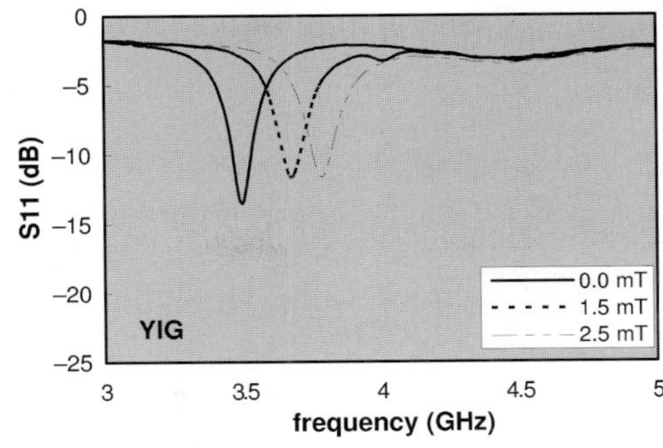

FIG. 2. RETURN LOSS CHARACTERISTICS FOR LiTi & YIG BASED ANTENNA

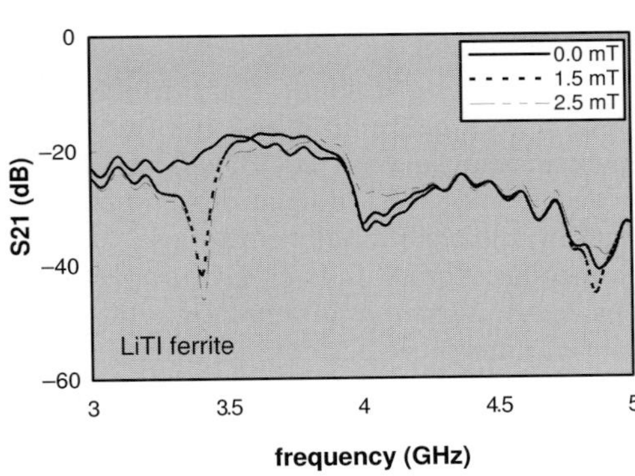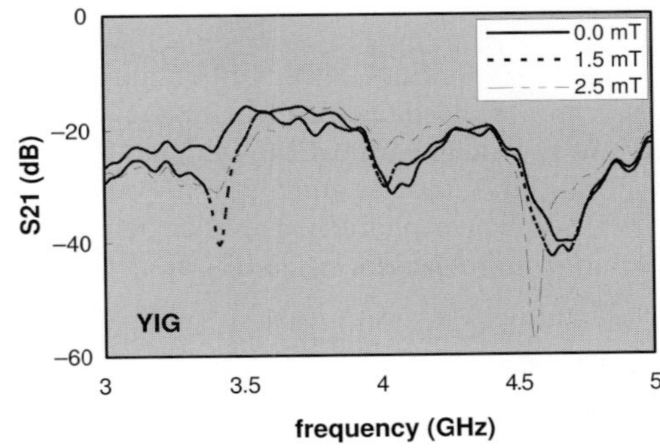

FIG. 3. TRASMITTING POWER CHARACTERISTICS FOR LiTi & YIG BASED ANTENNA

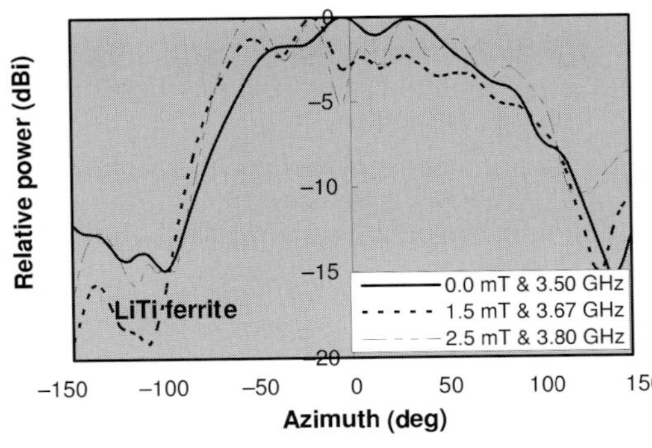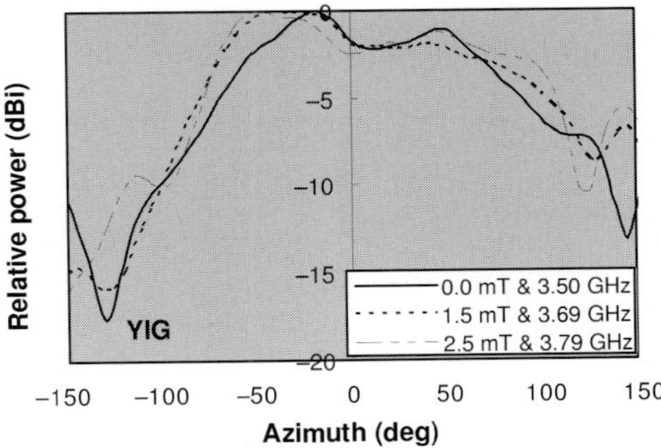

FIG. 4. RADIATION PATTERNS FOR LiTi & YIG BASED ANTENNA

Session F10

Applications

Chair: V.K. Varadan

Microcoil, carbon nanotubes, fiber composites, FSS and coatings for electromagnetic absorption and shielding

Vijay K. Varadan

Distinguished Alumni Professor of Engineering Science and Electrical Engineering

&

Professor of Neurosugery, College of Medicine

Pennsylvania State University, University Park, PA 16802 USA

ABSTRACT

The control of non-ionizing Electromagnetic (EM) radiation is becoming significant in many EM absorption and shielding applications in engineering and medicine. In this paper, design and development of novel composite materials and coatings employing distributed materials such as microcoil, chiral materials, ferromagnetic liquids, nonlinear metal oxides, ferroelectric tunable materials, voltage controllable dielectric and conducting polymers and microencapsulated balloons containing microcoil and ferromagnetic liquid are presented. One of the principal objectives of this effort is to develop both passive and smart (active) composites and coatings, with fast control architecture using Neural Networks, to control EM radiation over a very wide band range from very low frequencies -- where the wave is primarily magnetic in character-- to very high frequencies in the GHz range.

Results on both passive and active coatings on simple planar samples and complex 3D objects are presented which shows absorption in the order of -25 to -30 dB and EMI shielding effectiveness in the order of -60 to -110 dB for a wide range of frequency. The primary focus of the smart coatings is to employ Electroactive Polymeric (EAP) materials that would interact with the electromagnetic radiation, and perform a signature modification and evasion operation that has many applications

1.   INTRODUCTION

In many applications one is interested in coating materials, surface modification (smart skin) to reduce RCS and control IR emissivity and radiation. In the measurement arena, one may be interested in broadband absorbers for antenna shroud, anechoic chambers, and EMI shielding materials for hospital instruments. The traditional approach is to employ metal flakes and fibers, micro-balloons, semi-conducting polymers, carbon, ferrite, carbonyl iron, etc. The biggest drawback of the existing absorbers is: 1) they are fragile (e.g. carbon filled foam used in anechoic chambers), 2) they are relatively heavy because of the fillers such as magnetic powders, etc., 3) they do not provide broadband absorption, 4) they are not independent of all angles of incidence of the EM radiation. In this paper, an innovative and novel design and development of composite materials and coatings that mitigate many of the problems mentioned above is presented. The innovative approach to the design and development of composite EAP materials, microcoils and microcapsules with chiral and ferromagnetic materials, carbon nanotubes, coatings and actuators not only overcomes many of the shortcomings of conventional absorbers, but they provide them with other desirable attributes as well, such as, considerable material strength (due to carbon nanotubes) despite their light weight, ease of manufacturing, and cost effectiveness. Yet another salutary feature of such EAP composites is that they can be tailor-made with a combination of novel chiral materials, nonlinear metal oxides, ferroelectric tunable materials, voltage controllable dielectrics, and conducting polymers that can control the launching of the surface waves typically induced on the absorber for certain polarizations of the incoming wave.

Radar Cross Section (RCS) is the 'size' of an object seen by radar and is directly related to the scattering efficiency of the target. RCS depends upon many factors like shape, orientation, frequency, the angle at which the target is viewed, and the polarization of the transmitting and receiving antennas. A large number of papers dealing with simple and complex shaped targets have been published in literature [1-3]. RCS of complex shaped objects can be reduced to a limited extent by modifying the target shape and material. But for large reductions in the RCS of a target over a broad range of radar frequencies, it is necessary to coat the target with Electromagnetic Absorbing Material (EAM) that will absorb microwave radiation. An ideal electromagnetic coating for RCS reduction should be thin, light and easily applied and have the prescribed reflectivity reduction characteristics over a wide frequency range.

The possible use of microcoil composites for the reduction of electromagnetic reflection from metallic surfaces have been already investigated and established that, they are efficient microwave absorbers [4-11]. Chiral materials for the control of EM radiation may include polymers, non-linear metal oxides and some semiconductors (for details refer to the U.S. Patent of Drs. Varadan [2]). The wave-material interaction in such a medium involves both the left and right circularly polarized (LCP and RCP) fields. When an electromagnetic wave travels through a medium that contains chiral materials, it is forced to adopt the handedness of the chiral microstructure. One consequence of this is that an incident wave with one type of polarization, say RCP, generates scattered fields that have both the RCP and LCP characteristics, and traverse longer paths as a result of this mode conversion and scattering.

2. Composition of the EM absorber and the results

In the absorber discussed above, the medium is characterized by dielectric properties: permittivity, ε, permeability, μ, and chirality parameter, β. The right and left circular polarizations propagate at different velocities that depend on β. It is well known that the relative permittivity and permeability must be equaled to each other in order for an absorber to have ideal, non-reflecting characteristics. It has been demonstrated, both theoretically and experimentally, that it is possible to customize a medium that comes close to satisfying this ideal condition. The chiral non-linear metal oxides, e.g., carbon, SiC, SiN, TiC, etc., whose sizes range from a few microns to millimeters along with ferromagnetic liquid are ideal candidates for such coating. It is obvious that the thickness of the coating should be many orders less than the wavelength of the interrogating electromagnetic wave. This could be achieved with chiral materials provided sufficient surface area is introduced and also convert the plane waves into surface acoustic waves. For this purpose, we employ micro-encapsulation of microcoils in suitable polymer and distribute them in high dielectric host materials. For lower frequencies where magnetic effect is important, we introduce ferromagnetic liquid along with microcoil and carbon nanotubes in the microcapsules. The resulting microcapsules with microcoil and ferromagnetic liquid are shown in Figure 1a and 1b. The chiral non-linear metal oxides have been synthesized by using a microwave CVD techniques developed by the author. These chiral materials have excellent mechanical and elastic properties that render them suitable for integration with the EAPs. The change in the dielectric properties can be enhanced with tunable ferroelectric materials, which are either embedded in the EAP or are synthesized concurrently. One of the stable tunable materials that have been tested successfully is the barium strontium titanate (BST), synthesized at the paraelectric region.

With the advent of chiral materials, frequency-dependent and voltage-tunable dielectrics, we now have considerably more freedom than we did before in designing "sandwiches" to provide the desirable absorbing and shielding characteristics over a wide range of frequencies. This has been demonstrated in [12], where we have shown that a multi-layer 'sandwich' can be designed using the above approach.

Frequency Selective Surface (FSS) is a two-dimensional periodic array of conducting patches or aperture elements used in microwaves and optics, and is generally patterned on, or, embedded in a dielectric sheet. The frequency filtering property of FSS comes from its planar periodic structure. The elements reflect the incident microwave of a specific frequency range according to their shapes, periodicity and the dielectric property of the substrate. It is contended that the change in the absorption properties is because the FSS launches an infinite discrete set of inhomogeneous plane waves into the layer, the sum of which makes up a periodic wave. These guided waves inside the structure are attenuated within due to the losses within the layer.

In the engineering of microwave absorbers, many property aspects apart from the EM wave absorbing performance have to be considered, for example, physical strengths including tensile strength, bending strength, shear strength, compressive strength, temperature resistance, chemical resistance, etc. For the making of absorbers, the most important step is to design an optimized formula of the ingredients. The first consideration of the formula design is to obtain a high performance absorber, which can cover a certain frequency region. The absorbing performance depends on the additives. The reinforcing components like glass fibers may be considered after that of the absorbing components, because reinforcing components have very less effect on the absorption performance in the X and the Ku-Bands.

Polyurethane is used as the matrix/binder for all the samples. The other fillers added include carbon fibers, glass fibers micro-balloons and microcapsules with microcoil/carbon nanotubes and ferromagnetic liquid. It is important to select the proper carbon fiber length and concentration to achieve the proper conductivity in addition to the permeability due to the ferromagnetic liquid. Homogeneous dispersion of the carbon fibers in the matrix is also an important issue in the making of the EM absorbers. The micro-balloons offer homogenous distribution of the absorbing components and also provide uniform sized air bubbles. Using different sizes of micro-balloons at an optimal volume fraction, we can make composites with sufficient compressive strength at the lowest density. Although the glass fiber has no absorbent effect on the EM waves, it plays a key role in reinforcing the absorber. EM absorbers with different dielectric properties and thickness were made in order to investigate the impact of FSS and CNT on these absorbers. EM absorption of well over -20 dB has been obtained for a wide frequency band with a thin absorber of about 3mm.

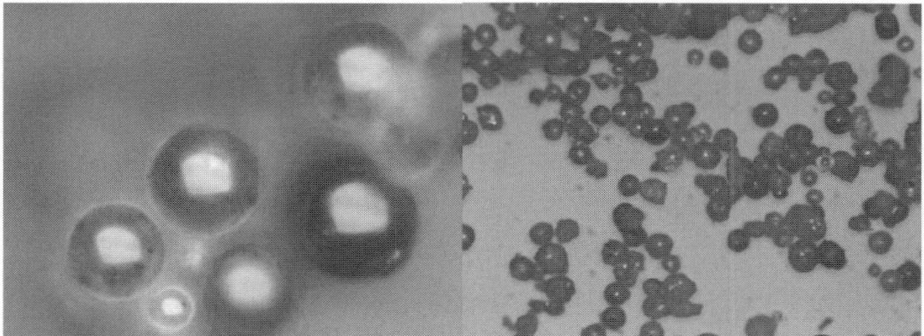

Figure 1a. Microcoil/carbon nanotube encapsulated in PMMA; Figure 1b same with ferromagnetic liquid

At Penn State, it has been proven that with a composite EAP with conducting polymer and tunable high permittivity materials, it is possible to make the absorber smart and adaptive with a dc bias field applied. Two approaches are being investigated: 1) embed the micro-chiral material in the EAP; apply dc bias field that deforms EAP and thus in turn changes the geometrical parameters of the chiral inclusions. It has been shown that the artificial magnetization produced in the chiral materials depends on the dielectric properties ε, while μ and the chiral parameter β, are in turn dependent on the diameter and length of helix structure [6]. The deformations in EAP thus activate these parameters and change the absorption and shielding effectiveness of the coating materials. The high dielectric materials will make the coating thinner and thus one could make 'walk-on' thin absorbers even for low frequency (in the order of 10 cm to 12 cm instead of 2m carbon cone absorbers currently being used in anechoic chambers).

Characterization of a fabricated EAM and its components is of importance both for the assessment of their performance and for the quality assurance. Material that is tested is usually in three forms: 1) finished absorber panels, 2) thin sheet used as components for EAM under development process, and 3) the uniform bulk materials. A number of measurement methods are available. Most commonly performed methods are based on free space techniques, such as: 1) NRL arch method, 2) RCS ranges method, and 3) HVS free space method [13]. Once the final EAM is made, its final evolution of the absorption characteristics (S11 and S21 and S12 measurements) can be carried out by HVS Free Space Method, NRL Arch method or in RCS range or Compact range.

We considered an object as shown in Figure 2 for evaluating the EM absorber. The object is placed at a distance of 5.18 meters from the transmitting and receiving antennas in the quiet-zone of the anechoic chamber. The anechoic chamber is 6.1 meters long, 3.5 meters wide and 3.5 meters high. The object support is a low loss, low reflecting polyform column. Measurements were carried out at each angle of incidence for 801 frequency points in 8-12.4 GHz frequency band with averaging and error correction. RCS measurements were made in the frequency range of 8-12.4 GHz. A flat metal plate is used as RCS standard.

A typical chiral coating with 2 to 3 mm thickness gives around 30 dB in S11 with metal backing. This type of coating will provide an excellent RCS reducing coating on complex targets. Reflection reduction of over -30 dB has been achieved for thin, metal backed coatings over a broad frequency band (2 to 110 GHz) and around -60 dB for thin and moderately thick 'walk-on' type absorbers for anechoic chambers. Some thin samples have been fabricated with a return loss of 20 to 28 dB over the frequency range of 8 to 18 GHz.

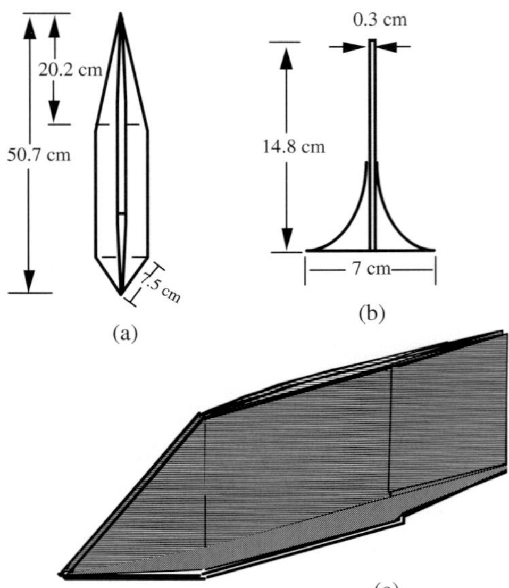

Figure 2 : Object considered

3. References

[1]   G.T.Ruck, Radar Cross Section Hand Book, Plenum Press, New York, 1970.

[2]   Radar Cross Section of Complex objects, W. Ross Stone (Ed.), IEEE Press New York, 1990.

[3]   E.F.Knott, J.F.Shaffer and M.T.Tuley, Radar Cross Section, Artech House Inc. MA. 1985

[4]   A. Lakhtakia, V.K.Varadan and V.V.Varadan, "Scattering and absorption characteristics of lossy dielectric, chiral, non spherical objects," Appl. Opt. 1985, 24, pp. 4146-4154, 1985.

[5]   A. Lakhtakia, V.V.Varadan and V.K.Varadan, "A parametric study of microwave reflection characteristics of a planar achiral-chiral interface," IEEE Trans. on Electromagn. Compat., Vol. EMC-28, pp. 90-95, 1986.

[6]   V. K.Varadan, V. V.Varadan and A. Lakhtakia, "On the possibility of designing anti-reflection coatings using chiral composites," Jour Wave-Material Interaction, Vol.2, 1, pp. 71-82, Jan. 1987.

[7]   R. Ro, V.V. Varadan and V.K.Varadan, "Electromagnetic activity and absorption in microwave chiral composites," IEE Proc-H Vol.39, 5, pp. 441-448, Oct. 1992.

[8]   Varadan, V.K. and Varadan, V.V., Electromagnetic Shielding and Absorptive Materials, US Patent #4,948,922, 1990

[9]   D. Ghodgaonkar, V.V.Varadan and V.K.Varadan, "A free space method for measurement of dielectric constants and loss tangents at microwave frequencies," IEEE Trans. Instrum. Meas. Vol. IM-38, 3, pp. 789-793, 1989.

[10] A.K.Bhattacharyya, "Control of radar cross section and cross polarization characteristics of an isotropic chiral sphere," Electronics Letters, Vol. 26, 14, pp. 1066-1067, Jan. 1990.

[11] A.K.Bhattacharyya, "Radar cross section reduction of a flat plate by RAM loading," Microwave and Opt Tech. Letters, Vol. 3, pp. 324-327, Sept. 1990.

[12] Y. Ma, V.K. Varadan, V.V. Varadan, "Frequency-selective Devices," Proc. SPIE 1558, 132-137, San Diego, (1991).

Effect of Temperature on LiZnTi Polycrystalline ferrite based Microstrip Tunable Ring Resonator

G.S.Tyagi, Rajeev Pourush, G.P.Srivastava and Ashish Dubey
Department of Physics & Computer Science,
Dayalbagh Educational Institute,
Dayalbagh-Agra (U.P)-282005

And

P.K.S.Pourush
Department of Physics,
Agra College, Agra,
(Dr. B.R.A. University, Agra)
Agra (U.P)-282001
E-mail- pourushpks@rediffmail.com

Abstract: *A tunable microstrip ring resonator was designed and fabricated at resonance frequency 9.30 GHz in the X Band using the thin metallic film on the bulk polycrystalline ferrite substrate. Temperature stability of fabricated ring resonator have been studied by keeping it in the room temperature range.*

1. Introduction

Resonators form the basic design elements in many circuits and components including filters, oscillators, couplers and antennas. Planer resonators offer several advantages over other conventional resonators including size, weight and cost. In microstrip planer resonators, two types of resonators are generally used namely linear and ring resonator. The ring resonators are better because they have the advantage of freedom from open-end effects.

For a microstrip ring resonator, the resonance condition is given by:

$$l\,\lambda = \pi\,(2a+w)$$

where l is number of azimuthal full-wave variations of the field, 2a is the inner diameter of the ring and w is line width.

2. Estimated Design Parameters and Device Fabrication

All the required parameters for the designing of tunable microstrip ring resonator on the ferrite substrate were estimated on the basis of different available expressions [5]. The estimated values are follows:

Effective dielectric constant of ferrite substrate (ε_{eff}) =10.399
Width of microstrip line (w) = 0.69 mm
Thickness of microstrip line (t) = 5.0 μm

Guided wavelength (λ_g) = 10.0 mm
Mean radius of the ring (R) = 4.77 mm
Inner radius of the ring (R_i) = 4.43 mm
Outer radius of the ring (R_a) = 5.12 mm
Capacitive coupling gap (δ) = 0.35 mm

The obtained dimensions of the tunable microstrip ring resonator has been fixed for 25 mm X 25 mm X 1mm, which is the size of the substrate available with us. Calculated dimensions of the device were put into AUTO CAD software in a personal computer and the plottings have been taken on a high quality glaze paper using GRAPTEC plotter. These plottings were photographed using high quality, high-resolution camera. Then the negative has been redeveloped as a positive photograph on a high-resolution glass plate. The mask was properly examined through a microscope for qualifying its use especially at the capacitive gaps of the device. To fabricate the desired metallic pattern, a thin film of Cr-Cu-Au was deposited by vacuum evaporation, which involves three main steps:

(a) Evaporation of very thin chrome layer on to the surface of the substrate.
(b) Evaporation of very thin layer of copper metal on the chrome layer.
(c) Electroplating the copper metal on to the layer produced by the step (b).

Similar steps have been followed for the deposition of ground plane on the other side of the substrate. Steps (a) and (b) provide, respectively, mechanical and electrical foundation layers on which to electroplate good quality bulk copper. After deposition of thin films a conventional photolithography procedure has been used to get the good quality desired metallic pattern on the polycrystalline ferrite substrate.

3. Experimental Procedure

For observing the influence of temperature over the performance of tunable microstrip ring resonator, the ring resonator was placed in an oven and the temperature of oven was varied up to 60 ^{0}c from the room temperature (25^0c). The magnetic tuning was measured by changing the strength as well as the direction of the applied dc baising field. The variation of loaded quality factor (Q_l) of the device with the external baising field in various directions is also measured. It is found that Q_l varies from 176 to 244. Of course these values are very low in comparison to the superconducting resonators but it is a good Q_l as far as ordinary conducting devices are concerned

4. Conclusions

The variation in characteristics of fabricated tunable microstrip ring resonator was studied by varying the temperature. It is concluded that the characteristics of fabricated microstrip ring resonator does not vary significantly in room temperature range while it gives the tunability of 31% by applying the external baising field [4]. Due to the use of metallic films and bulk polycrystalline ferrites, these devices are much cheaper [1] compared to the devices based on superconducting patterns and single crystals of YIG [2]. Of course the maximum loaded quality factor of this resonator is 244,whereas Q_l of the superconducting resonators (~1000) are much higher [3]. In many applications tuning of the circuit is more important than the Quality factor. These resonators can also be easily integrated as a sub-system in tunable oscillators and tunable filters. In fact such types of resonators are more useful in several microwave devices from the industry viewpoint, owing to their higher tunability and extremely low cost.

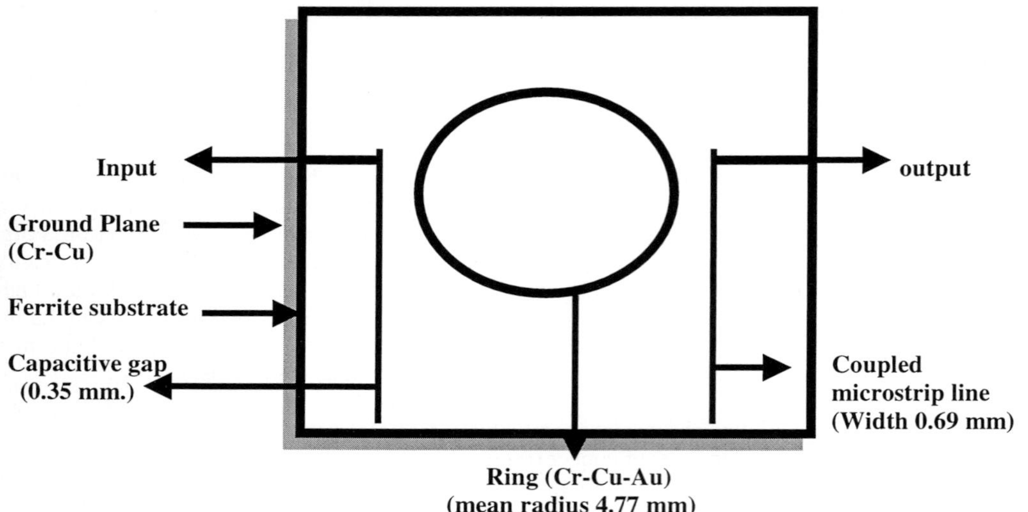

Figure 1: Geometry of microstrip ring resonator on polycrystalline ferrite substrate

References:

1. G.F.Dionne and D.E.Oates, IEEE Trans., MAG-33, pp.3421, (1997)
2. T.Fukusako and M.Tsutsumi, IEEE Trans., MTT-45, pp.2013, (1997)
3. D.E.Oates and G.F.Dionne, IEEE Trans., Appl.Supercond. -9,pp.4170, (1999)
4. Ashish Dubey, G.S.Tyagi, G.P.Srivastava, Electron.Lett. -37,pp.1296, (2001)
5. I.J.Bahl and R.Garg, IEEE Proc.-65,pp.1611, (1997)

Design of Interdigitial Transducers for Non-destructive Evaluation in Plates

J. Jin, S.T. Quek and Q. Wang
Department of Civil Engineering, National University of Singapore, Singapore
engp9892@nus.edu.sg

1. Introduction

Piezoelectric patches are widely used as transducers to launch elastic waves in substrates owing to their electro-mechanical behavior and can be employed to perform non-destructive evaluation (NDE) of the substrates. Interdigital transducers (IDT) are piezoelectric patches with two groups of interlaced electrodes deposited on their surface (see fig. 1). The electrode pattern of IDT controls the wave modes being excited and can be advantageously utilized to overcome the complication in signal interpretation arising from the existences of multiple modes in NDE of plates (Chang et al, 1994; Monkhouse et al, 1997; Wilcox et al, 1998; Varadan et al, 2000).

In this study, IDT were bonded onto plates to launch Lamb waves for NDE. The coupled structure of a plate bonded with IDT was modeled by considering both the strong electro-mechanical coupling effect and the interaction between the IDT and the substrate (Jin et al, 2002). The analytical model provides an efficient tool to investigate the effects of the parameters of IDT, namely, finger spacing, finger width, number of fingers and length of IDT.

2. Analytical Model

Consider the case where the length of the fingers of the IDT is long enough to neglect transverse effects (diffraction, guiding, etc.). Employing the plane strain assumption, the structure of IDT bonded onto a host plate may be represented by a 2-dimensional model (fig. 2). The IDT and the part of the host plate beneath IDT is considered as the near field, while the remaining part of the plate beyond the IDT is considered as the far field. Applying alternating voltages at high frequency between the two groups of metallic fingers of the IDT generate mechanical waves due to piezoelectric effect where electrical signals are transformed into mechanical strains. Hence, to model the near field response as accurately as possible, the interaction between IDT and the host plate must take into account the electro-mechanical coupling effect.

Wave solutions are obtained from electro-mechanical coupled governing equations using modal techniques. The amplitudes of the wave modes in the near field are obtained from a system of 10 equations by imposing the 10 boundary conditions. Spatial Fourier transform is employed in the wave propagation direction (x_2) to simplify the system of equations from 2-D into 1-D that is only dependent on the wavenumber. The results are then inversely Fourier transformed and integrated in the wavenumber (denoted as k) domain. Contour integral is performed around the poles which correspond to the wave modes of the coupled structure in the near field. The acoustic field is obtained as a combination of the modes in the near field,

$$T = \sum_{j=1}^{n} T_j = \sum_{j=1}^{n} \varphi_j \Gamma_j e^{ik_j x_2} \tag{1}$$

where T is stress field, φ_j is the amplitude of input electrical potential at wavenumber k_j, Γ_j is the transfer function of mode j.

Fig. 1 Interdigital transducer

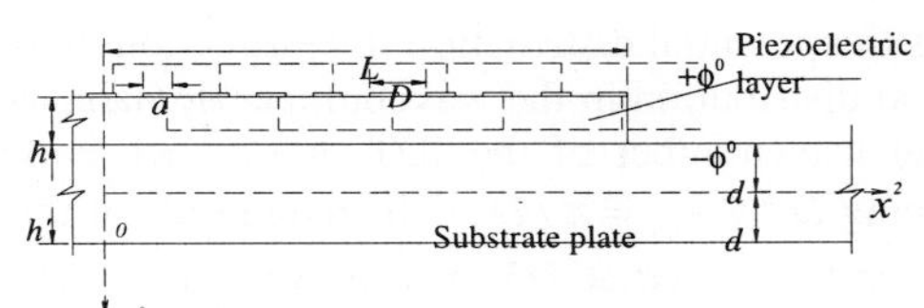

Fig. 2 Two-D model of IDT structure

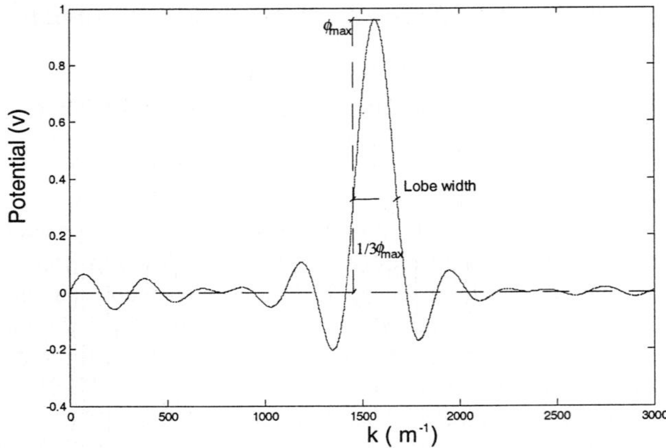

Fig. 3 Distribution of potential of IDT in wavenumber domain

The part of the host plate beyond the IDT is regarded as the far field, where the elastic waves propagate as free Lamb waves. From the solution of the excitation in the near field, the amplitudes of Lamb modes can be obtained by considering reciprocity relations and mode orthogonality,

$$a_q = \sum_{j=1}^{n} a_q^{(j)} = \sum_{j=1}^{n} (\varphi_j \frac{(v_{-q} \bullet \Gamma_j) \bullet \hat{x}_3 \big|_{x_3=-d}}{P_{(q)(-q)}} \frac{2\sin(i(k_j + k_{-q})l/2)}{i(k_j + k_{-q})e^{ik_{-q}l/2}}) \tag{2}$$

where a_q is the amplitude of Lamb mode q, subscript and superscript j refer to wave mode j in the near field, subscript q refers to Lamb mode q, v_{-q} is particle velocity field of Lamb mode $-q$ ($k_{-q}=-k_q$), $\bullet$ denotes the dot product of two vectors, d is half of the thickness of the substrate plate, l is the length of transducer and $P_{(q)(-q)}$ is given by

$$P_{(q)(-q)} = \int_{-d}^{d} (v_{-q} \bullet T_q - v_q \bullet T_{-q}) \bullet \hat{x}_2 dx_3 \tag{3}$$

Hence the displacement of the propagating Lamb mode can be expressed as

$$u(x_2) = \sum_q a_q u_q e^{ik_q x_2} \tag{4}$$

where u_q is the mode shape of Lamb mode q (Graff, 1975). Equation (2) can be alternatively expressed as

$$a_q = \sum_{j=1}^{n} \varphi_j Tr_j Rm_j \tag{5}$$

where $Tr_j = \dfrac{(v_{-q} \bullet \Gamma_j) \bullet \hat{x}_3 \big|_{x_3=-d}}{P_{(q)(-q)}}$ is a transfer function while $Rm_j = \dfrac{2\sin(i(k_j + k_{-q})l/2)}{i(k_j + k_{-q})e^{ik_{-q}l/2}}$ indicates

the mode conversion efficiency from mode j in the near field to Lamb mode q.

3. Design of IDT

In eq. (4), only φ_j is an implicit function of the geometry of IDT. The geometry of IDT determines the potential distribution in x_2 direction on the surface of the transducer. Figure 3 shows the potential distribution in the wavenumber domain, $\varphi(k)$, after applying spatial Fourier transform. The central wavenumber of the IDT, k_{idt}, which corresponds to the maxima, is associated with finger spacing D by $k_{idt} = \pi / D$. The magnitude of the maximum indicates the excitation strength. The lobe width, defined at 1/3 maximum of the main lobe, indicates mode selectivity of the IDT where a smaller width implies the IDT possessing better mode selectivity. Hence, for efficient Lamb mode excitation, the finger spacing must be designed such that the central wavenumber $k_{idt} = k_j$ at the excitation frequency, which ensures that φ_j achieves maximum value.

The maximum potential and lobe width against finger width (normalized by D) is plotted in fig. 4. It is quite understandable that the wider the electrode fingers, the larger maximum potential is obtained at k_{idt} because wider fingers receive electrical signals more efficiently. However, wider fingers results in wider lobe and hence reduces the mode selectivity. Hence, there is a trade-off when designing the finger width. Small width poses fabrication difficulty, provides inadequate electrical conductivity and are easily broken whereas large a/D may cause arcing when anti-phase signals are applied. Full metallization ($a/D = 1$) prevents the application of anti-phase electrical signals. However, fig. 4 shows that as a/D changes from 0.05 to 0.95, both the maximum potential and lobe width changes within a small range. Hence, considering the factors above, a/D in the region of 0.5 is recommended.

In the same manner, the effect of number of fingers can also be evaluated based on the maximum potential and lobe width for its excitation strength and mode selectivity. For the case of $a/D = 0.5$, the maximum potential and lobe width against the number of finger numbers is plotted in fig. 5. The maximum potential shows a linearly increasing trend as more fingers leads to more electrical energy dissipation in the piezoelectric layer resulting in stronger excitation. Mode selectivity improves as more fingers are used but in a nonlinear manner. The selectivity significantly increases from 1 to 4 fingers but shows minor improvements if the number is further increased. Both the excitation strength and mode selectivity requires larger number of fingers.

The number of fingers is related to the length of the piezoelectric layer used for a given finger spacing where the latter is fixed by matching the Lamb mode to be excited. In NDE application, pulse signals are commonly used. A long IDT patch may elongate the time span of the pulse which is undesirable when analyzing the arriving time of the pulse. Hence, the length of the IDT may be chosen to accommodate at least 4 fingers.

4. Experimental Results

A 20 mm x 10 mm x 0.5 mm thick piezoelectric patch with two groups of 5 fingers of 1 mm width and 4 mm spacing was bonded to a 2 mm aluminum plate for NDE purpose. Windowed pulses that consisted of several cosine cycles were sent to the IDT from a function generator at a frequency of 600 kHz. The two groups of metallic fingers were fed with voltages at 180 degrees phase difference. The maximum of the input alternating voltage was 10V. A wide-band piezoelectric sensor (8×8×0.5mm) was bonded on the aluminum plate 150 mm away from the IDT and connected to an oscilloscope.

When excited at 600kHz, two Lamb modes, namely A_0 and S_0, may propagate in the aluminum plate with wavenumber 1558m^{-1} and 720m^{-1}, respectively. For the finger spacing used, $k_{idt} = 2\pi / 0.004 = 1570m^{-1}$ which matched the wavenumber of A_0 well and hence A_0 was dominantly excited. From the group velocity of A_0, v_g=3.13km/s, and the flight-time of the peak in fig. 6, the propagation path of wave package was calculated to be 166μs×3.13km/s=518.8mm. This distance was found to be the wave package propagating from the IDT to the crack, and subsequently propagating from the crack to the sensor. Along with the excitation direction of the IDT, the location of the crack could be obtained accurately.

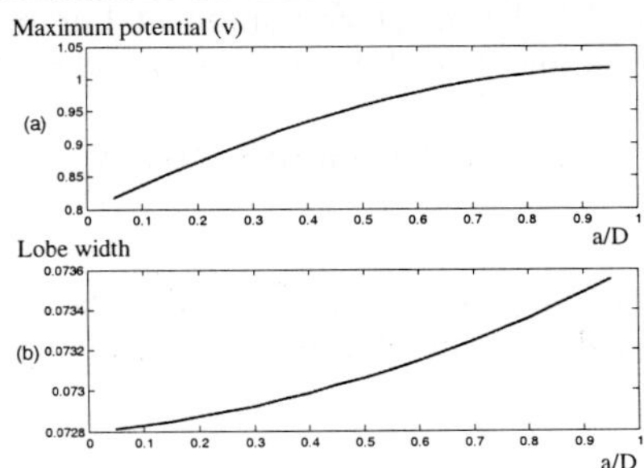

Fig. 4 (a) Maximum potential and
(b) lobe width against finger width

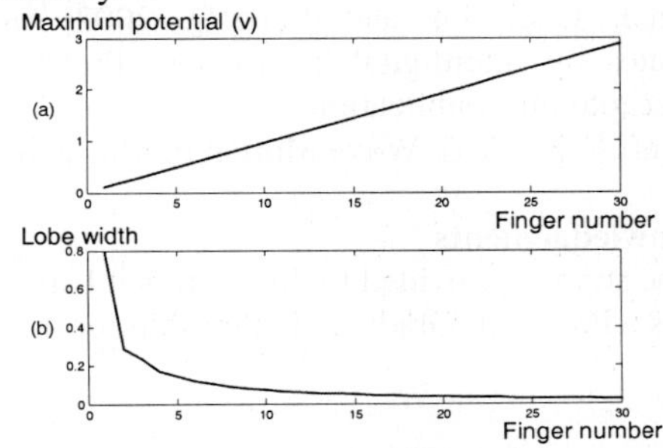

Fig. 5 (a) Maximum potential and
(b) lobe width against finger number

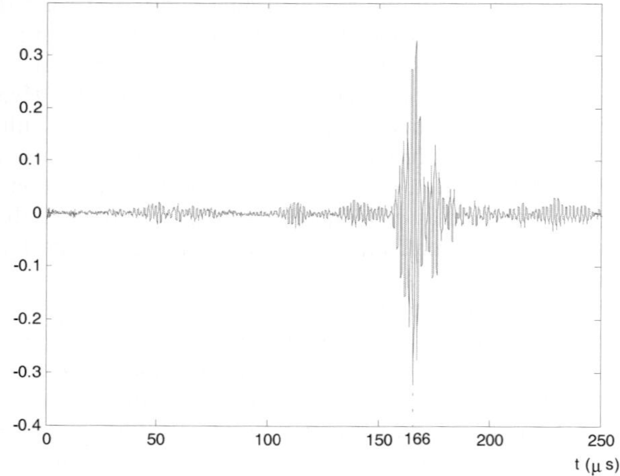

Fig. 6 Defects detection by IDT at 600 kHz

5. Conclusion

Analytical results show that the finger spacing controls the wavelength and is a fundamental design parameter. On the other hand, the finger width does not affect the excitation significantly. For fixed finger spacing, the IDT length and number of fingers are inter-related. They are designed so as to achieve sufficient mode selectivity and excitation strength while keeping the temporal width of signal package as small as possible for accurate flight-time measurement in NDE.

The IDT were fabricated in-house using commercially available piezoelectric patches. Experimental results based on the launching of single mode waves using the designed IDT agree well with those from the proposed analytical model. This technique was proved to be effective in conducting NDE in plates.

References

[1] Chang RE, Richie SM, Casey KJ, Malocha DC, 1994. Alternative nondestructive experimental approach to tangential static field detection in SAW structures imaging. Proceedings of the IEEE Ultrasonics Symposium; 1:341-345.

[2] Monkhouse RSC, Wilcox PD, Cawley P., 1997. Flexible interdigital PVDF transducers for the generation of Lamb waves in structures. Ultrasonics; 35:489-498.

[3] Wilcox P, Monkhouse R, Lowe M, Cawley P., 1998. The use of Huygens' principle to model the acoustic field from interdigital Lamb wave transducers. Review of Progress in Quantitative Nondestructive Evaluation; 1:915-922.

[4] Varadan VK and Varadan VV., 2000. Microsensors, microelectromechanical systems (MEMS), and electronics for smart structures and systems. Smart Materials and Structures; 9:953-972.

[5] Jin J., Quek S.T. and Wang Q., 2002. Analytical Solution of Excitation of Lamb Waves in Plates by Interdigital Transducer. Proceedings of the Royal Society of London, Series A. Accepted for publication.

[6] Graff K.F., 1975. Wave Motion in Elastic Solids. Dover Publications Inc., New York.

Acknowledgements

The support provided by the National University of Singapore in the form of a research grant as well as a President Graduate Fellowship for the first author for his PhD work is greatly appreciated.

Assessing the Performance of Ferrite Filters deploying Spread Spectrum modulation

S. B. Deosarkar[1]   T.R. Sontakke[2]   A.B. Nandgaonkar[2*]

[1,2*] Dr. Babasaheb Ambedkar Technological University, Lonere -402 103 (India)

[2] Shri. Guru Gobind Singh College of Engineering and Technology, Nanded - 431 602 (India)

Abstract

This paper presents analysis and behavior of ferrite beads with reference to the impedance, frequency, and spread spectrum modulation.

The first section of this paper will address the basic principles of EMI, ferrite chips, and will explain the characteristics of ferrites associated with EMI at low and higher frequencies. While the section two discusses reducing EMI with spread spectrum modulations.

Key words: Electromagnetic Interference (EMI), Ferrite filter, and Spread spectrum modulation.

1.Introduction

With the increasing emphasis placed on emissions and immunity by the FCC and other international regulatory agencies, design engineers are encouraged to prioritize EMC at the forefront of their designs. For this reason, there have been intense studies in the field of Electromagnetics which have provided break through in EMI filter technology, particularly ferrite chips / beads. This continuous effort for improvement has affected ferrite chips in areas such as size, performance, and popularity.

However, even with these trends of improvements, there still exists a sense of uncertainty on how to solve EMI problems using ferrite beads.

Ferrites are a class of ceramic ferromagnetic materials that can be magnetized to produce large magnetic flux densities in response to small-applied magnetization forces. Transfer of energy between the current and the magnetic field is effected through the "inductance" of the conductor. Placing a magnetically permeable material around the conductor increases the flux density for a given field strength and therefore increases the inductance.

Ferrite beads intended for EMI applications above 20 MHz are mixture of iron, nickel and zinc oxides that are characterized by high volume resistivity $(10^7$ ohm-cm) and moderate initial permeability (100 to 1500).

1.1 Ferrite impedance

The impedance of ferrite bead or core can be expressed as
$$Z = R\,(f) + j\,\omega\,L\,(f) \tag{1}$$
The frequency dependent loss term arises from the loss of energy incurred as a result of oscillation of microscopic regions with the ferrite. The loss and ferrite impedance can be expressed in terms of a complex permeability as
$$Z = K\,\{j\,\omega\,\mu o\,[(\mu'(f) - j\,\mu''\,(f)]\}$$
$$= K\,\omega\,\mu o\,\mu''\,(f) + j\,K\,\omega\,\mu o\,\mu'\,(f) = R\,f) + j\,\omega\,L\,(f) \tag{2}$$
Where,

$\mu'(f)$ and $\mu''(f)$- the real and imaginary component of the frequency dependent series complex relative permeability respectively.

K - a constant corresponding to number of windings and the core dimensions.

μo – free space permeability, ω - radian frequency $= 2\pi f$

The loss tangent tan of a ferrite material can be defined as the ratio of the imaginary part to the real part of the materials permeability.
$$\mathrm{Tan}\,(\delta) = \mu''\,(f)\,/\,\mu'(f) \tag{3}$$

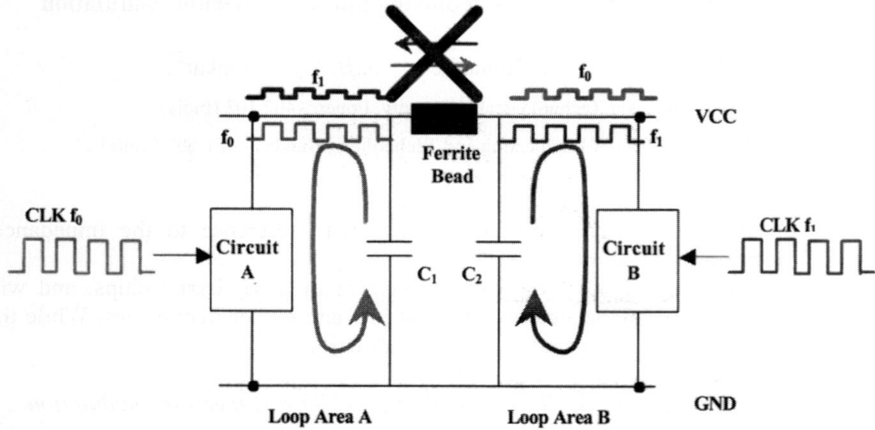

Figure 1: *A ferrite bead isolates high frequency currents between the circuit sharing the same Vcc rail*

As is true with the permeability, the loss tangent is frequency dependent. Thus from the knowledge of loss tangent one can easily estimate associated impedance and frequency.

In fact ferrites are often used in high frequency domain to prevent or to reduce unintended high frequency oscillations. The unique high frequency noise suppression performance of ferrites can be traced to their frequency dependent complex impedance as shown in the **Figure 4**. At low frequencies (below 10 MHz), a Murata type chip bead (*BLM-10A221SG) presents a small, predominately inductive impedance of less than 100 ohms. At higher frequencies, the impedance of the bead* increases to over 300 ohms, and becomes essentially resistive above 100MHz. When used as EMI filters, ferrites can thus provide resistive loss to attenuate and dissipate (as minute quantities of heat) high frequency noise while presenting negligible series impedance to lower frequency intended signal components.

1.2 Ferrite beads and frequency isolation [2][5]

Ferrite beads are used in local circuit loops. They are used either to confine local high frequency currents to the primary loop, or to prevent external high frequency current form entering the local loop. This is shown in **Figure 1.**

Furthermore, commercial grade ferrites suffer form magnetic saturation and leakage problems that also limit their performance. Because of the relatively low impedance, ferrite beads cannot be counted upon to completely prevent cross-loop current contamination in circuits sharing the same VCC rail. **Figure 1** shows the cross-contamination currents as waveforms of reduced amplitude (frequency f_1 and f_0 in loops A and B respectively). Cross contamination causes currents in primary loop A to leak into the adjacent loop B and vice versa.

1.3 E-field radiation

It is well known that differential mode E-field radiation at a distance of R meters is proportional to the loop area according to Equation (4) [1] [3]

$$\mathbf{E_{DM}} = 1.316 \mathrm{X} 10^{-14} f^2 \mathbf{I_d} \mathbf{A} / \mathbf{R} \qquad \text{Volts/m} \qquad (4)$$

Where **f** is the frequency, $\mathbf{I_d}$ is the amplitude of the current and **A** is the loop area.

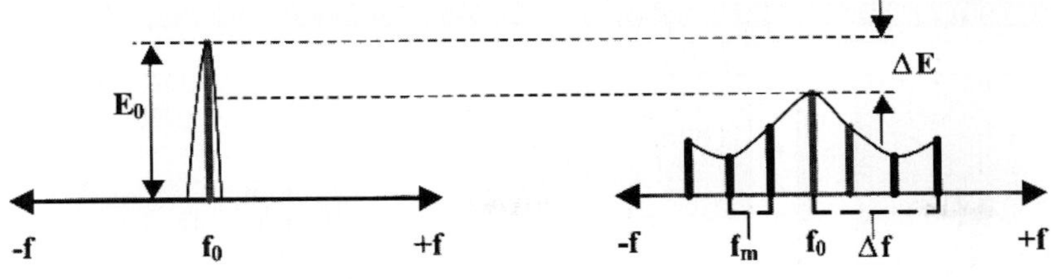

Figure 2: E-Field spectrum of clock; Fundamental frequency with and without FM

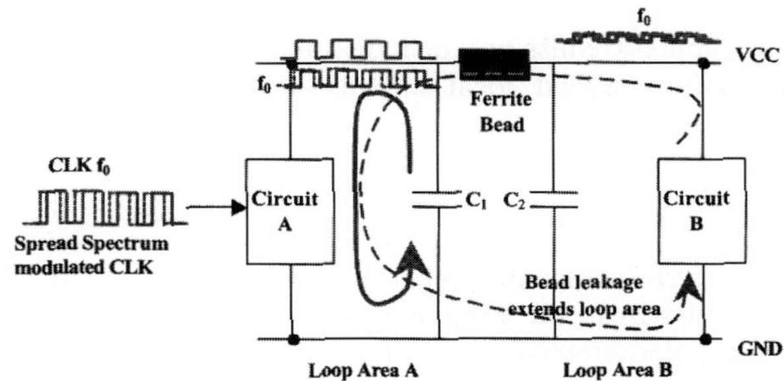

Figure 3: Spread spectrum modulation on clock in a two-loop circuit

Ferrite beads can therefore lower differential radiation by confining the current to the primary circuit. For proper usage however, one also needs to understand the limitation of these devices. Because ferrite beads have limited impedance cross-loop contamination that can easily occur, and adding more circuits on the Vcc supply rail extends the effective loop area over which these contamination currents can circulate. This is because the current simply leaks into more circuits, thereby circulating over a larger physical area. According to **Equation (4)**, a larger loop area translates into higher radiation.

2. Reducing EMI with Spread Spectrum modulation [2][4][5]

Spread Spectrum modulation is another word for Frequency Modulation (FM). EMI reduction by Spread Spectrum modulation makes use of the dispersion of energy from the target frequency over a wider band of frequencies. The width of the frequency band over which the energy is redistributed is known as the frequency spread (Δf).

Figure 2 shows the spread spectrum modulation, the peak amplitude E_0 of the fundamental frequency f_0 is suppressed by an amount equal to ΔE. The amount of suppression is a function of both the modulation rate f_m and the frequency Δf. The energy suppressed at the fundamental is distributed over the side bands generated by the modulation.

Figure 3 shows spread spectrum modulation on clock in a two-loop circuit. A key benefit of spread spectrum modulation is that even if the modulated signal is transmitted to another circuit on a different Vcc rail, or to an independent system on a completely separate power supply, the receiving circuit also benefits from the initial modulation. Furthermore, when spread spectrum modulation is applied to the input of a clock circuit that drives multiple output clock buffers, the modulation propagates through all the buffers, multiplying the effect and benefiting the entire system.

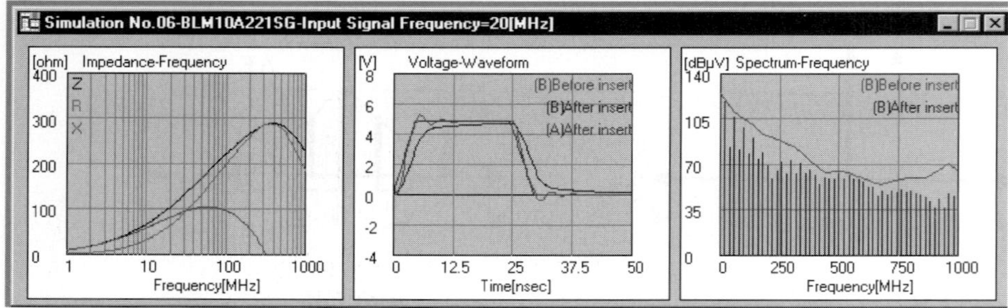

Figure 4: Simulation results showing the effect of ferrite bead insertion
(Simulated with Murata EMI Filter Selection Simulator BLM10A221SG)

Whereas ferrite beads only help suppress emissions in individual circuit loops, spread spectrum modulation extends EMI suppression beyond local circuits and provides system wide EMI reduction

2.1 Simulation results

The unique high noise suppression performance of typical ferrites can be traced to their frequency dependent complex impedance, as shown in the **Figure 4.**At low frequencies, below 10 MHz, a Murata EMI Filter presents a small, predominately inductive impedance of less than100 ohms, as shown **in Figure 4.**At higher frequencies, the impedance of the bead increases to over 300 ohms, and becomes essentially resistive above 100 MHz. The before and after insertions are also presented in the simulation graphs of Voltage waveforms and spectrum frequency.

3.Conclusion

Ferrite beads are used to isolate local high frequency current loops and lowers differential mode emissions. However, because the impedance of the ferrite bead is a function of frequency, and also because magnetic saturation and leakage tend to reduce its impedance, a ferrite bead cannot completely prevent cross-loop current contamination. Consequently, ferrite beads cannot efficiently suppress radiated emissions over a very wide range of frequencies.

Spread spectrum modulation help to suppress the amplitude of the target frequency by redistributing the energy over a wider frequency band. This technique not only lowers the radiation from the primary loop, but also in equal proportions, it suppresses the radiation from any extended current loops that may already exist. Thus, spread spectrum modulation is a powerful technique for EMI suppression, and extremely effective even if the circuits do not share the same power supply.

Obviously, the best results are achieved when spread spectrum techniques are combined with proper filtering implemented by strategically locating ferrite beads and bypass capacitors in the system. Spread spectrum modulation also help to prevent wasteful and indiscriminate use of ferrites along with significant amounts of shielding materials.

References

1. Clayton R. Paul: *IEEE Transactions on Electromagnetic Compatibility*, Vol. 31, No.2, May 1989, pp 189-192
2. White Paper: *EMI Reduction using Spread Spectrum Techniques*; PulseCore Semiconductor Inc, Santa Clara, CA.
3. G.K. Mithal and R. Mittal: *Radio Engineering*; Khanna Publishers, Delhi India.
4. Narvin K Simpson, Jim K.Omura,Robert A. Scholtz , Barry K.Levitt: *Spread Sprectrum Communications Handbook*,McGraw- Hill, Inc.,1985
5. Mills, J.P,Electromagnetic *Interference Reduction in Electronics Systems*,PTR Prentice – Hall Inc.,pp82-102,1993

Magnetocontrolled Elastic Composite Preparation and Characteristics

G.V.Stepanov, L.V.Nikitin*,A.I.Gorbunov, L.S. Mironova*, E.F.Levina
State Research Institute of Chemistry and Technology of Organoelement Compounds.
(GNIIChTEOS) Russia, Moscow. E-mail: stepanovgv@cityline.ru
*Moscow State University, Physical department, Moscow

At the present time it is actually developed the scientific-practical direction to the materials formation, which properties are changed under the external influence. Such materials are available to react on the temperature, pressure, moisture, electric and magnetic fields changes and so on. All of these are related to the kind of the so called "smart materials". One of such materials is the magnetoelastic (ME), which are available to change the form, dimensions and the elastic properties under the external magnetic field influence [1,2], to them are connected and the polymeric magnetic gels [3-6].

On the basis of such materials can be made the magneto-elastic thin-wall sensors, which can be considered as the magnetic analogs of the surface acoustical sensors, changing their properties in the function of the different influences of the external environment parameters. For example, for the distance pressure measurement, temperature, viscosity of the liquid and so on [7,8].

Experiment

In this report it is investigated the elastic properties of the ME in the function of the consistence and the magnetic field magnitude. The ME was getting by the dispersion of the ultra-dispersed magnetic particles (50-5000 nm) at the elastic polymeric matrices. It is investigated the ME, which was made on the silicon, isoprene and urethane rubbers. Silicone rubber whose elastic properties vary within a wide rage of 1 kPa – 1 MPa is most suitable. Ferrites and metal powders of 20nm – 2 μm size were used as magnetic fillers.

The main quantity of the tests was made with the use of the silicone polymeric elastomer (the SIEL mark of the GNIICHTEOS production). This resin composite was created by the high-temperature polymerization methods of the vinyl-containing silicone and the hydride-containing siloxane on the platinum contained catalyst.
Structure of silicone oligomer:
Deformation of magnetoelastic on external nonuniform magnetic field.

There are few works concerning the investigation of composite polymer materials that respond to a noticeable extent to an external magnetic field by changing their shape. Of these works, the studies on magnetic gels comprising chemically cross linked polymers filled with magnetic particles of 10 nm size are of primary interest [3-6]. In the cited works, it was shown that the shape of a magnetic gel in a non uniform magnetic field depends on the strength and configuration of the field. Deformation rating achieved 40% at low material tensile strength.
On the fig.1 it is shown scheme deformation of the ME at the external nonuniform magnetic field. Such ME is capable of elongation by hundreds percent in external magnetic field. Elongation value is determined by elasticoplastic properties of rubber. The ME elongation increases when Young's module value drops, magnetic powder content raises and external magnetic field increases. Ultimately, the material may break down under external magnetic field effect.

Such material property – to elongate (deform) under external magnetic field effect is most promising in respect of new devices design.
Deformation of magnetoelastic on external homogeneous magnetic field.
We have investigated the ME behavior in the homogeneous magnetic field. The cubic shape sample is positioned between the two poles of electromagnet. The magnetic field has changed between 0 till 3500 Oe. The ME deformation researches at the homogeneous magnetic fields has shown, that the magnetic-deformation elion effect in the homogeneous magnetic fields is absent.

The sample conserved the shape, given by the external mechanical force. The "plastic deformation" took place under the external mechanical force. The sample ME is become from the one side the more solid, hard, but from the other side the more plastic as the plastiline, that is from it you can made any different figures, but in the limits of those deformations, which permit the original magnetoelastic.

This moment the "plastic deformation" is observed. The "plastic deformation" magnitude is functioned form the MF size and is increased with the field growth from 1000 till 3500 Oe. After the MF turning-off the sample form is restituted (the original shape). By such way, the magnetoelastic has the "shape memory effect" (SME). The similar thermoelastic properties and super plasticity of shape memory alloys are observed [9].

On the fig.2 it is shown the material deformation at the magnetic field. The magnetoelastic at the form of the parallelepiped (A), is located between the magnet poles. After the magnetic field turn on at the magnitude till 350 mT the sample is not deformed in the practice (the elongation less than 5%)(B). However, by the mechanical deformation the sample become and conserve the given shape (C, D, I). After the magnetic field turn off the ME is restore the initial form (F).

Thus we observe effects of "plastic deformation" and "shape memory effect" in magnetoelastic.

<u>Viscoelastic properties of magnetoelastic on external homogeneous magnetic field.</u>

The ME elastic properties researches have been made as in longitudinal field, as in transverse magnetic field. The ME slice by the dimensions of 3x3x30 mm was placed in a lengthwise or in perpendicular direction to the magnetic field lines.

It is necessary to note, that the increase of elasticity of a sample in a longitudinal variable magnetic field by value 165 мT was insignificant up to 20 %. In a perpendicular constant magnetic field by the magnitude of 375 mT the sample elasticity could increase in 10 times. Detailed research of change of a stress in the sample at its stretching in various magnetic fields are submitted in a fig. 3. On these pictures the straight line 0 – the elastic stress deformation has got for the ME at the absence of the magnetic field. In this case the Young's module was calculated as the relation of change of a stress (F) from value of relative deformation (ε) and is equal 30 kPa.

The curves 1a, 2a, 3a, 4a reflects the stress change in the sample from the deformation at the magnetic fields 90, 165, 245, 335 mT in the magnetic field turn on with the next example extension, so called the initial cycle of the deformation curve.

The lines 1,2, 3, 4 – the examples deformation curves at the suitable magnetic fields after the load removal and the subsequent application (the weight-equal curve). The initial cycle of the sample deformation curve at the magnetic field is described by the deformation lag effect by the charge increase till 20 kPa.

For the calculation of the Young's module we have used the so called the weight-equal curves 1- 4. The curve beginning origins is not from the zero point on the abscise axe, but it is begin from the some rest deformation, as after the removal of the charge the example, located at the magnetic field, is not compressed in the whole, but it has the rest deformation. For the shown example by the field change from 0 to 335 mT, the Young's module of the elasticity is changed in 5 times from 30 till 154 kPa.

Let's consider dependence of change of size of the ME Young's module placed in a magnetic field from size of the initial Young's module. The lower the size of the ME initial Young's module is, the more increase of it in a magnetic field. For our samples we observed effect of the Me Young's module increased on ten times at increased size 3.5 kOe of the magnetic field and initial Young's module equal 8 кPa.

The dependence elastic from size of a magnetic field has not linear, parabolic character. That it is possible to explain by effect of an attraction magnetic particles and change spacing on centers, between magnetic particles.

The main criterion, describing the magneto-reological properties of the magnetoelastic, can be the exponent of the magneto-elastic susceptibleness. This exponent, showing the elastic change (dE) with the magnetic field change (dH), we can see from the picture- the elasticity as function of the magnetic field. So for a sample considered on a fig. 1,2 parameter of magneto-elastic receptivity for a field 335 мT is equal: $Е_н = \Delta E / \Delta H = 70кPa/T$ or 90H/A•м.

<u>Magnetoelastic preparation in the external magnetic field.</u>

These materials possess by the magnificent differences in the elastic module in the function of the deformation direction.

This compound was prepared by filler powder milling to liquid silicone rubber with the following orientation of the composite film in the magnetic field and the composite polymerization. The elastic-deformation properties were investigated of these received samples of the parallepiped shape. The magnetic field influence on the received samples elastic properties in these series was not great, but the elastic and deformation anisotropy effect was sufficient.

This composite Young's module is very differenced from the deformation direction. The sample elasticity, cluttered out in the longitudinal (0 degree) and the perpendicular (90 degree) direction relatively to the orientation axis is differed of 5 times.

From fig.4 we can see that the function curve of the strain (F) from the relative deformation (X) has two strictly expressed regions: the initial region- the low elasticity (LE) and the second- the high elasticity (HE), high tension stiffness. The elasticity Young's module (E) in the function of the sample deformation direction ME is 5 times changed for the low elastic region, that is E(0 degree)= 100 kPa and E(90 degree)= 23 kPa

The sample deformation magnitude (X) in the low-elastic region is much changed from the deformation direction. The borders of two deformation regions correspond to strain (F=20kPa). This deformation for the (LE) region sample (0 degree) is equaled 20 % (X = 0,2), but to the sample (90 degree) – 100 % (X = 1).

If the direction of deformation of a sample not considerably deviates a direction of axis magnetic orientation, the elasticity is sharply decreased.

For the composite on the basis of the rubber such high anisotropic elasticity is the new one.

CONCLUSION

Material properties studies demonstrated its ability to elongate by tens and hundreds percent in magnetic fields. Elasticity of the material placed into magnetic field raises up to 10 times. Depending on production technique it possesses significant elasticity anisotropy. Thus, I the obtained samples Young's modulus of elasticity 5 – fold differed on longitudinal and lateral strain. The analysis strain performance of ME located in homogeneous magnetic field demonstrated that the material is distinguished by "shape memory effect". It retains shape caused by external mechanical effect and return to primary state when magnetic field is switched off.

This kind of material can be find the application by the devices development in the electrotechnical branch, at the aviation and the others branches.

References
1. L.V. Nikitin, L.S.Mironova, G.V.Stepanov and A.N.Samus. Polymer Science, vol.43, no. 4,(2001), p.443.e, Ser.A
2. E.F.Levina, L.S.Mironova, L.V.Nikitin, G.V.Stepanov. Russian patent N 2157013/2000.
3. M.Zrinyi (HU); J.Gacs (HU); C.Simon (HU) Patent WO9702580 1997-01-23 H01F1/00; H01F1/44
4. M. Zrinyi, L. Barsi, and Duki, J. Chem.Pys., 104, no.21 (1996) p. 8750.
5. M. Zrinyi, L. Barsi, and D. Szabo J. Chem.Pys., 106, no.13 (1997) p. 5685.
6. M. Zrinyi, D. Szabo L. and H. Kilian Polym. Gels Networks, 6 (1998) p. 411.
7. C.A.Crimes, K.Loiselle, P.G.Stovanov and F.Tefiku., Smart Mater. Struct. 8 (October 1999), 639-646
8. D.Kouzou dis and C.A.Grimes, Smart Mater. Struct. 9 (December 2000), 885-889
9. A.A. Cherechukin, V.V.Khovailo at al. Proc. 18 Workshop on New Magnetic Materials for Microelectronics. Moscow (2002), p.863-865

This work is represented on the grant conference SfP-977998

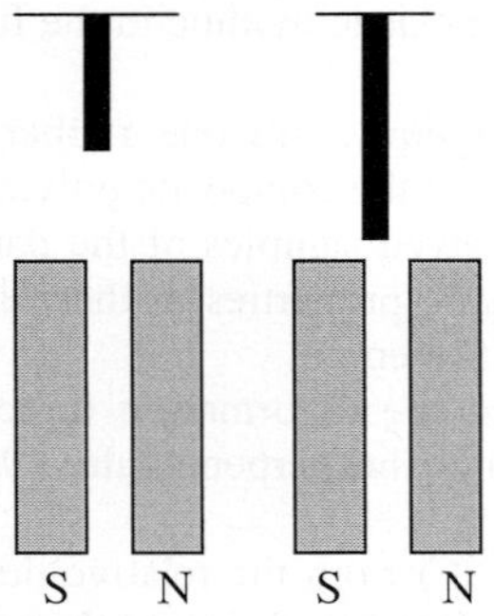

Fig. 1. Deformation of magnetoelastic in a nonuniform magnetic field.

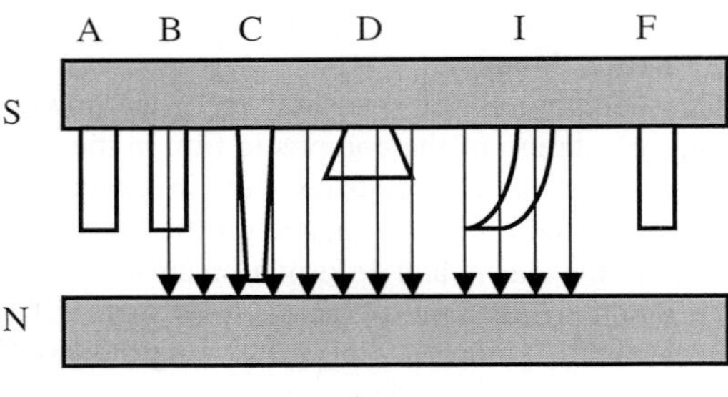

Fig.2. The circuit of deformation magnetoelastic in a homogeneous magnetic field

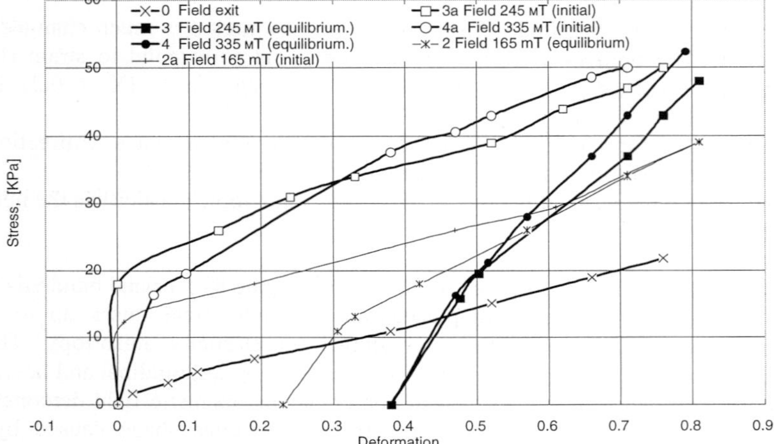

Fig. 3 Dependence of stress in the magnetoelastic from size of deformation for various magnetic fields.

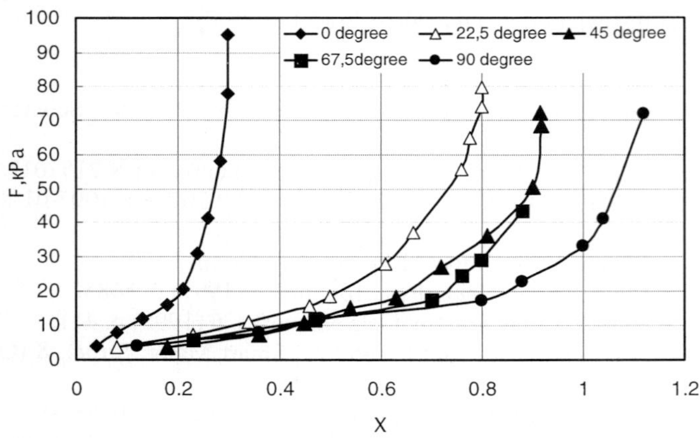

Fig.4 Dependence of the elastic stress F on the relative elongation X function in the oriented magnetoelastic. (0degree - direction elongated coincide with axis magnetic orientation)

Experimental Investigation on Piezoelectric Actuated Beams

Gupta V.K.[1]   **Seshu P.[2]**   **Issac K.K.[2]**

guptaeck@yahoo.co.in   seshu@iitb.ac.in   kurien@me.iitb.ac.in

1. Introduction

Piezoelectric materials are normally assumed to be linear and the actuation strain is modeled like thermal strain. The constitutive relations are based on the assumption that the total strain in the actuator is the sum of the mechanical strain induced by any stress, thermal strain due to any temperature and the controllable actuation strain due to electric voltage. Various theoretical and finite element models have been developed by researchers to study effect of piezoelectric actuation on beam, plates and shells (Crawley and deLuis (1987), Crawley and Anderson (1990), Park and Chopra (1993)). Few experimental works have been reported in the literature to verify theoretical and Finite Element models of piezoelectric actuation. While different theoretical and finite element models are available for different structures, experimental work is confined to straight structures like beams and plates with single actuators. In the current work, experimental results are presented for verifying the end-moment model of Crawley and Anderson (1990) using single and multiple actuators on a straight beam.

2. Tip Deflection of a Piezoelectricity Actuated Cantilever Beam

Simple relations are developed for calculating tip deflection of a cantilever beam and the applied voltage across the piezoelectric actuators based on Crawley and Andersons (1990) Bernoulli-Euler strain transfer model. Fig. 1 shows such a configuration and its equivalent representation for the beam with bonded piezoelectric actuators.

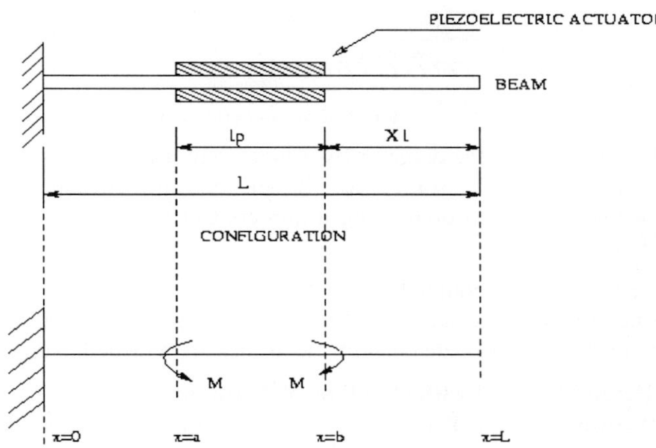

Fig 1 Cantilever Beam with Bonded Piezoelectric Actuators

The whole beam can be divided into three regions from x=0 to x=a, x=a to x=b and x=b to x=l. On application of the governing equation for the deflection of beam under the action of moment, we have

$$EI\frac{d^2y}{dx^2}=0 \ \Big|_{x<a} \ +\frac{M}{EI} \ \Big|_{x<b} \ -\frac{M}{EI} \tag{1}$$

[1] Sr. Lecturer, Mech. Eng. Dept., Engg. College, Akelgarh, Kota (Raj.), India - 324009
[2] Associate Professor, Mech. Engg. Department, I.I.T., Powai, Mumbai, India - 400076

Solving equation 1 and applying boundary conditions we get the deflection of the as -

$$Y_{tip} = \left[\frac{12(T+1)d_{31}V_{app}l_p}{\{(6+\psi)T^2+12T+8\}t_p^2} \right]\left[x_1 + \frac{l_p}{2} \right] \tag{2}$$

where, d_{31} is the piezoelectric coefficient of the actuator material, V_{app} is the applied voltage, $T=t_s/t_p$, and, subscript s represent structure and p represent piezoelectric actuator.

3. Experimental Investigation

In the previous section, a brief detail of analytical model for the analysis of piezoelectric actuated cantilevered beam has been discussed. Experiments are conducted to verify the theoretical prediction and to study different aspects of piezoelectric actuation of structures. Fig. 2 shows the experiment setup used for the static analysis of the beam.

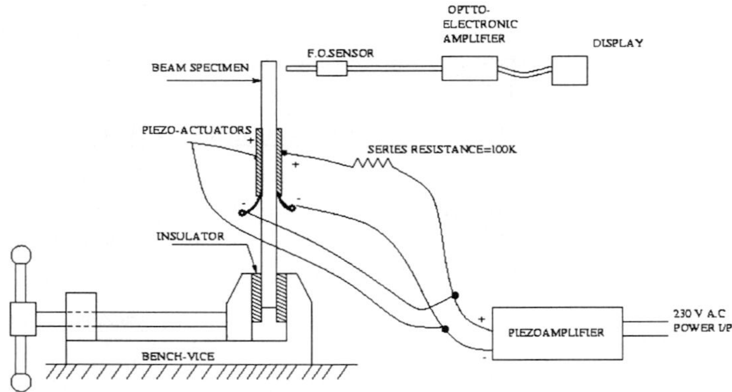

Fig. 2. Experimental Setup

Electric field is supplied to the actuator by a piezo-amplifier. A piezo amplifier (Piezo System Inc., USA) is a drive power source for the piezoelectric actuators. The deflection due to piezoelectric actuation is measured using a non-contact Fiber Optic (FO) displacement sensor (Philtec Inc., USA).

Piezoelectric actuators are mounted on top and bottom side of the straight beam with a non-conductive adhesive. Both the surfaces of the actuators are fully electroded. To access the lower (inner) surface of the actuators, metal shim electrodes are attached using conductive epoxy.

3.1 Straight Beam Mounted with One Pair of Actuators

Piezoelectric actuator is generally modeled as "end-moment actuator" [Crawley and Anderson, 1990] i.e. it exerts a moment at its ends. Experiments are first conducted to analyze deflection behavior along the length of the beam ($E=0.638\times10^5$ N/mm^2) (Fig. 3) and to examine the end-moment model. Material properties for this beam are.

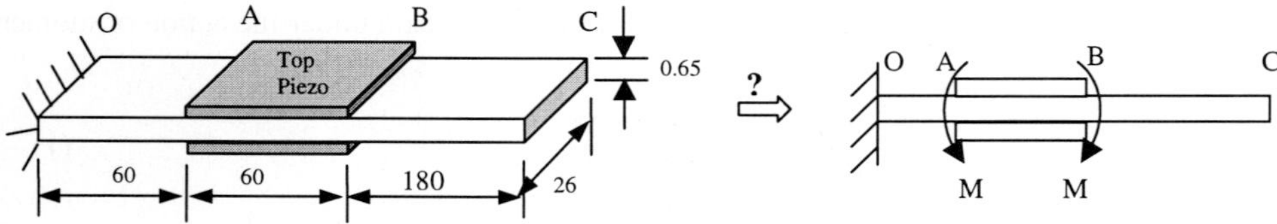

Fig. 3. Straight Beam for the Study of End-Moment Actuator Model

The actuators used are PSI-5A-S4-ENH (Piezo System Inc., USA), with $d_{31}=-190 \times 10^{-12}$ m/V, $E=0.66 \times 10^5$ N/mm^2, t=0.267mm. Deflections are measured at different points along length of the beam by moving the FO probe.

First FO probe is moved along the length of beam without actuation and then with actuation. Difference in the two values (without actuation and with actuation) gives the deflection at a particular point. The results for an applied voltage of 90V are plotted in Fig. 4.

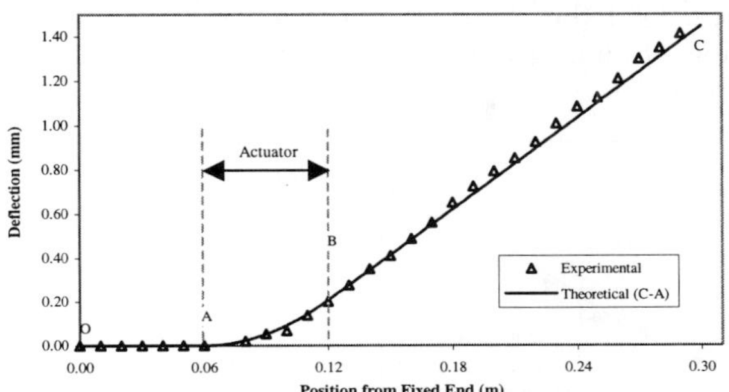

Fig. 4. Deflection along the Length of Cantilever Beam under Piezo Actuation

Location of actuator is shown by region AB on the deflection curve. Theoretical results by Crawley and Anderson (1990) are also plotted for comparison. Deflection in the region OA is zero as no actuator effect comes into picture in this region. Deflection in the region AB is non-linear due to the presence of the actuator. Region BC (starting from the farther end of actuator to the tip of beam) exhibits linear variation. Deviation of 10% is observed between theoretical and experimental values at the tip.

3.2 Straight Beam Mounted with Multiple Actuators

In order to further analyze the deformation pattern under multiple actuators, experiments are conducted on a straight beam (E=0.638 $\times 10^5$ N/mm^2, t=0.25mm) mounted with two pairs of actuators (SP-5H, Sparkler Ceramics, India with $d_{31}=-265 \times 10^{-12}$ m/V, $E=0.476 \times 10^5$ N/mm^2, t=0.25mm) at different locations (Fig. 5). The poling directions of both the actuators are same.

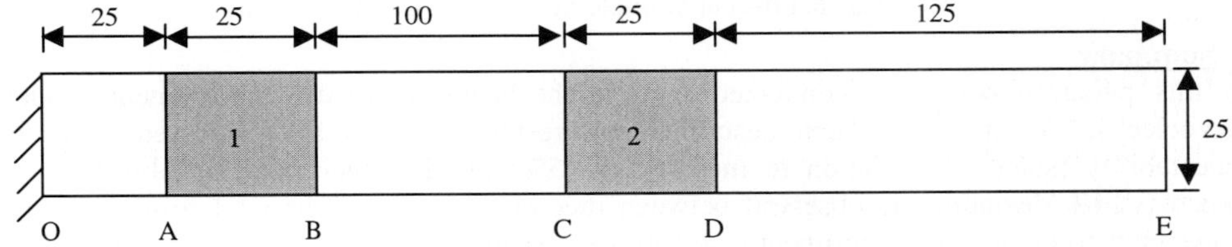

Fig. 5. Straight Beam with two Pairs of Acuators

The Crawley-Anderson model discussed in section 2 is extended here to consider multiple actuators (Actuator1 is supplied V_1 and actuator2 V_2 volts). Deflection at any point at 'x' distance from root is given by

$$y = \frac{12(T+1)d_{31}}{2((6+\psi)T^2+12T+8)t_p^2} \left(V_1\left((x\text{-}OA)^2 - (x\text{-}OB)^2 \right) + V_2\left((x\text{-}OC)^2 - (x\text{-}OD)^2 \right) \right) \qquad (3)$$

Three cases are considered. Deflections along the length are measured following the same procedure as in previous case. Deflections with respect to distance from the fixed end are plotted in Fig. 6. It is observed from the theoretical predictions that the starting region where no actuator is present (OA) has near zero deflection while in the region where actuators are present i.e. AB and CD have non-linear variation of deflection. Region after 2nd actuator (i.e. DE) has linear variation in deflection as expected in the end-moment model. It is also observed that deflection in region OABC is the same in all the three cases. The pattern in the region CDE is different due to effect of second actuator. When both actuators are supplied same voltage (Fig. 6(a)) the second actuator's effect is aiding the effect of first actuator. When both actuators are supplied same voltage but in opposite direction (Fig. 6(b)), second actuator nullifies the effect of first actuator and the deflection curve is almost horizontal. When second actuator is supplied half the voltage supplied to actuator 1 (Fig. 6(c)), as expected, the deflection pattern is in between the first two cases in region DE. It is also observed that the theoretical and experimental results are in close match (at worst 14%).

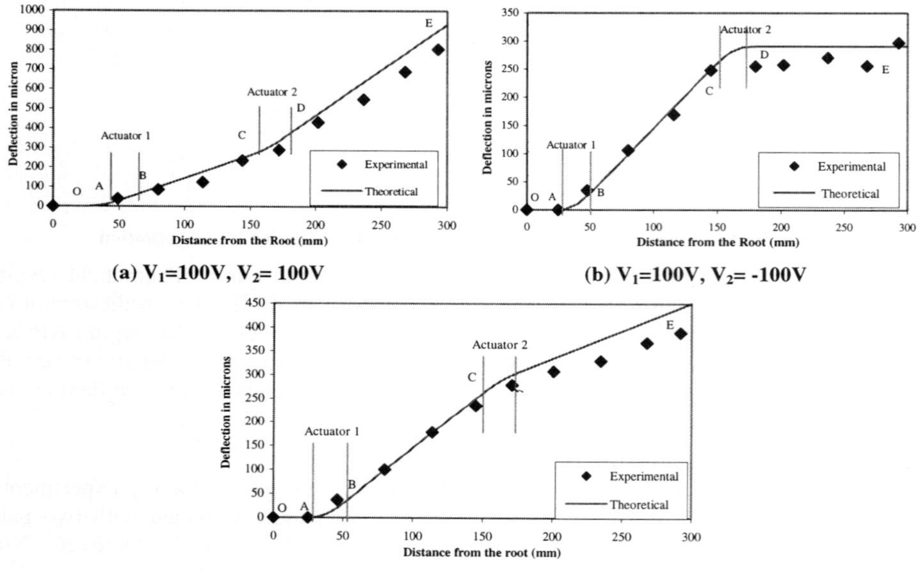

(a) V$_1$=100V, V$_2$= 100V

(b) V$_1$=100V, V$_2$= -100V

(c) V$_1$=100V, V$_2$= -50V

Fig. 6. Effect of Multiple Actuators on Cantilever Beam

4. Summary

In this paper, experiments conducted on straight beam to verify end-moment model of piezoelectric actuation has been described and results obtained are discussed. Very good repeatability (standard deviation to mean ratio <5%) of the experiments is observed. Good match (<14% deviation) is observed between theoretical and experimental results. Deviation between theoretical and experimental results can be due to unmodeled effects like adhesive bonding layer, error in measurement, non-linear behavior of piezoelectric coefficient at higher electric field, etc. It appears possible to obtain desired shape of the beam, by using multiple actuators and adjusting voltages supplied.

5. References

Crawley, E.F. and Anderson, E.H., 1990, "Detailed Models of Piezoceramic Actuation of Beams", Journal of Intelligent Material Systems and Structures, **1**, pp. 4-25.

Crawley, E.F., and deLuis, J., 1987, " Use of Piezoelectric Actuators as Elements of Intelligent Structures", AIAA Journal, **25(10)**, pp. 1373-1385.

Park, C., Walz, C. and Chopra, I., 1993, "Bending and Torsion Models of Beams with Induced Strain Actuators", SPIE Proceedings of Smart Structures and Intelligent Systems Meeting, Albuquerque, NM, **1917**, pp. 192-216

Magnetic Polymers

Sh. Tuichiev, V. Rashidov, A. Esfidary
Department of Physics, Tajik State University,
pr Rudaki 17, Dushanbe 734025, Tajikistan

Abstract – Magnetic polymers on the base of epoxy composition and polyethylene within magnetic powder were obtained. Different forms of magnets can be obtained from these materials, while magnetic field may be concentrated outside or within sample forms. The obtained polymer magnets were elastic and strong. They had high adhesion, and stability to influence of seawater, petroleum, etc.

INTRODUCTION

At the present time, magnetic polymers are used basically to filter liquids and gases in various areas of industry vastly. However, such materials are not studied sufficiently to be used in production of filter elements for petroleum derivatives, seawater, etc. since these materials deteriorate in such environments and are finally destroyed [1-4]. Therefore the issue of producing magnetic polymers and filter materials that resist such environments appears. The purpose of the present work is producing magnetic polymer compositions to be used in manufacturing of magnetic filters.

EXPERIMENTAL

As the samples of the investigations the following materials were used: epoxy, placticizer, stabilizer, and filler -- magnetic powder with high magnetic power (coercivity force Hc=1200 Э with the remnant magnetic induction B=0.37 T). Applying the ordinary technology of combination, filler entered the not so adhesive environment of epoxy composition. The size of the particles of the magnetic powder was 1-2 microns. The mixture of epoxy composition and the magnetic filler entered a special magneto-former and was spread. The magnetoformer produced plate samples with width of 2 mm. Changing the ratio of epoxy composition and the magnetic powder, 8 samples were obtained and their physical and technical properties were investigated. The mechanical properties of materials under one-dimensional tensional stress in standard equipment, with constant tension velocity of 10 mm/s were studied and the values of Young's modulus E, strength σd and the tearing deformation εd were determined. Moreover, applying the relative differential calorimetric principals, heat transfer coefficients (λ) of the samples were measured. As the environments, ordinary water, petrol AI-76, and technical oil type MV-1 were used. The samples were exposed to these environments, and the influences were studied for more than five months. The degrees of expansion of the samples in these environments were obtained from the relative changes in the mass of the samples   All measurements were done in temperature of 20-25°C, and humidity of 45-60%. In all the mechanical and thermal studies the number of the samples was 5-10 ones and the average values are given. The errors of experiments in mechanical and thermal studies did not exceed 5%.

RESULTS AND DISCUSSION

It is clear that the physical properties of polymer systems are determined by the filler – its characteristics, concentration, and its chemical reactions with the polymer [1-5]. In order to remove the chemical influences, chemical principles of selecting the composite couple-polymer were applied. Not only the process of technical magnetization determines the magnetic characteristics of magnetic polymers, but also the physical condition of the filler in the polymer. Apparently, this condition depends on the adhesion of the polymer at the first place. In order to produce a polymer environment with a specific adhesion, the proportion of epoxy, placticizer, and the stabilizer are changed. In the applied samples the polymerized compositions had very high Young's modulus properties and were not easily broken. The results of the studies showed that upon increase of concentration (c) of the magnetic filler, the value of magnetization of the remainder increases and while c = 0.8 it rises to the maximum level (Fig.1).

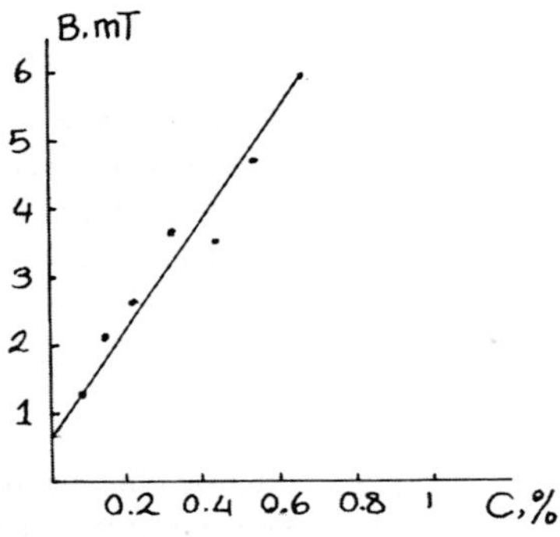

Fig.1. The dependence of magnetic induction to the concentration of magnetic powder in epoxy composition

In Table I the obtained composition samples and the magnetic property B for various compositions are given. It should be noted that the value of magnetic induction is B = 4.5 mT at the surface, and B = 0.5 mT at the height of h = 5 mm from the surface. The results in table 1 show that the proposed material samples have B = 1.4 – 22 mT while the strength is σd = 110 – 130 Mpa, and they are applicable as filtering materials. During the using process of epoxy composition, it encounters the influence of temperature, mechanical pressure, aggressive fluid environments, etc. and they change the mechanical and technical properties of the materials. Remarkable changes depend on absorption-diffusion events and the influence of aggressive environments on the magnetic polymers.

Sample Systems	Ratio of Components, Weight %		Characteristics	
	Epoxy Composition	Magnetic Powder	Magnetic Induction B, mT	Strength σ_d, MPa
1	80	20	1.4	110
2	70	30	3.2	115
3	60	40	3.6	120
4	55	45	4.7	120
5	50	50	4.1	130
6	45	55	5.3	124
7	40	60	6.7	118
8	10	90	22	70

Table I. Some characteristics of magnetic polymers.

In figure 2 dependence of strength σd and tearing deformation ed to concentration c of magnetic powder in epoxy composition is shown. It is seen that the value of σd is practically constant until c = 60% and at this level it starts to decrease. At the same concentration the tearing deformation decreases too. The observed changes in σd and εd can be explained as follows. While increasing the concentration of magnetic powder the mass of epoxy composition decreases relatively, and the modification of the filler does not occur; it loses its effectiveness in the material, and consequently the strength and deformation of the material decreases. It should be remarked that each sample of epoxy composition has its own individual properties. All the 8 samples were exposed to fluid environments (water, petrol AI-76, and technical oil type MV-1), and then their mechanical properties were measured. Results of mechanical investigations and absorption capacities of

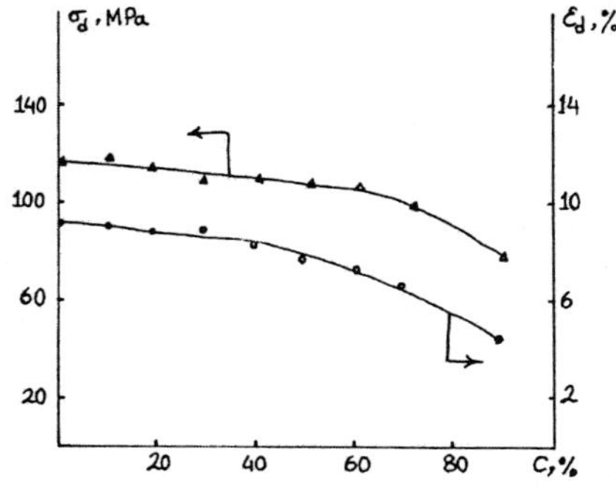

Fig.2. The dependences of strength σd and deformation εd to the concentration c of magnetic powder in epoxy composition.

some of the magnetic polymer samples are given (Fig 3, 4). Observations showed that after exposure to liquid environments changes in σ (t), ε (t), E (t) and α (t) occur. For studied samples, the changes in these values are similar to each other. For system 1, upon increase of exposure time (t), a not very large increase of ε, E, and σ appears. Analysis of these values indicates that system 1 is resistant to these environments. Diffusion and absorption of environment for system 1 is very low, however, the proportion of water absorption (α1) to petrol and technical oil absorptions (α2 and α3) is very large. Accounting the measurement errors, we can consider the magnitude of α for all environments to be the same. Liquid environment has little influence, or generally no influence on polymer magnetic system 1. In system 3 the level of changes in the magnitudes of σ, E and ε in the environments are not so different from each other. It means that system 3 is inert in these environments. The absorption curves show that α3 < α1 < α2. For system 5 the character of changes in σ, ε and E are different in various environments. The changes in σ and ε in these environments are anomalous: σ and ε decrease in water; σ increases and ε decreases in petrol; σ and ε increase both in technical oil. Such changes in α are observed as well: α3 <α2 < α1.

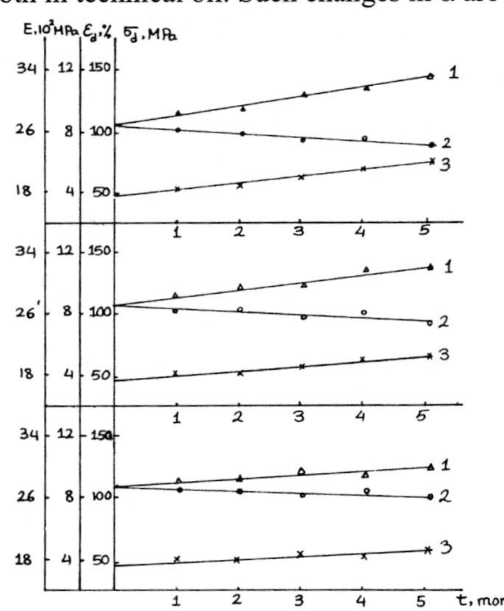

Fig.3. The change of strength σd (1), relative tearing deformation εd (2) and Young's modulus E (3) during time in liquids for system 1.

Up: water
Middle: petrol AI-76
Down: technical oil MV-1

For system 7, σ and E decreases in water and ε is constant. In petrol E and σ are constant and ε increases; in technical oil E, σ and ε increase with time t but α3 <α2 <α1. For all studied systems, heat transfer coefficients λ were measured and results are given in table 2. Note that λ is given for systems 1-8 when exposed to the environments for t = 1 month. The results show that λ changes in these environments differently. In water and technical oil λ tends to decrease and in petrol its decrease is observed. Increase of λ while the samples are exposed to petrol depends partially on the density of the structural defects [5]. The changes in the heat transferring of polymer magnets while exposed to fluid environments indicate that these changes are basically determined by the nature of mutual influences between the particles of the materials. On the base of the achieved results from magnetic polymers, filters for technical oil that contains magnetic particles were designed and built. The efficiency of the filters was obtained from the relative change of concentration of the mixtures during the filtration time [1] of technical oil (see Fig. 5). In this figure the efficiency of filters according to [1] (curve 1) and the efficiency of the filter we produced are given. The results of measurements showed that the filtration capacity of the filter we produced is more than two times larger than the other typical filter, and its efficiency is about 99%.

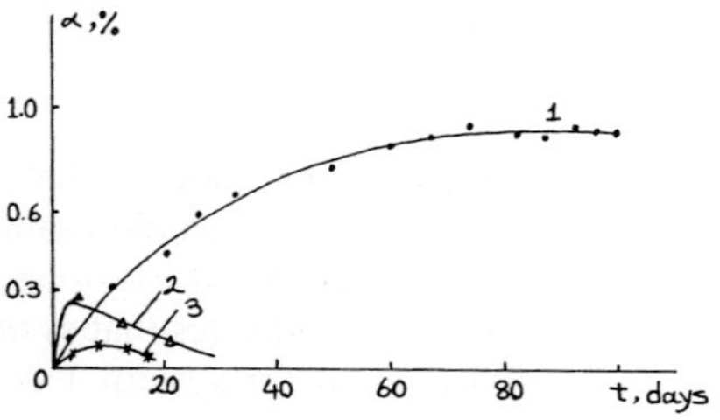

Fig. 4. The dependence of expansion coefficient α to exposure time t in liquids for system 1.
1. Water
2. Petrol AI-76
3. Technical oil MV-1

The proposed magnetic filter based on magnetic polymers, viz. epoxy composition is very resistant to petroleum and can be used repeatedly in various areas of petroleum industry. Moreover, geometrical changes in the construction of such filters make them appropriate for vast application in electro-technical and automobile industry.

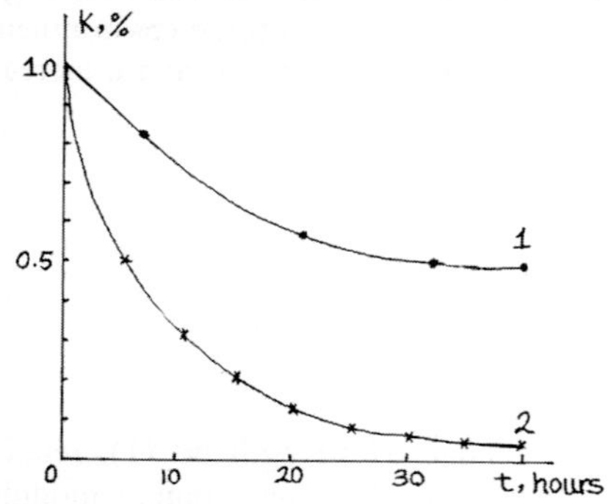

Fig. 5.The efficiency of filtration (K) of ferromagnetic particles of technical oil during filtration time (t).
1.Filter [1]
2. The proposed filter

System Numbers	Heat Transfer Coefficient λ, Wt/MK			
	Initial	Water	Petrol AI-76	Technical oil MV-1
1	0.130	0.092	0.118	0.110
2	0.174	0.133	0.250	0.144
3	0.179	0.140	0.254	0.150
4	0.164	0.148	0.204	0.212
5	0.168	0.159	0.209	0.205
6	0.176	0.165	0.215	0.220
7	0.193	0.178	0.263	0.233
8	0.207	0.149	0.209	0.258

Table 2. Heat transfer coefficient of magnetic polymers λ (Wt/MK)

REFERENCES

1. V. K. Shishkin et al Cert. No: 1330666A1, Bulletin No 1, 1985, USSR.
2. V. K. Shishkin et al Cert. No: 1294479A1, Bulletin No 9, 1985, USSR.
3. V. V. Shezhkov et al Cert. No: 677017, Bulletin No 28, 1979, USSR.
4. P. P. Edel'man et al Cert. No: 1360798A1, Bulletin No 47, 1989, USSR.
5. Yu. K. Godovskii, Teplophysica Polymerov, Moscow, Khimiya, 1982

Session F13

HTS and Thin Films (1)

Chair: X. Yao

Ferroelectric and Ferromagnetic
Glass Ceramics for Microwave Applications

Yao Xi
Functional Materials Research laboratory
Tongji University, 200092, Shanghai China
yaoxi@fmrl.ac.cn

Conventional ferroelectric and ferromagnetic ceramics are crystalline ceramics. The amount of crystalline phase is almost closed to hundred percent. The introduction of a small amount glass phase or other second phase will much reduce the dielectric permittivity and the magnetic susceptibility of the materials. However, in many cases such as microwave applications, high permittivity and high susceptibility are not preferable to the requirements. In many cases, very complicated and comprehensive requirements of electric, magnetic, mechanic behaviors are of much importance to the specific applications. The processing requirements to the materials such as co-firing of the material with other materials and/or electrodes, sometimes dominates the selection of materials. A new group of microcrystalline ferroelectric and ferromagnetic glass ceramics is introduced to meet such special requirements. Most major ferroelectrics and ferromagnetics can be prepared as glass ceramics by using sol-gel technology. The glass content of the glass ceramics can be adjusted in very wide range from less than a few percent to more than ninety percent. The size of ferroic crystallites can be controlled in the range from nanometer sizes to micrometer sizes. The electric and magnetic behaviors of the glass ceramics will mostly inherit from the crystalline components, while the mechanical and thermal processing behaviors will be mostly depending on the glass components. The electric, magnetic, elastic, and thermal behaviors of the glass ceramics can be tailored in much wider range compare to the crystalline counter parts. The ferroic glass ceramic can be prepared in the form of bulk ceramics, thick and thin films. There are a lot of advantages of glass ceramics, such as low densification temperatures, fine and dense microstructures, low specific density, low acoustic impedance, low and adjustable thermal expansion, good mechanical strength and toughness, better temperature and frequency stability of dielectric/magnetic behaviors. Discuss of the idea to develop the ferroelectric and ferromagnetic glass ceramics will be given in this presentation. Examples of the processing and microwave applications of the materials will be also given.

Improvement of dielectric properties of $Ba_{0.5}Sr_{0.5}TiO_3$ thin films

K. B. Chong[1], Linfeng Chen[2], L. B. Kong[2], X. S. Rao[2], C. Y. Tan[1], C. K. Ong[1] and T. Osipowicz[3]

[1]Centre for Superconducting and Magnetic Materials , department of Physics, National University of Singapore.

[2]Temasek Laboratories, National University of Singapore.

[3]Research Centre for Nuclear Microscopy, Department of Physics, National University of Singapore.

ABSTRACT

Ferroelectric thin films like $Ba_{0.5}Sr_{0.5}TiO_3$ are useful in several high-density electronic and tunable microwave device applications , because of its high dielectric constant and low loss. The reported values for dielectric constant and dielectric loss of $Ba_{0.5}Sr_{0.5}TiO_3$ thin films were varied from different research groups. We address to improve the dielectric properties of $Ba_{0.5}Sr_{0.5}TiO_3$ (BST) thin films by doping with other low loss oxide. We choose Al_2O_3 as dopant due its low microwave dielectric loss. The Al_2O_3-BST thin films were fabricated by pulsed laser deposition technique. The Structural phase composition were determined by X-ray diffraction (XRD). The characterization of dielectric properties for Al_2O_3-BST thin films were carried out by using non-destructive microstrip dual resonator. The dielectric loss are significantly reduce with the Al_2O_3-doping. Consequently, the figure of merit for Al_2O_3-BST thin films were founds to improvement with the content of Al_2O_3. Thus the Al_2O_3-BST thin films showed the potential to be exploited for tunable microwave devices applications.

The ferroelectric barium strontium titanates thin films , $Ba_{1-x}Sr_xTiO_3$ have been investigating as a most promising candidates for microelectronic application like high-density dynamic random access memories and infrared detector, because of its high dielectric constant (ε_r) values, low dielectric loss, good high frequency characteristics and thermal stability. In addition, the nonlinearity dielectric property of $Ba_{1-x}Sr_xTiO_3$ thin films with respect to applied dc voltage make it attractive for tunable microwave devices such as phase shifters, high-Q resonators, tunable filters and varactors[1-2]. The Curie temperature, T_C of $Ba_{1-x}Sr_xTiO_3$ decreases linearly with increasing strontium concentration. Thus the transition temperature, electrical and optical properties of $Ba_{1-x}Sr_xTiO_3$ can be tailored over a broad range to meet the requirements of various electronic applications. Usually, ferroelectrics thin films with T_C below operating temperature are employed in practical device applications, since the ferroelectric materials has lower dielectric loss in paraelectric states due to the disappearance of hysteresis. Therefore the reason why $Ba_{1-x}Sr_xTiO_3$ with composition around $Ba_{0.5}Sr_{0.5}TiO_3$ (BST) has been widely used for room temperature applications[3].

Tunable ferroelectric devices showed the advantages compared to ferrites and semiconductor, i.e. low losses, low operation voltage, and fast switching times. Ferroelectrics thin films offer the additional advantages of lightweight, compactness and lower processing temperatures. The size of the tunable microwave devices can be effectively reduced with using high dielectric constant materials because wavelength (λ) in dielectrics is inversely proportional to $\sqrt{\varepsilon_r}$ of the wavelength (λ_o) in vacuum $\left(\lambda = \lambda_0 / \sqrt{\varepsilon_r}\right)$. Beside that, the inverse of the dielectric loss $(Q = 1/\tan\delta)$ is required to be high for achieving prominent impedance matching and stability in microwave transmitter components. Thus high dielectric tenability and low dielectric loss are the basic requirement for tunable microwave devices.

One of the critical issues for practical device application of ferroelectric materials is the reduction of the dielectric losses[2-4]. Recently, many efforts have been put-in to reduce the dielectric losses, and found that doping of other oxides with low dielectric losses into BST is an effective way to reduce the dielectric losses. Previous works tried to reduce the dielectric loss of $Ba_{0.5}Sr_{0.5}TiO_3$ ceramic, thick and thin films, though adding MgO and ZrO_2 and reported that MgO was the best dopant in terms of reducing dielectric losses, which, therefore, led to extensive studies on the doping effect of MgO on the dielectric tunable properties of $Ba_{1-x}Sr_xTiO_3$. However, less information is available as to whether complex oxides can be used as an alternative additive like MgO to improve dielectric properties of microwave tunable ferroelectrics. Unfortunately, the dielectric loss $(\tan\delta)$ of the BST is normally high, which is around 0.03 for $x=0.5$ in the present study on pulsed laser deposited thin film. The reported values of $\tan\delta$ for BST from different research groups are varied.

In this study, we chose to address the problem of dielectric loss by doping the BST with Al_2O_3. Al_2O_3 is chosen as dopant due to its low microwave dielectric loss. The dielectric properties of doped $Ba_{0.5}Sr_{0.5}TiO_3$ (BST) film were measured by a home-made microstrip dual resonator near 7.7 GHz.

The BST target with diameter 2.5 cm was prepared using $BaTiO_3$, $SrTiO_3$ powders with ratio 1:1, via the conventional ceramic processing. A KrF excimer pulsed laser with pulse energy 250 mJ and repetition rate of 5 Hz was employed to deposited the Al_2O_3 doped BST thin films on (100) $LaAlO_3$ single crystal substrates with size of $10\times5\times0.5$ mm^3. In order to get the different dopant amount in doped BST films, BST target with Al_2O_3 plates on its surface was employed in the film deposition. The optimization deposition condition has been investigated and found that high quality epitaxial Al_2O_3 doped BST thin films can be achieved with the substrate temperature of 650 °C and an oxygen pressure of 0.2 mbar. The depositions were carried out for duration 45 minutes, and a post-deposition *in-situ* annealing was done in the PLD chamber right after the deposition process at the same temperature but with a higher oxygen pressure of 1 atm. The as-grown films were measured around 500nm thick by using SEM. And also the optimum distance between substrate and target was 4.5 cm for this temperature and pressure. The doped BST films with different Al_2O_3 content are deposited by different coverage area of Al_2O_3 over the BST target. The films produced from the targets with 1/10, 1/5, and 2/5 Al_2O_3 coverage of the BST target were abbreviated as BST1, BST2 and BST3. BST film without doping was also deposited for comparison. The relative concentration of Al_2O_3 content in BST1: BST2: BST3 were found to be 1: 4: 9 by using Rutherford Backscattering Analysis.

A Philips PW 1729 type X-ray diffractometer with Cu K_α radiation was employed for determine the structural phase composition and crystallization of the Al_2O_3 doped BST thin films. The dielectric properties of Al_2O_3 doped BST films in terms of ε_r and $\tan\delta$ were measured by a home-made non-destructive microstrip dual-resonator method at room temperature and microwave frequency near 7.7 GHz. The microstrip dual-resonator consisted of two planar half-wavelength resonators coupled through a gap of 36 μm. The film under test was placed on top of the microstrip circuit and covering the gap between two microstrip resonators. Finally the value for ε_r and $\tan\delta$ of the films derived from the resonant frequencies f_1, f_2 and the corresponding quality factor Q_1, Q_2, of the microstrip dual-resonator[5]. In order to study the electric field dependence of the ferroelectric thin films, a maximum dc voltage of 2.1 kV was applied through two electrode pads on the microstrip circuit board across a gap of about 2.6 mm, corresponding to a maximum electric field of ~8.1 kV/cm.

XRD patterns of the deposited BST films shown in Fig. 1 and display that the intensity peak was broadened with increasing Al_2O_3 dopant. This means the grain size of the Al_2O_3 doped BST films decreased with increasing Al_2O_3 content. The XRD does not detect Al_2O_3 because relatively small amount of Al_2O_3 compared with BST. However, two unidentified peaks were observed (marked ♦ as in Fig. 1) where are attributed to a second phase caused by the excessive Al_2O_3 content for the BST3 film.

Fig. 2 shows the dielectric constant of the doped BST films as a function of applied electric field. The ε_r of Al_2O_3 doped BST films decrease with increasing Al_2O_3 content, which pure BST films (ε_r=1622), BST1(ε_r=1387), BST2(ε_r=1311) and BST3(ε_r=870). All thin films are epitaxially c-axis oriented, which are indicated by (100) and (200) peaks in XRD patterns, and thus produced high ε_r values since highly c-axis oriented films provide a strong polarization direction and tends to form a concentrated polarization which results in higher ε_r compared to randomly oriented samples. The $\tan\delta$ of Al_2O_3 doped BST films as a function of the applied electric field showed in Fig. 3. The $\tan\delta$ have significantly decreased as 0.030 to 0.011, corresponding to the samples BST and BST3, respectively.

Tunable microwave device applications require a high dielectric constant and low dielectric loss. The tunability, loss tangent and T_C of pure BST thin films are directly proportional to the barium/strontium ratio. We have calculated dielectric tunability and reported that dielectric tunability was reduced from 22.0% (BST thin film) to 15.9% (BST3). In order to compare the quality of ferroelectric films for tunable microwave devices application, usually a parameter namely figure-of-merits (K) will be employ, where the K defined as:-

$$K = \left[\frac{tunability}{\tan\delta}\right] = \left[\frac{\left(\varepsilon_{r_0} - \varepsilon_{r_b}\right)/\varepsilon_{r_0}}{\tan\delta}\right] \tag{1}$$

with ε_{r_0} is the dielectric constant value at zero applied electric field and ε_{r_b} represent the dielectric constant for the case maximum applied electric field. By using Eq. 1, we found that K factors is increased with Al_2O_3 content, with K=7.33 (BST) and K=14.5 (BST4).

In summary, highly c-axis oriented Al_2O_3 doped BST films have been deposited on $LaAlO_3$ single-crystal substrate, by PLD technique. We reported that K factors of BST films were improved significantly by doped Al_2O_3 into BST films. Thus ε_r, $\tan\delta$ and dielectric tunability of the Al_2O_3 doped BST thin films have decreased due to increasing Al_2O_3 content. We conclude that dielectric properties of BST ferroelectric thin films be improved for microwave device applications with Al_2O_3 doping.

References

1. C. L. Chen et al., Appl. Phys. Lett. **75**, 412 (2001).
2. L. C. Sengupta and S. Sengupta, IEEE Trans. Ultrason. Ferroelect. Freq. Control. **44**, 792 (1997).
3. C. L. Huang et al., Mater. Res. Bull. **37**, 449 (2002).
4. M. W. Cole et al., J. Appl. Phys. **93**, 9218 (2003).
5. C. Y. Tan et al, Rev. Sci. Instrum. (submitted).

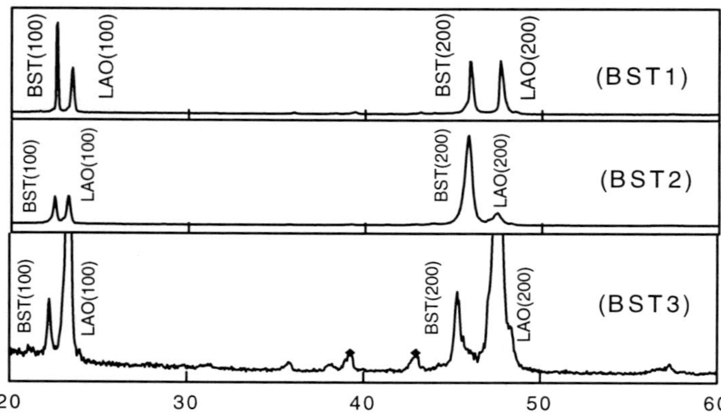

FIG. 1 XRD patterns of the Al_2O_3 doped BST films on $LaAlO_3$ substrates

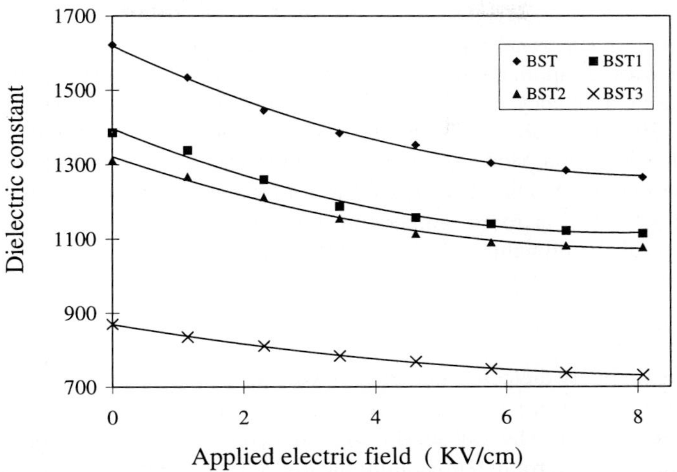

FIG. 2 Dielectric constant of the doped BST films as a function of applied electric field.

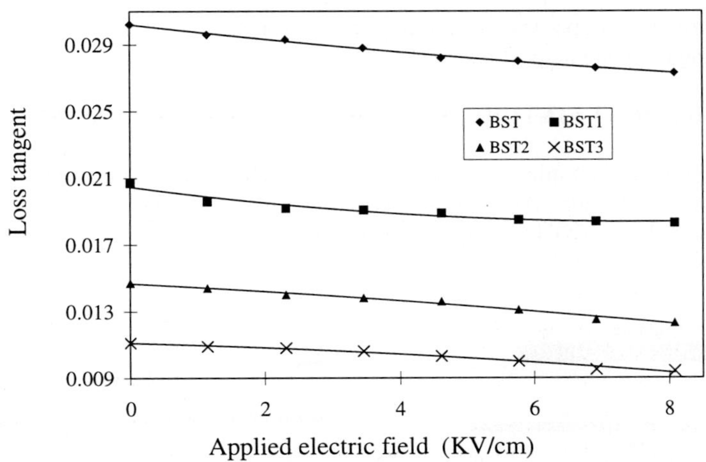

FIG. 3 Loss tangent of the doped BST films as a function of applied electric field.

Temperature Dependence of Relative Permittivity and Loss Tangent of Novel Transparent Substrate Material at Microwave Frequencies

J. Mazierska, M. V. Jacob, T. F. Au Yeung

Electrical and Computer Engineering, James Cook University, Townsville, QLD 4811, Australia

Abstract — Recently a transparent dielectric substrate based on $Ba(Sn,Mg,Ta)O_3$ material has been developed for microwave applications. We measured the real relative permittivity of this transparent substrate at frequency of 3.2 GHz using a split post dielectric resonator technique at temperatures from 230 K to 300K, and a high resolution cryogenic post resonator at 9.2GHz from 38K to 88K. The Transmission Mode Q factor Technique was used for data processing to ensure high accuracy of the measurements. Measured values of ε_r' and $\tan\delta$ of transparent BSMTO turned out to be 25.2 and 3.45×10^{-5} at room temperature respectively. The $Ba(Sn,MgTa)O_3$ transparent substrate may be useful for miniaturised planar antenna arrays and other microwave applications where transparent substrates are needed.

I.   INTRODUCTION

In the past twenty years many dielectric materials have been developed with the dielectric constant (ε_r') varying from 1.03 to 110 for diverse microwave applications like dielectric filters, substrates for MMICs and antennas [1]. The $Ba(Sn,MgTa)O_3$ material has been developed as a dielectric resonator in the eighties [2]. At room temperature the dielectric exhibited ε_r' of 24.2, $\tan\delta$ of 3.9×10^{-5} at frequency of 6.5GHz and τ_f between 0 and 6ppm/°C [3]. Recently a transparent substrate has been developed based on the $Ba(Sn,MgTa)O_3$ dielectric resonator material [4].

Precise knowledge of microwave properties of any dielectric used is necessary for efficient design of microwave systems containing these dielectrics. Usually the real part of the complex permittivity (ε_r'), loss tangent $(\tan\delta)$ and temperature dependence of the resonant frequency $(\Delta f_r/f_r$ or $\tau_f)$ is given to describe usefulness of a given dielectric material [2, 3, 5]. Measurements of ε_r', $\tan\delta$ and $\Delta f_o/f_o$ of low loss planar dielectrics can be performed using the cavity perturbation and the dielectric resonator (post or split post) perturbation techniques [6].

In this paper microwave properties of the transparent substrate based on $Ba(Sn,MgTa)O_3$ at varying temperatures using a cryogenic post resonator and a split post dielectric resonator are presented.

II.   MEASUREMENT FIXTURE, SYSTEM AND DATA PROCESSING METHOD

The transparent substrate sample under test (75mm diameter and 0.61mm thickness) was inserted into a microwave resonator. As mentioned earlier two dielectric resonators were used for measurements, namely a 9.2GHz cryogenic post resonator and a split post structure resonating at 3.2GHz, and as shown in Fig. 1.

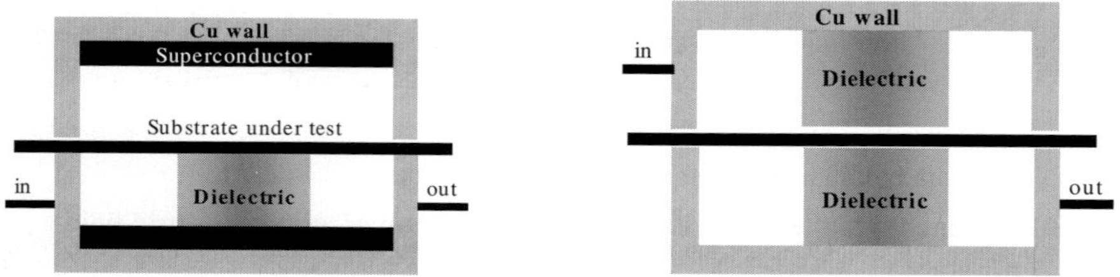

Fig. 1 Schematic diagrams of (a) Cryogenic Post resonator and (b) Split Post dielectric resonator

The resonators were especially designed to achieve high sensitivity. Additionally the post resonator has endwalls made of High Temperature Superconducting thin films (DyBa$_2$Cu$_3$O$_7$ films of 700 nm thickness on MgO substrate) [7] to increase the resolution of the measurements. The real part of permittivity of the substrate under test is calculated as [8]:

$$\varepsilon_r = 1 + \frac{f_0 - f_r}{h f_0 K_\varepsilon (\varepsilon_r{}', h)} \tag{1}$$

where f$_0$ and f$_r$ are resonant frequencies of the resonator without and with the sample, h is thickness of the sample, K_ε is a function of ε_r' and h. Details of calculation of the different coefficients and permittivity is described in reference [8]

The loss tangent of a tested substrate was determined from the measured unloaded Q$_o$-factor of the resonator with and without the sample. For Split post-dielectric resonator, the loss tangent is calculated using the equation:

$$\tan\delta = (Q_o^{-1} - Q_{DR}^{-1} - Q_c^{-1}) / p_{es} \tag{2}$$

and for cryogenic post resonator, the loss tangent is calculated using the equation [8]:

$$\tan\delta = (Q_o^{-1} - Q_{DR}^{-1} - Q_{c1}^{-1} - Q_{c2}^{-1}) / p_{es} \tag{3}$$

where p_{es} is the electric energy filling factor of the sample, Q$_{DR}$ is the Q-factor of the post dielectric, Q$_c$ or Q$_{c1}$ and Q$_{c2}$ represent the conductor losses of metal and superconducting parts of the resonator. Equations to calculate Q$_{DR}$, Q$_{c1}$ and Q$_{c2}$ are given in [7, 8].

To determine the resonance frequency and Q-factor of the resonator, a measurement system consisting of Network Analyser (HP 8722C), closed cycle refrigerator (APD DE-204), temperature controller (LTC-10), vacuum Dewar, a PC and one of the two resonators was used. To ensure accurate values of the Q-factor, we have used the TMQF technique [9] to remove noise, crosstalk and uncalibrated cables and adaptors from the measured data. Obtained results of ε_r, tanδ and $\Delta f_r / f_r$ of the Ba(Sn,MgTa)O$_3$ based transparent substrate are given below in Section III.

III. MEASUREMENT RESULTS OF THE TRANSPARENT SUBSTRATE

Computed values of ε_r and tanδ based on measurements with the Split Post dielectric resonator at frequency of 3.2 GHz in the temperature range from 230K to 293K are shown in Fig.2. Measured dielectric constant and loss tangent at frequency of 9.2 GHz using the high resolution cryogenic post resonator in the temperature range from 38K to 86K are shown in Fig. 3.

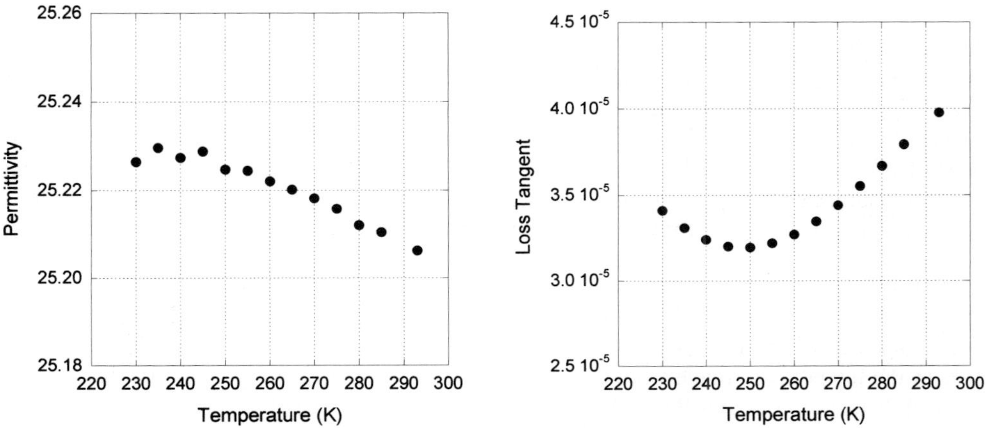

Fig. 2. Real relative permittivity and loss tangent of transparent BSMTO measured at 3.2GHz

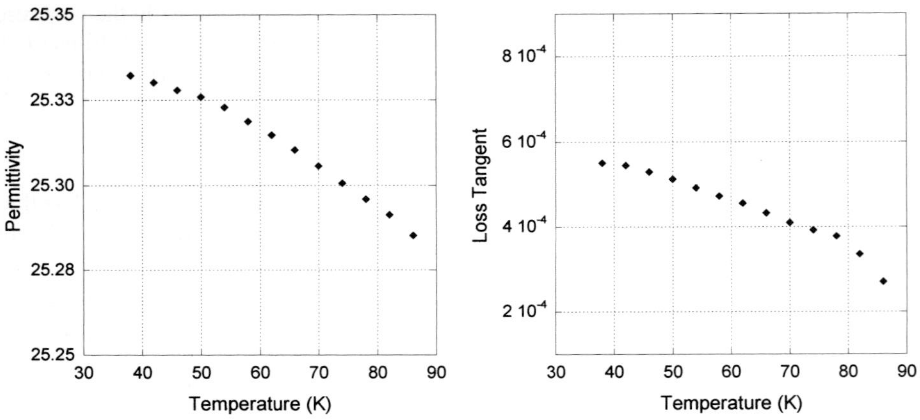

Fig.3 Real part of relative permittivity and loss tangent of transparent BSMTO measured at 9.2 GHz

Fig. 4 presents combined results of the real relative permittivity and loss tangent of the transparent substrate. Loss tangent values measured at 38-86K and 9.2 GHz were scaled to 3.2 GHz using the linear frequency dependence.

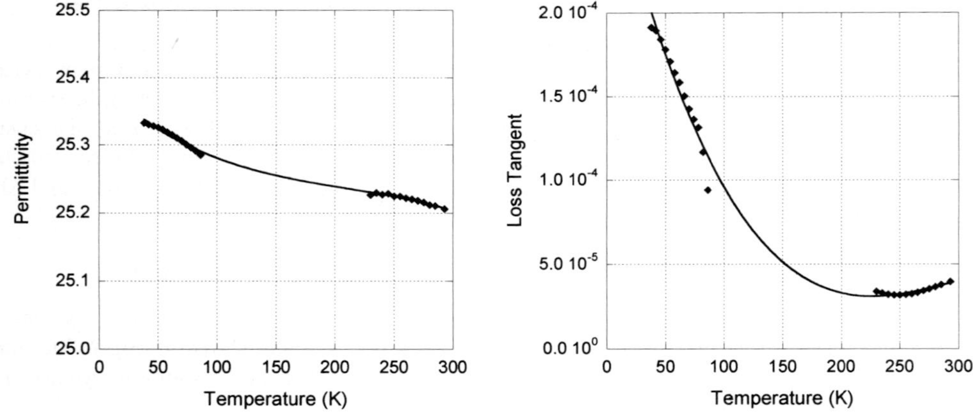

Fig. 4 Combined measured ε_r and $\tan\delta$ of transparent BSMTO at 3.2GHz

When comparing measured properties of the transparent substrate under test with values of the standard dielectric resonator $Ba(Sn,MgTa)O_3$ material [2] one can find a close similarity in values of the real part of relative permittivity. The room temperature values of $\tan\delta$ are also similar. However the temperature dependence of the loss tangent is very different; for the transparent BSMTO $\tan\delta$ decreases strongly with the increase in temperature at cryogenic temperatures while for the resonator BSMTO loss tangent increases with temperature. Also values of $\tan\delta$ at very low temperatures are much higher for the transparent substrate than for the resonator material (less than 1×10^{-4}).

A measured dependence of the resonant frequency for both test resonators containing the transparent substrate under test is shown in Fig. 5. The parameter $\Delta f_r/f_r$ is approximately 12.6 ppm/K in the temperature range from 230K to 300K, and 7 ppm/K at cryogenic temperatures. For the typical dielectric resonator $Ba(Sn,MgTa)O_3$ material, τ_f is between 0 and 6ppm/K [3].

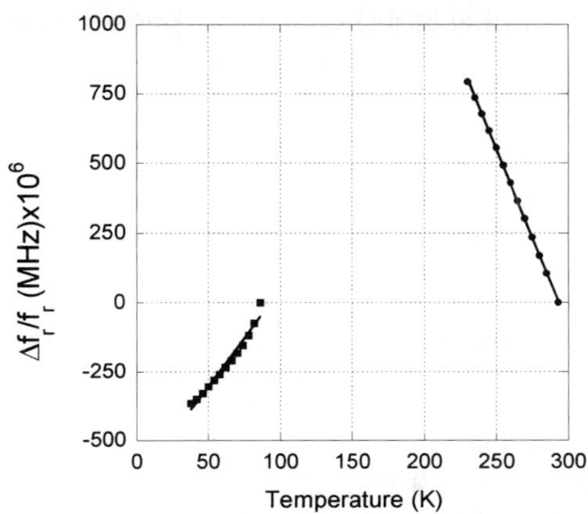

Fig. 5 Frequency change of BSMTO based transparent substrate as a function temperature.

IV CONCLUSIONS

A recently developed transparent substrate based on $Ba(Sn,MgTa)O_3$ dielectric has been tested at microwave frequencies as a function of temperature. Measured values of ε_r' of the transparent BSMTO vary from 25.34 to 25.21 in the temperature range 38-300K. The loss tangent is the highest at lowest temperatures (1.9×10^{-4} at 38K) and decreases to 3.45×10^{-5} at room temperature. The temperature dependence of the resonant frequency of the cryogenic post resonator with the substrate under test is approximately 7ppm/K and 12.5ppm/K for the split post resonator. This low loss medium permittivity transparent substrate may be very suitable for several microwave applications.

Acknowledgements: The authors would like to thank Dr. H.Tamura and Murata Mfg. Ltd. for supplying samples and information. Also the support of ARC Large grant A00000799 and JCU CRIG fellowship is acknowledged.

REFERENCES
[1] [online] M.T. Sebastian and N. Mc N Alford, "List of microwave dielectric resonator materials and their properties", www.sbu.ac.uk/dielectric-materials, accessed on 10th August 2003.
[2] H. Tamura, H. Matsumoto and K.Wakino, "Low temperature properties of microwave dielectrics", *Japanese Journal of Applied Physics*, vol. 28, pp.21-23, 1989.
[3] Y. Higuchi and H. Tamura, "Recent progress on dielectric resonator materials with their applications from microwave to optical frequencies", *Journal of European Ceramic Society* (in press).
[4] H. Tamura, MURATA Mfg Ltd, private communication, 2003.
[5] A. N. Luiten, M. E. Tobar, J. Krupka, R. Woode, E. N. Ivanov and A. G. Mann, "Microwave properties of a rutile resonator between 2 and 10K", *Journal of Physics D: Applied Physics*, vol. 31, pp. 1383-1391, 1998.
[6] J. Krupka and M. D. Janezic, "Nondestructive permittivity measurements of low loss substrates", *IMAPS, Workshop on Ceramic Applications for Microwave & Photonic Packaging*, Providence, Rhode Island, April 2002.
[7] M. V. Jacob, J. Mazierska, J. Krupka, "Cryogenic post dielectric resonator for precise microwave characterization of planar dielectric materials for superconducting circuits" (to be published).
[8] J. Krupka, A. P. Gregory, O. C. Rochard, R. N. Clarke, B. Riddle, J. Baker-Jarvis, "Uncertainty of complex permittivity measurements by SPDR resonator technique" *Journal of the European Ceramic Society*, vol. 21, pp. 2673-2676, 2001.
[9] K. Leong and J. Mazierska, "Precise measurements of the Q-factor of transmission mode dielectric resonators: accounting for noise, crosstalk, coupling loss and reactance, and uncalibrated transmission lines", *IEEE Trans. on Microwave Theory and Technique*, vol 50, pp 2115-2127, 2002.

Deposition and characterization of BaZr$_x$Ti$_{1-x}$O$_3$ ferroelectric thin films for microwave tunable applications

L. B. Kong[1], L. Yan[2], K. B. Chong[2], C. Y. Tan[2], L. F. Chen[1] and C. K. Ong[2]

[1]*Temasek Laboratories, National University of Singapore,*
10 Kent Ridge Crescent, Singapore 119260
[2]*Center for Superconductor and Magnetic Materials, National University of Singapore,*
Lower Kent Ridge Crescent, Singapore 119260

Abstract

BaZr$_x$Ti$_{1-x}$O$_3$ (BZT, x=0.20, 0.25, 0.30 and 0.35) ferroelectric thin films were deposited on Pt/Ti/SiO$_2$/Si and (100) LaAlO$_3$ substrates via a pulsed laser deposition (PLD). The deposited BZT thin films were characterized by X-ray diffraction (XRD), scanning electron microscopy (SEM) and electrical measurement. The BZT thin films were of single phase perovskite structure. The films deposited on the Pt/Ti/SiO$_2$/Si substrate were polycrystalline, with a zero field dielectric constant of 641, 507, 485 and 419, and a maximum dielectric tunability of 59%, 68%, 65% and 53%, for the BaZr$_x$Ti$_{1-x}$O$_3$ films of x=0.20, 0.25, 0.30 and 0.35, respectively. The films deposited on the LaAlO$_3$ substrate were epitaxial. BaZr$_{0.25}$Ti$_{0.75}$O$_3$ thin film on LaAlO$_3$ possessed a zero field dielectric constant of 413, a dielectric loss of 0.01 and a maximum dielectric tunability of 18%.

1. Introduction

Ferroelectric materials have been acknowledged to be promising candidates for microwave device applications, such as tunable oscillators, phase shifters and varactors, due to the strong dependence of their dielectric constants on the electric field, and fast responses [1-3]. Barium strontium titanate (Ba$_{1-x}$Sr$_x$TiO$_3$ or BST) has been extensively investigated for tunable device applications [4-6]. There were also other perovskite ABO$_3$ type materials, such as KTaO$_3$ [7], K$_{0.5}$Na$_{0.5}$NbO$_3$ [8] CaTiO$_3$ [9] and BaZr$_x$Ti$_{1-x}$O$_3$ [10], with high tunability and reasonably low dielectric loss. Among these materials, BaZr$_{0.3}$Ti$_{0.7}$O$_3$ was reported to be the most promising candidate in terms of microwave tunable properties. However, those properties were only available for BaZr$_{0.3}$Ti$_{0.7}$O$_3$ ceramic. Although BaZr$_x$Ti$_{1-x}$O$_3$ thin films were prepared by sol-gel [11] and pulsed laser deposition (PLD) [12] methods, less information is available on their dielectric tunable characteristics. In this paper, preparation and characterization of BaZr$_x$Ti$_{1-x}$O$_3$ (BZT, x=0.20, 0.25, 0.30 and 0.35) ferroelctric thin films, via a pulse laser deposition, will be reported.

2. Experimental procedure

BaZr$_x$Ti$_{1-x}$O$_3$ (x=0.20, 0.25, 0.30 and 0.35) targets with a diameter of about 2.5 cm, were prepared using commercial BaCO$_3$, ZrO$_2$ and TiO$_2$ powders, via the conventional ceramic processing. The powders were thoroughly mixed and calcined at 1300°C for 4 h. The calcined mixture was crushed, compacted and finally sintered at 1400°C for 4 h. All targets were characterized by X-ray diffraction to be of single phase perovskite structure (result not shown).

BaZr$_x$Ti$_{1-x}$O$_3$ thin films were deposited on Pt/Ti/SiO$_2$/Si and (100) LaAlO$_3$ single crystal substrate, via a PLD with a KrF excimer laser at 5 Hz repetition frequency, with an energy of 250 mJ/pulse. The deposition was conducted for 45 minutes, at a substrate temperature of 650°C and a chamber oxygen pressure of 0.2 mbar.

The deposited BaZr$_x$Ti$_{1-x}$O$_3$ thin film films were characterized by X-ray diffraction (XRD) using a Philips PW 1729 type X-ray diffractometer with Cu K$_\alpha$ radiation. For electrical measurement, Au top electrode with a diameter of 0.2 mm was deposited to form a sandwich structure, using a JEOL magnetic sputtering equipment. Low frequency (100 kHz) dielectric properties of the BZT thin films on the Pt/Ti/SiO$_2$/Si substrate were measured using an HP4194A LCR analyzer. A sandwich structure with a Au top electrode was made for the low frequency dielectric characterization. Microwave dielectric constant of the films on LaAlO$_3$ substrate was measured using a non-destructive microstrip split resonator method at ~7.9 GHz, which was based on the original design of Galt *et al.* [13-15]. The microstrip split resonator, formed by a straight microstrip line with a 30 μm gap in the center, was patterned on a TMM10i microwave substrate. The films were placed on top of the microstrip line, covering the gap. The dielectric constant ε and loss tangent $tan\delta$ of the film were derived from the odd-mode resonant frequency f_0, and quality factor Q, of the microstrip split resonator. For the study of the electric field dependence of the LAO-BST thin films, a maximum dc voltage of 2.4 kV is applied through two electrode pads on the microstrip circuit board across a gap of about 2.6 mm, corresponding to a maximum electric field of ~9.2 kV/cm.

3. Results and discussion

Fig. 1 shows the XRD patterns of the BZT thin films deposited on Pt/Ti/SiO$_2$/Si substrate. It is noticed that the samples are all polycrystalline of single phase perovskite structure, which is evidenced by the strongest diffraction peak of (110). All diffraction peaks coming from BZT thin films are indexed in terms of pseudocubic structure. Peaks from the Pt electrode and substrate are also indicated in the figure. By carefully examining the (110) diffraction peak (inset in Fig. 1), one can notice that it shifts to lower 2θ angle. The peak shift to lower diffraction angle indicates an increase in unit cell. This is expected since the ionic radius of Zr^{4+} (0.86 Å) is larger than that of Ti^{4+} (0.745 Å) [11]. Therefore, replacement of Ti^{4+} with Zr^{4+} should result in an increased unit cell of the BZT thin films.

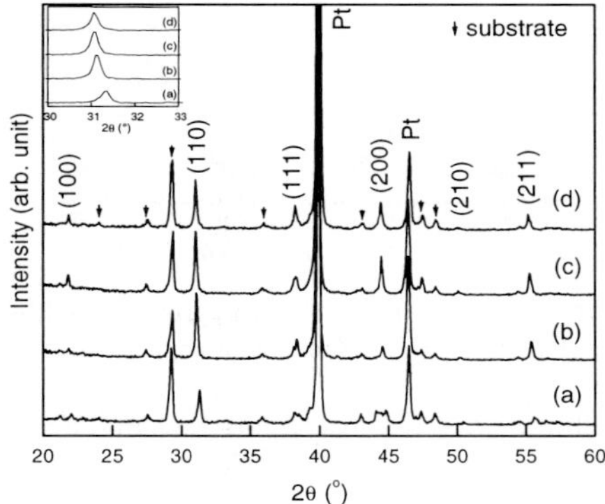

Fig. 1 XRD patterns of the BaZr$_x$Ti$_{1-x}$O$_3$ thin films on Pt/Ti/SiO$_2$/Si: (a) 0.20, (b) 0.25, (c) 0.30 and (d) 0.35

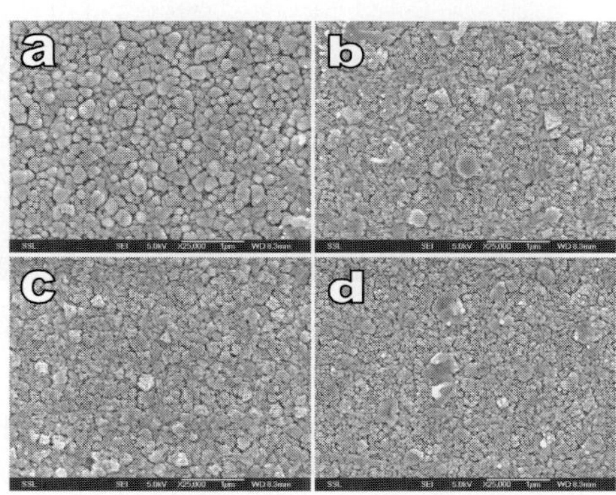

Fig. 2 SEM images of the BaZr$_x$Ti$_{1-x}$O$_3$ thin films on Pt/Ti/SiO$_2$/Si: (a) 0.20, (b) 0.25, (c) 0.30 and (d) 0.35.

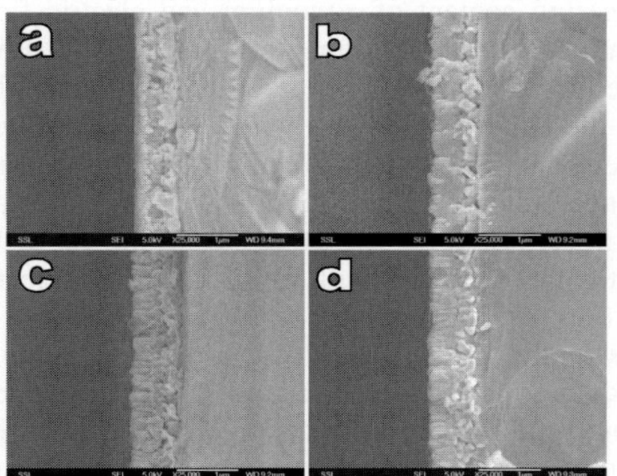

Fig. 3 Cross-sectional SEM images of the BaZr$_x$Ti$_{1-x}$O$_3$ thin films on Pt/Ti/SiO$_2$/Si: (a) 0.20, (b) 0.25, (c) 0.30 and (d) 0.35.

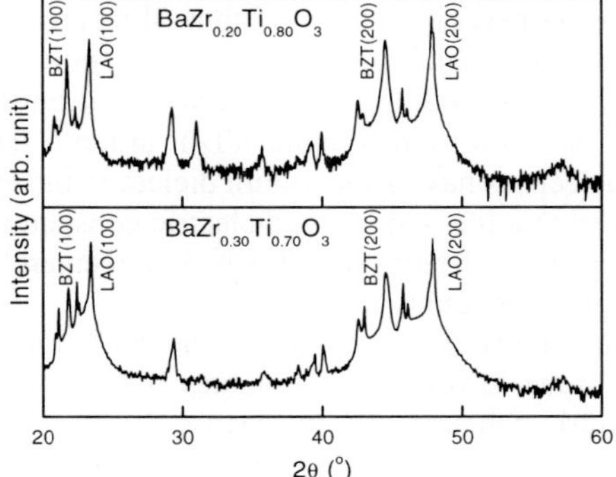

Fig. 4 XRD patterns of the BaZr$_x$Ti$_{1-x}$O$_3$ thin films on LaAlO$_3$.

SEM images of the BZT thin films are shown in Fig. 2. The grain size of the BZT thin film decreases slightly with increasing zirconium content. Their thicknesses, estimated from the cross-sectional SEM images shown in Fig. 3, are about 0.4 µm in all cases, indicating a similar deposition rate of the four compositions in the present study.

Representative XRD patterns of the BZT thin films ($BaZr_{0.20}Ti_{0.80}O_3$ and $BaZr_{0.30}Ti_{0.70}O_3$) grown on the $LaAlO_3$ substrate are shown in Fig. 4. It shows that the BZT thin films are expitaxially grown on the $LaAlO_3$ substrate. Our result is slightly different from that reported by Dixit *et al.* [11], who used sol-gel process to deposit BZT thin films on $LaAlO_3$ substrate. In that case, randomly oriented BZT thin films were obtained for the compositions of $BaZr_xTi_{1-x}O_3$ with x<0.3, while highly textured samples could only be achieved with x≥0.3. This difference in the preferred orientation of the BZT thin film may be attributed to the different deposition methods used in the present work and that reported. The microstructures of the BZT thin films on $LaAlO_3$ are significantly different from those of the films deposited on $Pt/Ti/SiO_2/Si$. SEM image of $BaZr_{0.25}Ti_{0.75}O_3$ film deposited on $LaAlO_3$ substrate is shown in Fig. 5. Uniformly distributed grains with narrow grain size distribution are observed in the epitaxially grown BZT thin films.

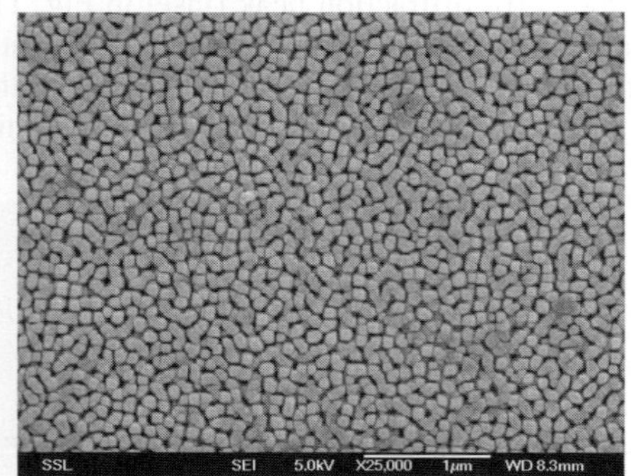

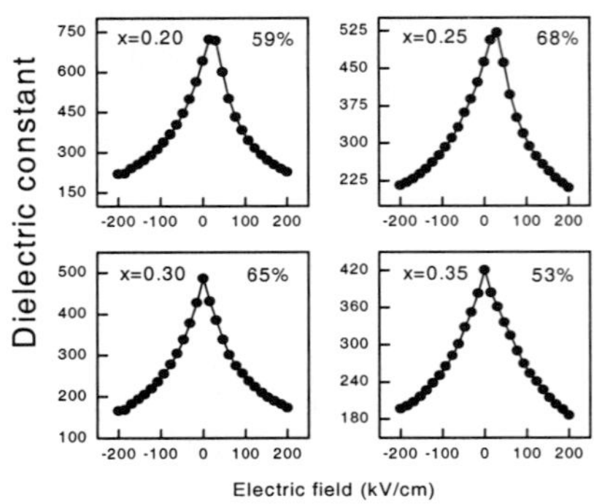

Fig. 5 SEM image of $BaZr_{0.25}Ti_{0.75}O_3$ thin film on $LaAlO_3$.   Fig. 6 Dielectric constant of the $BaZr_xTi_{1-x}O_3$ thin films on $Pt/Ti/SiO_2/Si$ as a function of applied field.

Low frequency dielectric constants, of the BZT thin films deposited on the $Pt/Ti/SiO_2/Si$ substrate, as a function of electric field, are plotted in Fig. 6. The zero field dielectric constants, of the $BaZr_xTi_{1-x}O_3$ films with x=0.20, 0.25, 0.30 and 0.35, are 641, 507, 485 and 419, respectively. Their maximum dielectric tunabilities are correspondingly 59%, 68%, 65% and 53%, at a maximum electric field of 200 kV/cm. The dielectric losses of all four thin films are around 0.02 at zero electric field, showing a slight decrease with increasing content of zirconium and electric field. These properties of the BZT thin films are comparable to that of barium strontium titanate (BST) based films, which have been widely reported in the literature [2-4]. The larger dielectric tunability observed for x=0.25 and 0.30 may be attributed to the fact that the films with these compositions have a Curie temperature (T_C) around measurement (room) temperature (~20°C), because ferroelectrics have a maximum dielectric tunability at around phase transition temperature. It is also noticed that the maximum dielectric constant is not at zero field for x=0.20 and 0.25, indicating the presence of a hysteresis. This is in agreement with the fact that BZT with x≤0.25 is in ferroelectric state at room temperature.

The dielectric constants of the BZT thin films in the present study are almost an order of magnitude lower than the values for BZT ceramics reported in the literature [10]. The reduced dielectric constant of ferroelectric thin films, as compared to their bulk ceramic counterparts, has been mainly attributed to the formation of a nonferroelectric interface layer between the film and substrate. However, other factors, such as defects, stress and small grain size, are also considered to be responsible for the reduction in dielectric constant.

The dielectric constant as a function of electric field of $BaZr_{0.25}Ti_{0.75}O_3$ film on $LaAlO_3$ substrate is shown in Fig. 7, with a zero field dielectric constant of 413, a dielectric loss of 0.01 and a maximum dielectric tunability of 18%. Since the dielectric tunability of the samples on $LaAlO_3$ substrate was measured at a maximum electric field of ~9.2 kV/cm, it is possible to achieve higher dielectric tunability sufficient enough for practical applications, if the electric field can be increased.

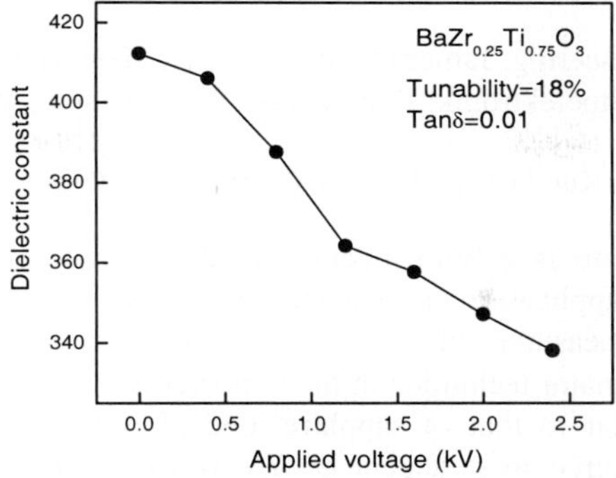

Fig. 7 Dielectric constant of BaZr$_x$Ti$_{1-x}$O$_3$ thin films on LaAlO$_3$ as a function of applied voltage.

4. Conclusions

BaZr$_x$Ti$_{1-x}$O$_3$ (BZT, x=0.20, 0.25, 0.30 and 0.35) ferroelectric thin films, with a thickness of ~0.4 μm, were deposited on Pt/Ti/SiO$_2$/Si and (100) LaAlO$_3$ substrates via a pulsed laser deposition (PLD). The BZT thin films were all of single phase perovskite structure. The films deposited on the Pt/Ti/SiO$_2$/Si substrate were polycrystalline, whereas those on the LaAlO$_3$ were highly textured. The films on Pt/Ti/SiO$_2$/Si had a zero field dielectric constant of 641, 507, 485 and 419, and a maximum dielectric tunability of 59%, 68%, 65% and 53%, for the BaZr$_x$Ti$_{1-x}$O$_3$ films of x=0.20, 0.25, 0.30 and 0.35, respectively. BaZr$_{0.25}$Ti$_{0.75}$O$_3$ thin film on LaAlO$_3$ possessed a zero field dielectric constant of 413, a dielectric loss of 0.01 and a maximum dielectric tunability of 18% at a maximum electric field of ~9.2 kV/cm and frequency of 7.9 GHz. It can be concluded that BaZr$_x$Ti$_{1-x}$O$_3$ thin film is a promising candidate for microwave tunable device applications.

References
1.   C. H. Mueller, R. R. Romanofsky and F. A. Miranda, IEEE Potentials **20**, 36 (2001).
2.   S. S. Gevorgian and E. L. Kollberg, IEEE Trans. Microwave Theory Tech. **49**, 217 (2001).
3.   D. S. Korn and H. D. Wu, Integr. Ferroelectri. **24**, 215 (1999).
4.   S. Gevorgian, E. Carlsson, E. Wikborg and E. Kollberg, Integrated Ferroelect. **22**, 765 (1998).
5.   F. Deflaviis, N. G. Alexepolous and O. M. Staffsudd, IEEE Trans. Microwave Theory Tech. **45**, 963 (1997).
6.   S. W. Kirchoefer, J. M. Pond, A. C. Carter, W. Chang, K. K. Agarwal, J. C. Horwitz and D. B. Chrisey, Microwave Opt. Tech. Lett. **18**, 168 (1998).
7.   C. Ang, A. S. Bhalla, R. Guo, L. C. Cross, Phys. Rev. B **64**, 184104 (2001).
8.   S. Abadei, S. Gevorgian, C. R. Cho, A. Grishin, J. Appl. Phys. **91**, 2267 (2002).
9.   J. Hao, W. Si, X. X. Xi, R. Guo, A. S. Bhalla, L. E. Cross, Appl. Phys. Lett. **76**, 3100 (2000).
10. Z. Yu, C. Ang, R. Guo, A. S. Bhalla, Appl. Phys. Lett. **81**, 1285 (2002).
11. A. Dixit, S. B. Majumder, A. Savvinov, R. S. Katiyar, R. Guo, A. S. Bhalla, Mater. Lett. **56**, 933 (2002).
12. S. Halder, S. B. Krupanidhi, Solid State Comm. **122**, 429 (2002).
13. D. Galt, J. C. Price, J. A. Beall and R. H. Ono, Appl. Phys. Lett. **63**, 3078 (1993).
14. D. Galt, J. C. Price, J. A. Beall and T. E. Harvey, IEEE Trans. Appl. Supercond. **5**, 2575 (1995).
15. C. Y. Tan, L. F. Chen, K. B. Chong, Y. T. Ngow, Y. N. Tan and C. K. Ong., Rev. Sci. Instrum. (submitted).

Microwave Properties of Yttrium Vanadate Crystals at Cryogenic Temperatures

Mohan V. Jacob[a], Janina Mazierska[a], Jerzy Krupka[b], Dimitri Ledenyov[a] and Seiichi Takeuchi[c]

[a]Electrical and Computer Engineering, James Cook University, Townsville, QLD 4811, Australia
[b]Instytut Mikroelektroniki i Optoelektroniki Politechniki Warszawskiej, Koszykowa 75, 00-662 Warszawa, Poland
[c] Tokyo Denki University, 2-2, Kanda-Nishiki-Cho, Chiyoda-ku, Tokyo, 101-8457, Japan

Abstract — Yttrium Vanadate is a birefringent crystal material used in optical isolators and circulators with potentials for application in cryogenic microwave devices. As microwave properties of YVO_4 are not known, we measured the complex permittivity at frequency of 25 GHz using the Hakki-Coleman dielectric resonator technique in the temperature range from 13 K to 80K. The ε_r of YVO_4 turned out to be similar to that of sapphire, one of popular dielectric materials used at microwave frequencies. Measured loss tangent of YVO_4 was of the order of 10^{-6} at cryogenic temperatures. As Yttrium Vanadate is easy to fabricate and machine, it may replace the expensive sapphire in some microwave applications.

I. INTRODUCTION

Yttrium Vanadate (YVO_4), is a birefringent crystal grown by the Czochralski method and is characterised by good mechanical and physical properties [1-3]. YVO_4 is considered very suitable for optical polarizing components due to its wide transparency range and large birefringence. The YVO_4 crystals are mainly used in optical components such as fiber optical isolators and circulators, beam displaces and other polarizing optical systems. Neodymium-doped yttrium vanadate ($Nd:YVO_4$) is one of the most promising commercially available diode-pumped solid-state laser materials for optical communication. The YVO_4 has better thermal stability and physical and mechanical properties than $CaCO_3$, more than three times larger birefringence than $LiNbO_3$, and lower hardness than rutile (TiO_2) what greatly reduces cost of fabrication. Also the crystal growth and fabrication YVO4 are easier than similar birefringent crystals.

Even though YVO_4 is well characterised at optical frequencies, there is hardly any data on microwave properties of this material. In this paper we present for the first time results of precise measurements of the permittivity and loss tangent of YVO_4 crystals fabricated by [3] at frequency of 25GHz and cryogenic temperatures from 13 K to 80 K. We have used a superconducting dielectric resonator technique combined with the multi-frequency Transmission Mode Q-Factor (TMQF) method [4, 5] for data processing to ensure high accuracy of calculated values of ε_r and $\tan\delta$.

II. DIELECTRIC RESONATOR MEASUREMENT METHOD

The dielectric resonator technique can be used for the microwave characterisation of dielectric materials [6-10]. The permittivity and loss tangent of the dielectric material under test can be calculated from the unloaded Q-factor and the resonant frequency of the resonator. In our work the Hakki-Coleman dielectric resonator containing yttrium vanadate under test, schematically shown in Fig. 1, was used as the resonating structure. The resonator consisted of the copper cavity of diameter of 9.5 mm and 3 mm height with High Temperature Superconducting endplates to increase sensitivity and to reduce uncertainty in loss tangent measurements. The YVO_4 sample was machined into a cylinder shape with the aspect ratio (diameter to height) equal to 1.61, with 3.09 mm height and 4.99 mm diameter. The sample was oriented and machined in such a way that its z-axis was parallel to the crystal optical axis with accuracy better than 0.5^0.

Measurements were carried out using the TE_{011} mode of the resonator for which the electric field is perpendicular to z-axis and hence the component of the complex permittivity of YVO_4

perpendicular to the optical axis was determined. The real part of relative permittivity ε_r was determined from measurements of the resonant frequency as the first root of the following transcendental equation [11] using software SUP12 [12].

$$k_{\rho1}J_0(k_{\rho1}b)F_1(b) + k_{\rho2}J_1(k_{\rho1}b)F_0(b) = 0 \qquad (1)$$

The loss tangent (tanδ) of YVO$_4$ was computed from the measured Q_0-factor of the resonator on the basis of the well known loss equation [11], namely:

$$\tan\delta = \frac{1}{\rho_e}\left[\frac{1}{Q_0} - \frac{R_{SS}}{A_S} - \frac{R_{SM}}{A_M}\right] \qquad (2)$$

where Q_0 is the unloaded Q-factor of the resonant structure with the sample, R_{SS} and R_{SM} are the surface resistance of the superconducting and the metallic parts of the cavity respectively, A_S and A_M are the geometric factors of the superconducting part and metallic parts of the cavity and ρ_e is the electric energy filling factor.

Geometric factors A_S, A_M, and ρ_e to be used in (2) were computed using incremental frequency rules [11] and the calculated values for the YVO$_4$ and Sapphire (used for measurement of R_{SS} and R_{SM}) are given in Table I. Details of surface resistance measurement is described in IIIA.

Table I Calculated geometric factors of the dielectric resonator with YVO$_4$ and sapphire rods.

Dielectric	YVO$_4$	Sapphire
f_{res}	24.4 GHz	24.65 GHz
A_S	19593	22319
A_M	291.6	280.6
ρ_e	0.97	0.97

III. MEASUREMENTS OF MICROWAVE PROPERTIES of YVO$_4$ CRYSTALS

The measurement system we used for microwave characterisation of the YVO$_4$ sample is shown in Fig. 2. The system consisted of Network Analyser (HP 8722C), closed cycle refrigerator (APD DE-204), temperature controller (LTC-10), vacuum Dewar, a PC and the Hakki-Coleman dielectric resonator in the transmission mode.

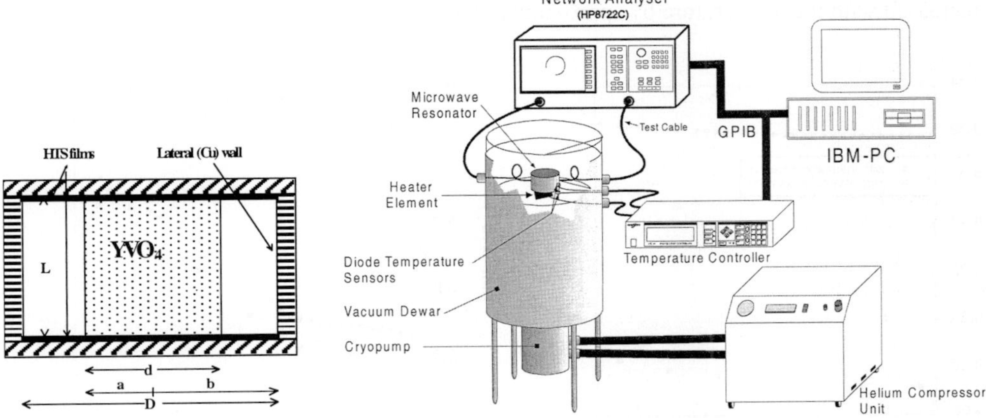

Fig. 1. Hakki-Coleman resonator   Fig. 2 Measurement system.

III.A Measurements of R_{SS} and R_{SM} of the Hakki-Coleman cavity

The surface resistance of the HTS thin films (R_{SS}) and copper walls (R_{SM}), necessary for calculations of tanδ of YVO$_4$ sample were measured in the same measurement system with the same dielectric resonator but with the sapphire rod instead of YVO$_4$. To obtain precise values of surface resistances we have measured S-parameters (S$_{21}$, S$_{11}$, and S$_{22}$) around the resonance. The measured data sets were processed with the Transmission Mode Q-Factor technique [4,5] to obtain the loaded Q-factor (Q$_L$), coupling coefficients and the unloaded Q-factor as mentioned in the Introduction. The TMQF method accounts for noise, delay due to un-calibrated transmission lines and its frequency dependence, and crosstalk in measurement data and hence provides accurate values of Q$_L$ and the coupling coefficients β_1 and β_2. The unloaded Q-factor was subsequently calculated using the exact equation [13],

$$Q_0 = Q_L(1 + \beta_1 + \beta_2)$$
(3)

Assuming loss tangent of the sapphire rod as 10^{-7}, the surface resistance of copper R$_{SM}$ and superconductor R$_{SS}$ were calculated using eq. (2) [11]. A comprehensive information on R$_s$ measurements using the TMQF technique is given in reference [5].

III.B Measurements of ε_r and tanδ of YVO$_4$

The Hakki-Coleman resonator with HTS endplates containing the YVO$_4$ sample was cooled from room temperature to approximately 13 K, and the resonant frequency of 24.4 GHz was obtained. The S$_{21}$, S$_{11}$ and S$_{22}$ parameters data sets around the resonance were measured as a function of increasing temperature from 13 K to 81 K, and the Q$_0$-factor and f$_{res}$ were calculated using the TMQF technique and eq. (3) as described in IIIA. The real part of relative permittivity ε_r of the Yttrium Vanadate sample was calculated from the measured resonant frequency using eq. (1) and taking the thermal expansion phenomenon into consideration. Values of the expansion coefficients (α) of YVO$_4$ for cryogenic temperatures were not available and hence we assumed the thermal expansion coefficients as for another optical material CaF$_2$ [14]. At room temperature α for CaF$_2$ is around 17×10^{-6}/K while it is 8.5×10^{-6}/K for YVO$_4$. Therefore our assumption of α is most probably higher than the real values.

Measured real part of relative permittivity ε_r of the YVO$_4$ sample measured at temperatures from 13K to 80K is shown in Fig. 3. The 'circles' represent values of ε_r computed assuming the thermal expansion coefficient of CaF$_2$, and the 'squares' represent the permittivity with the thermal expansion neglected. The difference between values of permittivity with and without thermal expansion considered is approximately 1%. The ε_r exhibited the magnitude of approximately 9.3 and increased with the temperature by approximately 0.1%; from 9.318 to 9.326.

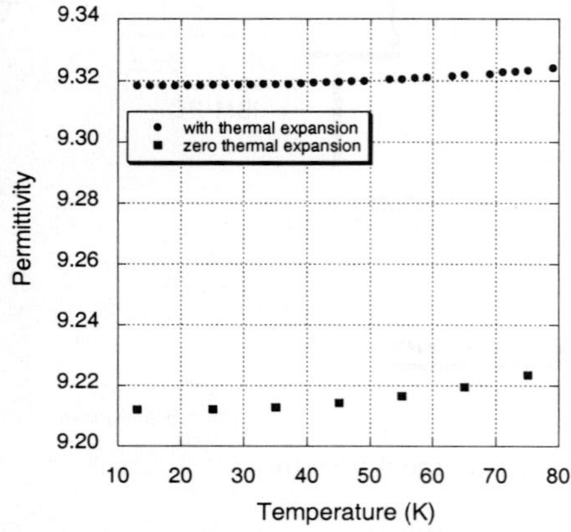

Fig. 3 The permittivity of YVO$_4$ crystal at 24.4 GHz as a function of temperature.

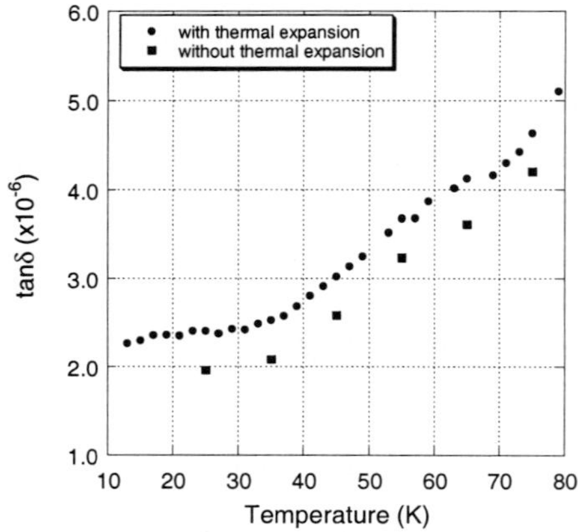

Fig. 4 The loss tangent of YVO$_4$ crystal at 24.4 GHz as a function of temperature.

Fig. 4 shows the measured temperature dependence of loss tangent of the YVO$_4$ sample at 24.4 GHz calculated using (2) from the measured unloaded Q-factor. The loss tangent exhibits an increase of 125% in the temperature range from 13 K to 80 K; the measured tanδ of YVO$_4$ were 2.265×10^{-6} and 5.1×10^{-6} respectively. Using a linear scaling, the calculated loss tangent of YVO$_4$ at temperature of 13 K and frequency of 10 GHz is only 9.2×10^{-7}.

IV CONCLUSIONS

The complex permittivity of YVO$_4$ birefringent crystal has been measured precisely at frequency of 24.4 GHz at cryogenic temperatures using the Hakki-Coleman dielectric resonator with superconducting endplates. Our measurements accounted for noise, crosstalk and uncompensated cables and adaptors and were based on accurate equations for the unloaded Q-factor. The YVO$_4$ was found to exhibit ε_r varying from 9.31 to 9.32, comparable to the permittivity of sapphire, and tanδ from 2.2×10^{-6} to 5.1×10^{-6} in the temperature range from 13 to 80 K. Our measurements have shown that YVO$_4$ is a very low loss material at cryogenic temperatures. Hence apart from the optical applications Yttrium Vanadate can be useful in cryogenic microwave circuits as a replacement for sapphire.

ACKNOWLEDGEMENT
This work is supported by ARC-Large grant (A00105170) and ARC-Linkage International (LX0242351). The first author acknowledges the James Cook University Post Doctoral Fellowship and the James Cook University MRG grant.

REFERENCES

1. [online]: YVO4, www.newphotons.com/YVO4.htm (November, 2002).
2. [online]: www.coteck.com/egscp.htm, (November 2002).
3. The Institute of Materials Research and Technology, Warsaw, Poland.
4. Leong K. and Mazierska J., *Journal of Superconductivity,* vol. 14, pp. 93-103, 2001.
5. Leong K. and Mazierska J., *IEEE Transactions on Microwave Theory and Techniques*, vol. 50, pp. 2115-2127, 2002.
6. Kobayashi Y. and Katoh M., *IEEE Trans. on Microwave Theory and Techniques,* vol. 33, pp. 586-592, 1985.
7. Krupka J., Geyer R.G., Kuhn M. and Hinken J., *IEEE Transactions on Microwave Theory and Technique*, vol. 42, pp. 1886-90, 1994.
8. Krupka J., Derzakowski K., Tobar M.E., Hartnett J., and Geyer R.G., *Measurement Science and Technology*, vol. 10, pp. 387-392, 1999.
9. Jacob M. V., Mazierska J., Leong K. and Krupka J., *IEEE Transactions on Microwave Theory and Techniques,* vol. 50, pp. 474-480, 2002.
10. Jacob M. V., Mazierska J., Ledenyov D. and Krupka J., Journal of the European Ceramics Society (in press).
11. Krupka J., Klinger M., Kuhn M., Baranyak A., Stiller M., Hinken J. and Modelski J., *IEEE Transactions on Applied Superconductivity*, vol. 3, pp. 3043-3048, 1993.
12. Krupka J. (Software in Fortran to calculate surface resistance, permittivity and loss tangent of dielectric materials, resonant frequency of differing TE and TM modes), 2000.
13. Ginzton E. L. "Microwave Measurements", *McGraw Hill Book Co.,* 1985.
14. Touloikian Y. S., Kirby R.K., Taylor R.E. and Lee T.Y.R., "Thermal expansion - Nonmetallic solids", *Thermophysical Properties of Matter Data Series*, vol 13, IFI/Plenum, New York, 1977

Session F14

HTS and Thin Films (2)

Co-chairs: C.K. Ong
                    S. Matitsine

Superconductor ($YBa_2Cu_3O_{7-\delta}$) and ferromagnetic ($La_{0.67}Ca_{0.33}MnO_3$) heterojunctions: Role of epitaxy-scheme

Sanghamitra Khatua[1,2*], D.C. Kundaliya[2], S. Wanchoo[3], R. Pinto[2], S.K. Malik[2]

1. Jai Hind College, Bombay, India – 400 020
2. Dept. of Condensed Matter Physics & Material Science, Tata Institute of Fundamental Research, Bombay, India – 400 005
3. R.D. National College, Bombay, India – 400 050

*Corresponding author, e-mail : skhatua@yahoo.com

Abstract

Good quality bilayers of ferromagnetic/superconductor (F/S) and superconductor/ferromagnetic (S/F) heterostructures were prepared by pulsed laser deposition. Detailed magnetization studies on these structures were carried out using a SQUID magnetometer. Notwithstanding, the identical growth condition, the epitaxial sequence seems to greatly influence the magneto-transport properties of these structures. The superconducting property of the first structure(F/S) is retained, while in the latter (S/F) structure, it is seriously impaired. We suggest that the contrasting behaviour of the two, otherwise similar, structures originates from the predominance of either of the two competing phenomena at the interface, namely, Andreev reflections and quasiparticle tunneling.

Introduction

Fabrication of hetero-epitaxially grown high temperature superconductor (HTSC) based, multilayer has recently attracted much attention due to the interesting physics underlying it and its vast potential for technological application [1]. The vortex dynamics of such mixed-state superconductors can be greatly affected by the layered heterostructure, especially in the limit of quantum tunneling and a strong interfacial scattering of spin-polarized/quasiparticles [2,3]. Yet it is fraught with many open questions as to why both enhancement and suppression of superconductivity are observed in generically similar materials. Whether, a non-equilibrium energy or a net quasiparticle density is the origin of the above observation, still remains an unresolved issue. Therefore, systematic studies of these colossal magnetic resistance (CMR) and HTSC hybrids are imperative, since they hold the key to understand and control the CMR/HTSC heterostructures that can lead to novel spin based devices. The decisive factors, obviously are the electronic and magnetic properties at the interfacial region and the thermal management of the structural entity. With this objective, we undertook the following study of $YBa_2Cu_3O_{7-\delta}$ (YBCO) and $La_{0.67}Ca_{0.33}MnO_3$ (LCMO) layered structures. These two particular materials were chosen because LCMO can be indexed as pseudo-cubic with a lattice parameter $\sim$ 3.86 Å which is close to that of YBCO a-b lattice parameter. This similarity, in principle, makes it possible to grow quality heteroepitaxy of LCMO and YBCO structures.

Experiment

Thin films of (S) YBCO (2000 Å), bilayers (F/S) LCMO (2000Å)/YBCO (2000Å) and (S/F) YBCO/LCMO, of identical sizes, were prepared *insitu* by pulsed laser desposition (KrF, 248nm) on $LaAlO_3$ substrates from stoichiometric targets of YBCO and LCMO, under optimum growth conditions. X-ray diffraction studies show all films to be c-axis oriented. The superconducting transition temperature (Tc) of (S) and (F/S) were found by a.c. susceptibility method. In both cases,

Tc ~ 85K with sharp transition (ΔTc ~ 2K). d.c. magnetization studies were carried out on all the three; S, F/S and S/F films using a SQUID magnetometer in a wide range of temperatures.

Results

A near identical superconducting-nonmagnetic(normal) transition temperature Tc in S and F/S films indicates that the introduction of a hetero (ferromagnetic) layer underneath does not do any damage to its superconducting properties [Fig.1]. This is noteworthy because in the reverse epitaxial growth i.e. S/F film the Tc was extremely low (<50K) without any particular, well-defined transition [Fig.1]. The remnant magnetization of S/F film drops to almost $1/10^{th}$ of the F/S value [Fig. 2]. The most significant feature is the emergence of a broad maxima at a still lower temperature (~ 25K) in the ZFC plot of S/F structure [Fig.2]. (This contrasting behaviour was reproduced in several S/F and F/S layers, which we grew later.)

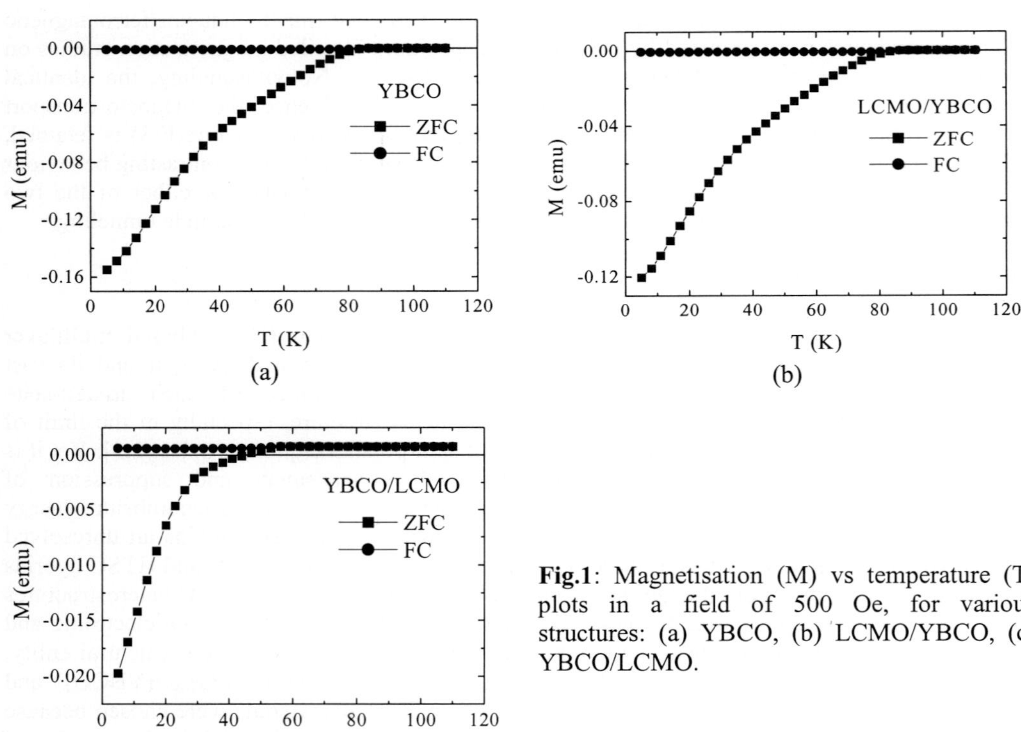

Fig.1: Magnetisation (M) vs temperature (T) plots in a field of 500 Oe, for various structures: (a) YBCO, (b) LCMO/YBCO, (c) YBCO/LCMO.

Discussion

We focus on the concept and the role played by the Andreev bound states (midgap states) on the transport properties at CMR and HTSC interface. The phonon mediated superconductivity of these Andreev states and the spin fluctuation via magnetic scattering as well,strongly influence the interplay of magnetism and superconductivity at the interface.

In the ferromagnetic region, the temperature dependent exchange interaction results in spin sub-band splitting and in a near perfect spin polarization (polarization factor ~1). In an hybrid system such as F/S structure, charged particles in the ferromagnetic region, having excitation energy (E) less than the superconducting gap energy (Δ) are reflected from the interface as holes (Andreev reflection)[4-7]. In superior quality junctions, this can result in a *non-equilibrium* spin proliferation/accumulation in the vicinity of the interface, thereby generating a local *interfacial resistance* formed, notionally at the spin relaxation length scale of the ferromagnet. This can further make the interface opaque to any charge transport and induce increased Andreev reflections. An enhanced/high degree of Andreev reflection therefore means, less probability of charge transport and consequently less Cooper pair breaking, thereby preserving the superconducting properties of the YBCO cap layer.

Another probable explanation is based on RKKY interaction. The long range RKKY exchange interaction in the S-layers at the interface predicts the formation of layered AFS (Antiferromagnetic superconductor) state at the interface [8]. In this AFS state, the magnetic order parameter is π phase shifted. The π phase shift, markedly suppresses the Cooper- pair breaking effect on the exchange field in the S – layers, hence retaining its superconducting properties. However, a decrease (~1/3) in the remnant magnetization could be the nullifying effect of AFS state on the lower ferromagnetic layers close to the interface [Fig.2].

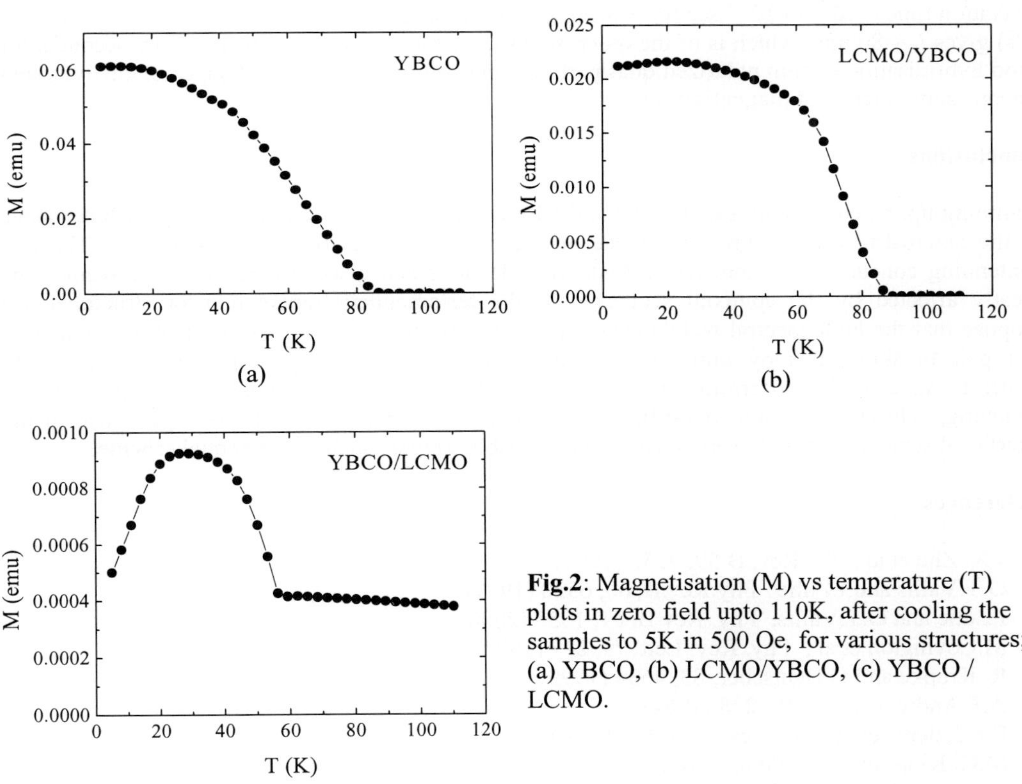

Fig.2: Magnetisation (M) vs temperature (T) plots in zero field upto 110K, after cooling the samples to 5K in 500 Oe, for various structures; (a) YBCO, (b) LCMO/YBCO, (c) YBCO / LCMO.

Next, we discuss the reverse hetero structure i.e. S/F bilayer. A predominant $d_{x^2-y^2}$ pairing symmetry in HTSC (YBCO) allows low energy excitation[9,10]. So, a fairly large population of quasi-particles exist even at low temperatures. Further, an impedance mismatch for thermal/electrical transport at the interface of two dissimilar materials causes a chemical potential E_μ to be developed. So if S and F share a common boundary, few thermally induced quasiparticales may *diffuse* from S to F. Since ferromagnetic materials have near perfect spin polarization, only those quasi particles whose spins are parallel to those of the majority carriers in the ferromagnet (F) can diffuse across the interface. Consequently, spin polarized quasi particle of those Cooper pairs are left behind. This can lead to an *autocatalytic/self-driven* process of further pair breaking and loss of superconducting properties. Another contributing factor is the vortex pinning energy of the YBCO layer [11]. A decoupling of YBCO layers due to proximity of the ferromagnetic layer can greatly impair its pinning ability, which is quintessential to high quality superconducting properties. The reduction in coupling strength is probably due to a competition between negative magneto-resistance in LCMO and the positive magneto-resistance of the flux motion in YBCO.

Moreover, magneto tunneling being primarily an interfacial phenomenon, an intrinsically large spin wave excitation can lead to an enhanced electron-magnon scattering. It is possible that it may cause some *spin-flip*. This can give rise to a signature of antiferromagnetic ordering, manifested as the broad maxima (T ~ 25K) in our result [Fig. 2(c)]. The spin-flip due to exchange interaction between the conduction electrons and magnetic moments causes depairing of Cooper pairs leading to a drastic reduction of transition temperature.

Finally a rough estimate of spin diffusion length l_s of the quasiparticles ($l_s = \overline{(l_0 V_F \tau_s)}$), τ_s = spin relaxation time ~ 10^{-12} to 10^{-13} sec for YBCO, l_0= mean free path ~ 20 nm, V_F = Fermi velocity 10^5 m/s) gives l_s ~ 80 nm, which is of the order of our layer thickness [12]. This probably accounts for good hybridisation of spin polarized quasiparticles and strong suppression of superconductivity and a depression in remnant magnetisation.

Conclusions

Summing up, heterostructures of LCMO and YBCO were grown epitaxially. The growth sequence of the heterostructure has great influence on the magnetic/superconducting properties. The two contending concurrent phenomena of Andreev reflections and quasiparticle tunneling seem to be greatly affected by the epitaxial sequence. In the ferromagnetic/superconductor structures, we propose that the high spectral weightage of Andreev reflections suppresses quasiparticle tunneling and pair breaking, thereby minimally affecting the superconducting properties of S-layer. In contrast, superconductor/ferromagnetic structure shows great predisposition to quasiparticle tunneling, setting in a chain of further pair breaking. The inherent local strain generated at the interfacial segment during the epitaxial growth possibly triggers such a preferential scheme.

References

1. J.X. Zhu et al , Phy.Rev. B 59, 9558 (1999)
2. O.T.Vallis and I.Zutic Phy.Rev.B, 60, 6320 (1999)
3. I.Zutic and O.T.Vallis, Phy. Rev B, 61, 1555 (2000)
4. M.Covington et al , Phy. Rev. Lett. 79, 277 (1997)
5. R. Krupke and G.Deutscher, 83, 4634 (1999)
6. A.F.Andreev, JETP 19, 228 (1964)
7. F.J. Jedema et al, Phy. Rev B 60. 16549 (1999)
8. M.G. Khusainov., JETP 82, 278 (1996)
9. A. Schmid and G.Schon, J.Low Temp. Phy 20, 207 (1975), H. Mongonery and G.P.Pells, J.Appl. Phys 14, 525 (1963)
10. J.Y.T.Wei et al, Phy. Rev. Lett 81, 2542 (1998), C.C. Tsuei et al, Phy. Rev. Lett 73, 593 (1994)
11. Li Q, Xi X, Inam A, Vadiamanali.S,, Phy Rev Lett 64,3086 (1999)
12. Y. Zha V. Barazkin and D.Pines , Phy.Rev B 54, 7561 (1996) and reference therein.

Optical Properties of Vacuum Evaporated amorphous CdTe Thin Films

R.Sathyamoorthy*, S.Lalitha, S.Senthilarasu, A.Subbarayan and K.Natarajan
Department of Physics, Kongunadu Arts and Science college, Coimbatore, Tamilnadu. India PIN:
641 029. * Author : 91-422-642095 E-Mail: rsathya59@yahoo.co.in.

Abstract

In this paper, we have reported the optical properties of vacuum evaporated CdTe thin films. Thin films have been prepared from the polycrystalline powder of CdTe on well-cleaned glass substrate under the vacuum of 10^{-6} Torr. The thicknesses of the films were calculated by the quartz crystal monitor (57nm to 575nm). The transmittance characteristics of films were studied by the spectrophotometer. The films were analyzed by an X-ray diffractometer. Lattice constants of the films were determined from the X-ray diffraction pattern using a Nelson-Relay plot. The optical constants (the refractive index n and the absorption constant k) and absorption coefficient α were measured on CdTe thin films in the wavelength range 190nm to 2500 nm. In these, we have observed direct band gap decreases with increasing thickness. The refractive index is found to be decreases of the wavelength of incident photon.

INTRODUCTION

Cadmium Telluride is considered to be a one of the best materials for the fabrication of thin film solar cells [1,2]. In the last decade, much attention has been paid to the development of low cost, high efficiency thin film solar cells Polycrystalline CdTe is a most suitable candidate for such solar cells due to its ideal bandgap, high absorption coefficient, and ease of film fabrication. The optical constants (the refractive index n and the absorption constant k) of thin film semiconducting materials may depend on the conditions of preparation and as the rate of evaporation and the substrate temperature during the deposition process. They may also depend on the film thickness especially in the first stages of deposition. The purpose of this work is to study the thickness depends on structural and optical properties of vacuum evaporated CdTe thin films.

EXPERIMENTAL

Cadmium Telluride thin films were deposited by thermal evaporation under the vacuum of 10^{-5} torr. The films were prepared on well cleaned glass substrates. First cadmium telluride was evaporated from the molybdenum boat. A constant rate of evaporation (0.1Å/sec) was maintained to prepare all the CdTe thin film samples. Rotary drive also used to maintain uniform thickness of the samples prepared. The rectangular samples of (dimension 3.75x2.5 cm) vacuum evaporated CdTe films were used for the structural and optical studies. The structural characterization was carried out by x-ray diffractometer with filtered CuKα radiation (λ= 0.15406 nm). The JASCO V – 570 spectrophotometer is used to record transmission spectra of CdTe thin films in visible and NIR region. Thickness of the samples are measured by quartz thickness monitor and verified by Multiple Beam Interferometry (MBI) technique.

RESULTS
1.Structure

Diffraction spectrum for bulk powder is shown in Fig (1). The spectrum of CdTe powder exhibits sharp peaks at 2θ equal to 23.7°, 39.3° and 46.6° which corresponds to diffraction from (111),(220)and (311) planes of the cubic phase respectively. Both the peak height and peak position are in good agreement with ASTM data (16-770) for cubic CdTe, whereas the as grown high thickness films showed preferential growth of film crystallites corresponding to textured (111) growth. The (111) direction is the close packing of the Zinc blende structure and this type of ordering is often observed in polycrystalline films [3,4]. Fig (2) shows the x-ray diffractogram of vacuum evaporated CdTe thin film of thickness (2060 Å). The featureless spectra indicate that the growned films are amorphous in nature. We have taken XRD pattern for films of various thickness

(<4000 Å). The results are clearly indicates that the films are amorphous in nature. Fig (3) represents XRD pattern for CdTe thin films coated at higher thickness (5466 Å). This spectra indicates that the growned films are polycrystalline in nature. Here we observed that the crystallite size increases with increase in film thickness [3] and thus the observed crystalline nature, for higher thickness thin films. Crystallite sizes were determined from x-ray diffraction data using the Scherer formula

$$D = 0.9 / B \cos \theta$$

Where D is grain size, is the wavelength of the x-rays and B is the width of the peak at full width half maximum. The value of 'D' for as grown film the calculated d-spacing and lattice parameter along with the probable hkl values were presented in table1.

θ	d	hkl	a=b=c (cubic)	crystallite size
11.300	7.824	-	-	127.91Å
19.55	2.309	220	6.531	120 Å
23.500	3.783	111	6.481	65.03Å

2.Optical

In order to calculate the optical parameters of the films from the transmittance spectra (Fig.4.) We applied the model developed by Perrion. In the model the substrate is taken to be non-absorbing with a thickness sufficiently large to allow the use of its average transmittance. Complete coherence in the CdTe film and complete incoherence in the substrate are assumed. According to this model, the transmission in the low absorption range is

$$T = \frac{t_0^2\, t_1^2\, t_2^2\, \exp(-4\pi kd/\lambda)}{1 - r_1^2 r_2^2 - 2r_0 r_1(1 - r_2^2)\exp(-4\pi kd/\lambda)\cos(4\pi nd/\lambda) - r_0^2(r_1^2 - r_2^2)\exp(-8\pi kd/\lambda)} \quad ----(1)$$

Where t_0, t_1 and t_2 are the transmission coefficients at the interfaces air-thin film. Thin film substrate and substrate – air respectively: r_0, r_1 and r_2 are the corresponding reflection coefficients at these interfaces: k is the extinction coefficient of the thin film, n its refractive index and d its thickness. The reflection and transmission coefficients may be expressed as functions of the various refractive indices through the Fresnel equations.

For the high absorption range the model leads to

$$T = \frac{16n_s(n^2 + k^2)\exp(-4\pi kd/\lambda)}{\{(n+1)(n+n_s^2)+k^2\}\{1+n)^2+k^2\}} \quad ----(2)$$

In this equation k, d and n have the same meanings as in equation (1) and n_s is the refractive index of the glass substrate. We examine the $\alpha^{\frac{1}{2}}$ Vs hυ, $\alpha^{2/3}$ Vs hυ and α^2 Vs hυ relations but only the last yields a linear dependence (Fig.7.). This indicates that absorption in the high absorption range takes place through direct inter band transitions [3] .The Table 2 gives the band gaps Eg obtained by extrapolation to zero absorption. We have observed two band gap energies for all the films.

Thickness (Å)	Band gap (Eg$_1$)	Band gap (Eg$_2$)
570	1.58 eV	2.2 eV
1100	1.56 eV	2.0 eV
2000	1.5 eV	1.8 eV
5450	1.48 eV	-

DISCUSSION

CdTe films of different thickness have been taken for the structural analysis. The structure of the films are found to be amorphous for certain films of thickness (<4000 Å) where as the crystallinity was observed for the films of thickness >4000 Å. The stress are also developed in the film due to the lattice misfit. However the stress has two components, thermal stress arising from the difference of expansion coefficient of the film and substrate, and internal stress due to the accumulating effect of the crystallographic flaws that are built into the film during deposition. All the CdTe films at room temperature (303K) thus the internal stress dominated in these evaporated films other than thermal stress. From the calculated structural parameters we have drawn Nelson-Riley plots (Fig.8 and Fig.9) with error function. Fig.8 clearly differentiates bulk powder and evaporated CdTe film (5450Å) The change in the lattice parameter is due to the strain developed in the grains of the deposited films.

A clear relation exists between the transmission spectra of the CdTe films and their thickness. A high transmittance and a well defined absorption edge are seen in films grown at high thickness. The decrease in the IR transmittance and the loss of definition of the absorption edge in the films having lower thickness must be attributed to absorption mechanisms other than ideal CdTe inter band transitions [4]. A primary consideration is the absorption by phases other than crystalline CdTe. The presence of free tellurium in some films could account for their non-ideal transmission spectra [5]. But we did not get any free tellurium peaks in XRD pattern. The grain boundary being highly disordered and a sink for impurities can have properties very different from those of the grain giving a marked contribution to the absorption for the films with low grain sizes. This contribution has been modelled by considering the grain boundaries as amorphous regions [6]. Whatever the process by which grain boundaries contributes to the crystalline imperfections, the lower will be the absorption. A high transmission and well defined absorption edge in the polycrystalline films with larger grain size and stronger orientation of the crystallites. The optical band gap (E_g) its narrowing in the films grown at higher thickness may be related with the existence with in the bandgap of a high density levels it energies near the bands which can give raise to band tailing as has been suggested for other polycrystalline materials [7]. Two band gap energies were obtained for all the films. These two band gap energies may be attributed due to the spin orbit splitting of the valance band. Similar observations have been made by earlier investigators [8,9] for CdTe thin films. The bandgap decreases with increase of film thickness.

CONCLUSION

The thermally evaporated CdTe thin films are amorphous and crystalline nature for lower and higher thickness respectively. The evaporated films are highly influenced by the internal stress results amorphous nature. The optical transitions are found to be direct and allowed. The band gap energy decreases with increase in film thickness.

REFERENCES

1. J.L.Loferski, J.Appl.Phys, 27 (1956) 777.
2. M.Rodot, Rev.phys.appl.12, (1977) 411.
3. J.I.Pankove, Optical Process in semiconductors, Dover Publications, New York, 1975, p.34
4. J.Aranda, J.I.Morenza, J.Esteve and M.Codina, Thin solid Films, 120 (1984), 23-30
5. M. Cardona, K.L. Shaklee and F.H. Pollak, Physics Rev. 154(1967), 696.
6. J.Szezyrbrowski and A. Czapla, Thin solid films, 460 (1977), 127.
7. L.Y.Sun, L.L.Kazmerski, A.H.Clerk, P.J. Ireland and D.W.Morton, J.Vac, Science Technology 225 (1978) 265.
8. G.K.M.Thutupalli and S.G.Tomlin, J.Appl.phy.D, 9 (1976) 128.
9. R.Sathyamoorthy, PhD Thesis, Bharathiyar University, 1991.

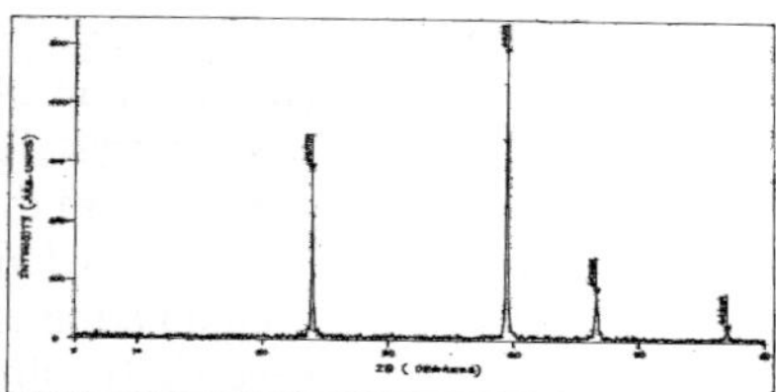

Fig.1. XRD pattern for CdTe bulk.

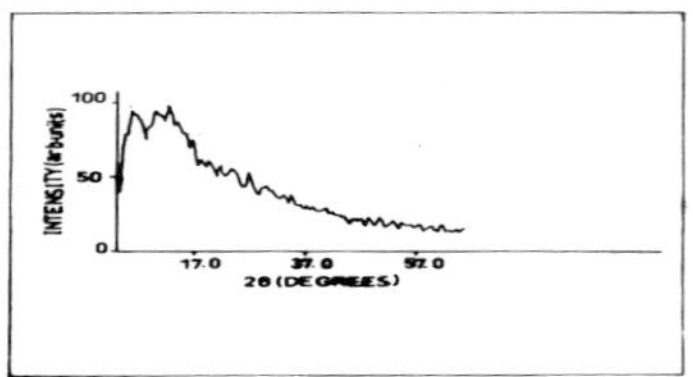

Fig 2. XRD pattern for the CdTe Film (d = 2060Å)

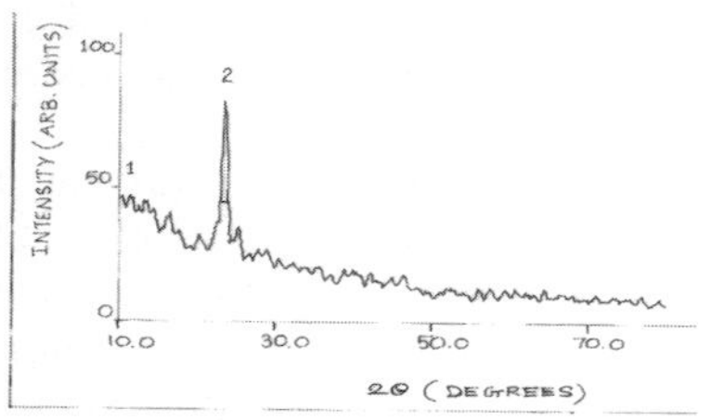

Fig.3. XRD pattern for the CdTe thin film (d =5466Å)

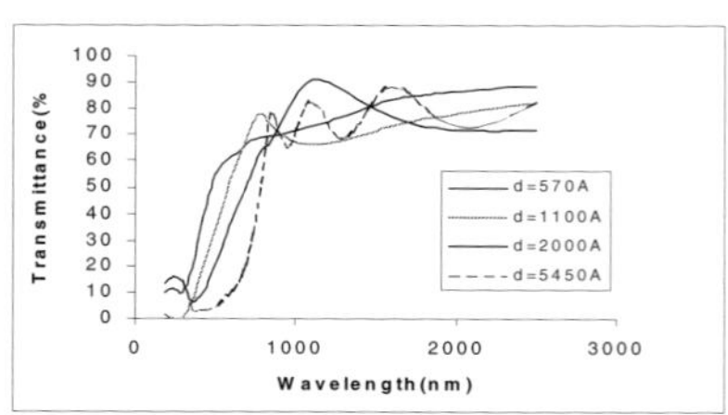

Fig.4. The optical transmittance spectra of vacuum evaporated CdTe thin films

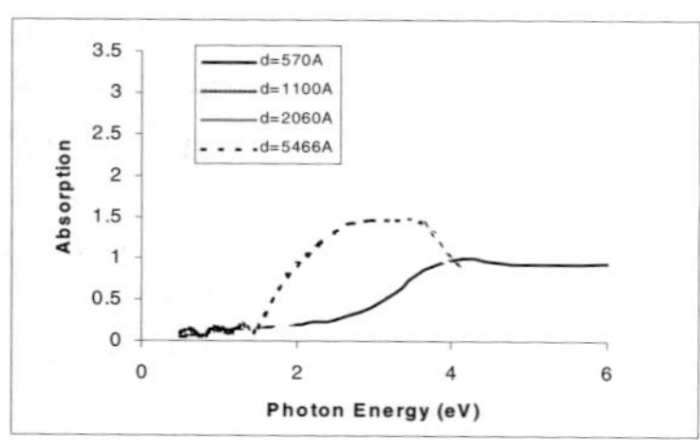

Fig.5. Photon Energy Vs Absorption

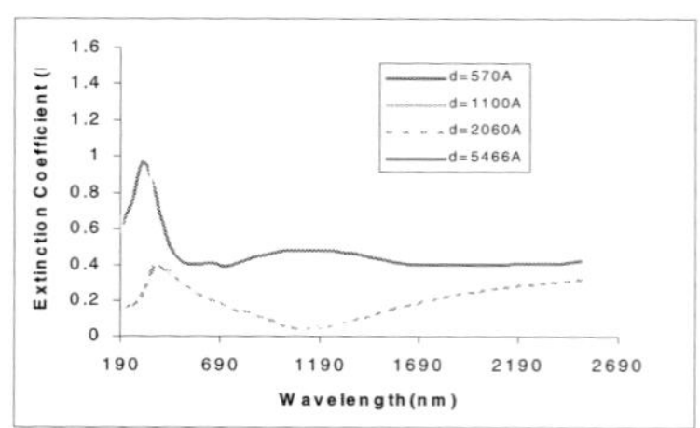

Fig.6. Wavelength Vs Extinction Coefficient (k)

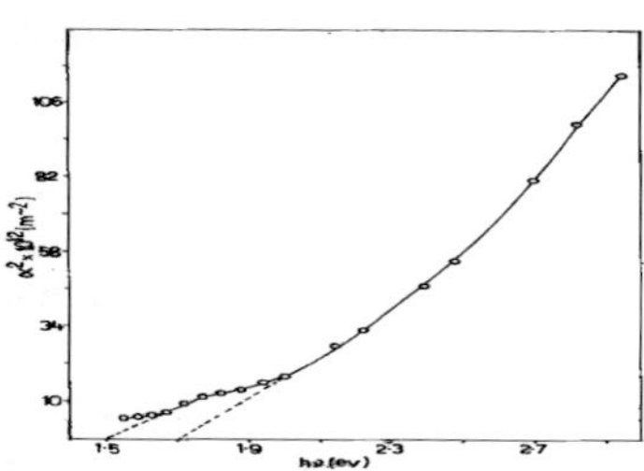

Fig.7. Variation of α^2 with hυ for CdTe thin film.

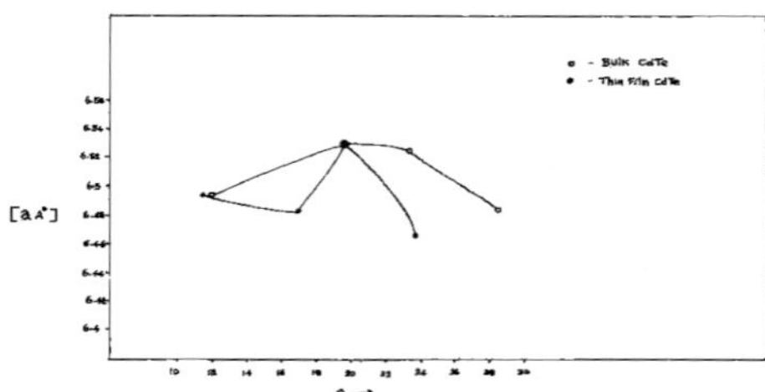

Fig.8. Variation of Lattice constant for Bulk CdTe and Thin film CdTe

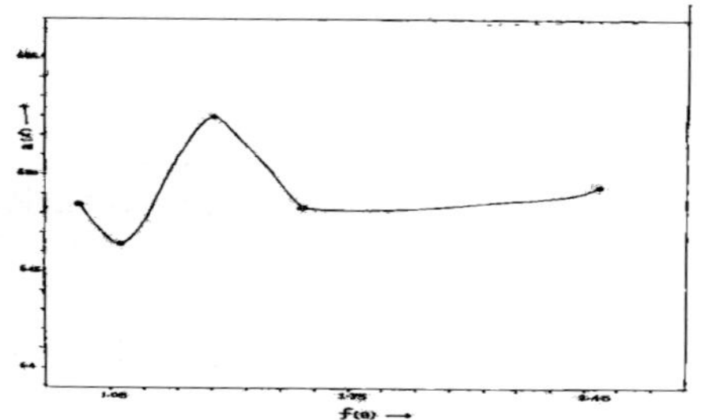

Fig.9. Nelson – Riley plot for CdTe films for determination of Lattice constant a: error function $f(\theta) = \frac{1}{2} (\cos^2\theta / \sin\theta + \cos^2\theta / \theta)$

Structural, Optical and Electrical Properties of Zinc Phthalocyanine (ZnPc) thin films

S.Senthilarasu, R.Sathyamoorthy*, K.Kanmani, S.Lalitha and A.Subbarayan
Department of Physics, Kongunadu Arts and Science College,
Coimbatore 641 029. Tamilnadu, India.
* Author : 91-422-642095 E-Mail: rsathya59@yahoo.co.in.

Abstract

In this article, we have reported the structural, optical and electrical properties of ZnPc thin films. The samples were prepared by thermal evaporation method .The thickness of the samples were measured by the quartz crystal monitor. The structural analysis were carried out by X-ray diffraction (XRD) and Scanning Electron Microscope (SEM). The XRD studies and SEM studies of the ZnPc thin films are reveals that the films are amorphous in nature. The optical properties in the transmission mode have studied by spectrophotometer in the visible region. The possible transition in these films is found to be direct and allowed. AC conduction mechanism and dielectric properties in these films (Al-ZnPc-Al structure) were studied by the LCR meter for various frequencies (12 Hz to 100 KHz) at different temperatures. The field dependence behaviour on activation energy and possible conduction mechanism in the ZnPc films under dc field have also been discussed.

INTRODUCTION

ZincPhthalocyanine (ZnPc) is a well-known organic semiconductor with large absorption coefficient [1,2], high thermal and photochemical stability [2] and for that reason they are good potential candidates in solar-to-electric energy conversion. In this paper, we have reported the Structural, Optical and Electrical properties of vacuum evaporated ZincPhthalocyanine(ZnPc) thin films.

The ZincPhthalocyanine (ZnPc) thin films are prepared by vacuum evaporation method under the pressure of 10^{-6} m.bar on to a well cleaned glass substrates. The thickness of the thin films has measured using quartz crystal thickness monitor. The films are prepared in various thicknesses. The structural characterization is studied by x-ray diffraction method. The optical transmittance and absorption spectra of the ZnPc thin films are studied by spectrophotometer. Al-ZnPc-Al structure has been made to study the dielectric and AC conduction studies of ZnPc films. AC conduction mechanism and dielectric properties in these films were studied by the LCR meter for various frequencies (12 Hz to 100 KHz) at different temperatures.

RESULTS AND DISCUSSION
1. SURFACE AND STRUCTURAL

The SEM photograph (d=2565Å) are depicted in fig.1. It is observed that the film is uniform, continuous and made of loosely packed in amorphous structure. No periodic features are observed in the SEM photographs. This may be lead to conclude the nature of the film to be loosely packed disorder structure leading to amorphisation. The X-ray diffractogram of ZnPc films of thickness 2565Å are given in the fig.2. The observed featureless spectra in the film leads us to understand the films are amorphous in nature, and thus confirms the glassy nature of the ZnPc thin film.

2.OPTICAL

The ZnPc layer obtained by thermal evaporation is blue in colour, highly transparent and firm. The transmittance spectra of the films in the visible region for the different thickness films are shown in the fig.3. It reveals that the transmittance decreases with the increase of film thickness. The band gap energy have been calculated from the plot α^2 versus photon for all the films and it was found to be 1.82 eV for the film of thickness 1875 Å (fig.4.). The band gap energy is found to be decreases with increase in film thickness (Table.1)

Thickness (Å)	Band gap energy (eV)
1140	1.83
1875	1.81
2565	1.79

3.DIELECRTIC PROPERTIES

The capacitance and loss factor (tanδ) are important scale factors to analyze the dielectric properties. In the present study the equivalent series capacitance (C_s) was measured and whenever necessary a series – parallel conversion was made. The dielectric constant of ZnPc films are calculated using the formula $\varepsilon' = C_s d / \varepsilon_0 A$ and it was found to be 0.02 for 1KHz at room temperature.

FREQUENCY AND TEMPERATURE EFFECT

The changes in capacitance with frequency at different temperature for ZnPc film capacitor formed by vacuum evaporation are represented in fig.5. From the figure it is seen that the capacitance decreases with increase in frequency at all temperatures. The large increase in capacitance towards the low frequency region may be attributed to the blocking of charge carriers at the electrodes. Actually, the charge carriers present in the film migrate upon the application of the field and because of the impedance to their motion at electrodes there is a large increase in the capacitance at low frequencies. The variations of tanδ with frequency at different temperature for ZnPc film are represented in the fig.6. It is observed that the value of tanδ increase in frequency for all temperatures. It is found that the presence of loss peaks in the lower frequency region (10-500Hz) shifts to higher frequency region with increasing temperature. Generally the deficiency or the imperfection in the solid state materials pave way to form dipoles which leads to occurrence of Debye type dispersion [3] such a dispersion could be expected in vacuum evaporated ZnPc thin films. The temperature coefficient of capacitance (tcc), temperature co-efficient of permittivity (tcp) for ZnPc film has also been evaluated and found to be 9870ppm/K for 50 Hz and 5869 ppm/K for 1KHz respectively.

A.C CONDUCTION

Fig.7 represents the dependence of a.c conduction on frequency at different temperatures for vacuum evaporated ZnPc thin films of thickness 2565Å. The conductance is found to increase linearly with increase of frequency in all the films in accordance with the relation $G_p \; \alpha \; f^n$. The values of exponent n have been estimated and found lie between 0.96 and 1.66. Many hopping system mostly disordered or amorphous materials exhibit this behaviour. The frequency dependence of conductivity is a general feature of hopping systems [4]. The value of n in general decreases with increase of temperature. The fig.8 display the temperature dependence of conductance for ZnPc film at different frequencies. The activation energy has been calculated from this plot and is found to be 0.248ev at 1KHz. The low value of activation energy suggests that the hopping conduction may be due to electrons rather than ions [5].

D.C CONDUCTION

Information about the conduction mechanism can be obtained from I-V characteristics at different temperature. LogI Vs LogV characteristics of ZnPc film of thickness 2565Å for different temperature are represented in fig.9. It is seen that the curve for each temperature exhibits two regions namely AB and BC. In the region AB at low voltage the conduction is ohmic ($I\alpha V$), indicating that the current is controlled by thermally generated carriers. In the region BC corresponding to approximately 6v to 13v, a traps square law region ($I\alpha V^2$) is obtained. The similar behaviour has been reported by G.D. Sharma et.al. [6] for Al-ZnPc-ITO system.Fig.10 represents the variation of log current with the square root of applied field. It is observed that in the high field region the ZnPc film exhibit linear current field characteristics for all the temperatures, this behaviour indicates that the conduction mechanism may be of either Schottky or Poole-Frenkel type.   By taking the dielectric constant ε' to be 0.02 and $\varepsilon_0=8.85\times10^{-12}$ the theoretical value of β_{PF} & β_S are calculated and given in the Table 2.

Temp in K	β Experimental x 10^{-4} (mv)$^{1/2}$	β Theoretical x 10^{-4} (mv)$^{1/2}$	
		Schottky	Poole-Frenkel
303	1.2020		
323	1.0707		
343	1.1090	5.368	2.68
363	0.7200		
403	1.1193		

From the evaluated values indicated in the table it has been found that the experimental value of β coincides with the theoretical value for the Poole – Frenkel conduction mechanism. Therefore the conduction mechanism may be of the Poole – Frenkel type. The graph is drawn between 1000/T (K^{-1}) Vs LogI is shown in fig.11. From this figure the activation energy has been determined at different voltages using the relation

$$I = I_0 \exp(-E / KT)$$

The activation energies calculated for different voltages are given in the table 4. From this it is evident that the activation energy decreases as the field increases. The fig.12 is plotted between voltage Vs activation energy. It was found that activation energy decreases as the voltage increases and the corresponding Zero field activation energy was found to be 0.50 eV.

CONCLUSION

ZincPhthalocyanine thin films were obtained by thermal evaporation method. The XRD studies and SEM studies of the ZnPc thin films are reveals that the films are amorphous in nature. The optical transition in these films is found to be direct and allowed. The dielectric studies clearly indicate that the Debye type of polarization exists in the films. The electrical conduction mechanism prevails in ZnPc films under ac and dc fields are found to be electronic hopping and Poole Frankel type respectively.

REFERENCES
1. D.Wrobel, A.Boguta, R.M.Ion, Int.J.Photoenergy 2 (2000) 87.
2. C.C.Leznoft, A.B.P.Lever, Phthalocyanines: Properties and Applications, VCH, New York, 1996.
3. P. Debye, Polar Molecules, Dover, New York, 1929.
4. A.K. Jonscher and M.H. Nathoo, Thin Solid Films, 12 (1972) 515.
5. W.S. Chan and A.K. Jonscher, Phys. Stat. Sol., 32 (1969) 749.
6. G.D. Sharma, S.G. Sangodkar, M.S. Roy, Materials Science and Engineering, B41 (1996) 222 - 227.

Fig.1. SEM Photograph for ZnPc thin film (d=2565Å)

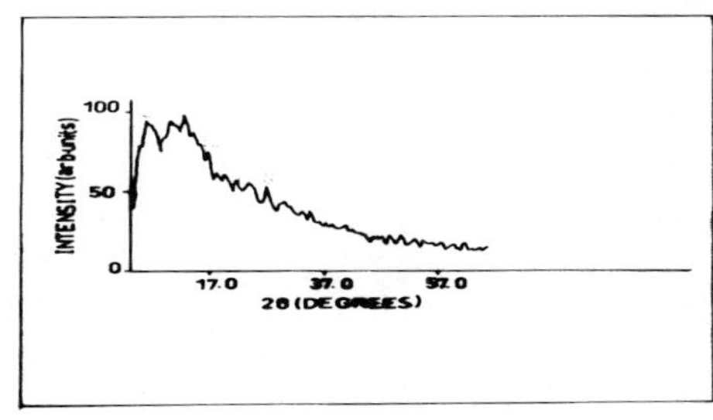

Fig.2. XRD pattern for ZnPc thin film(d=2565Å)

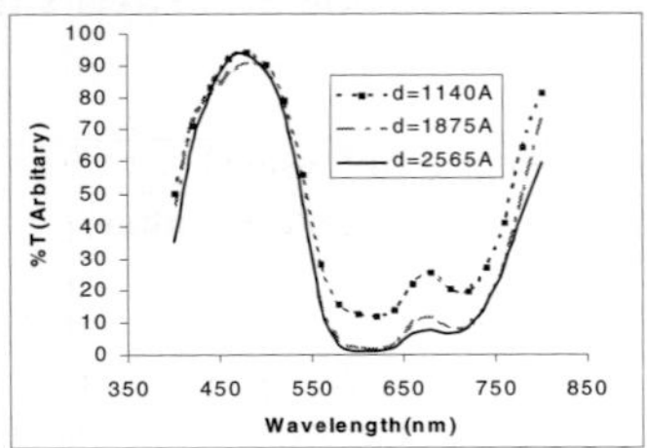

Fig.3. Transmittance spectra for ZnPc films at different thicknesses

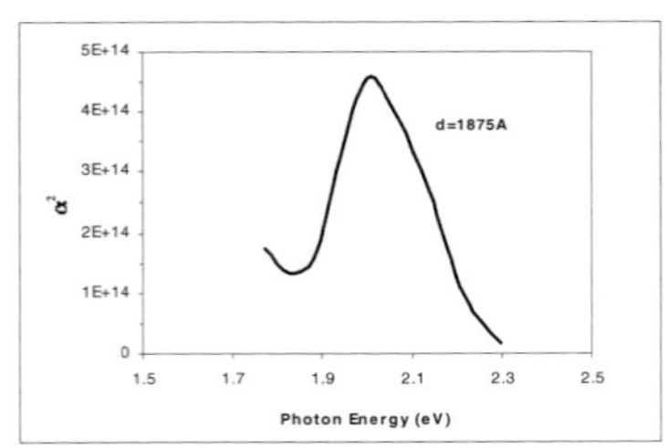

Fig.4. Variation of α^2 with incident photon energy

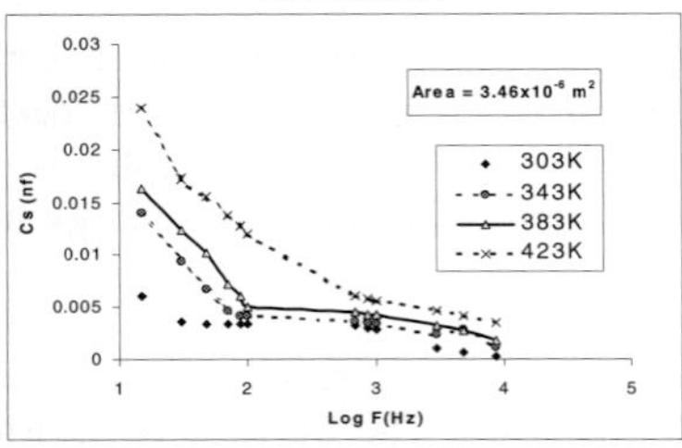

Fig.5. Change of Capacitance (Cs) with Log frequency at different temperatures.(d=2565Å)

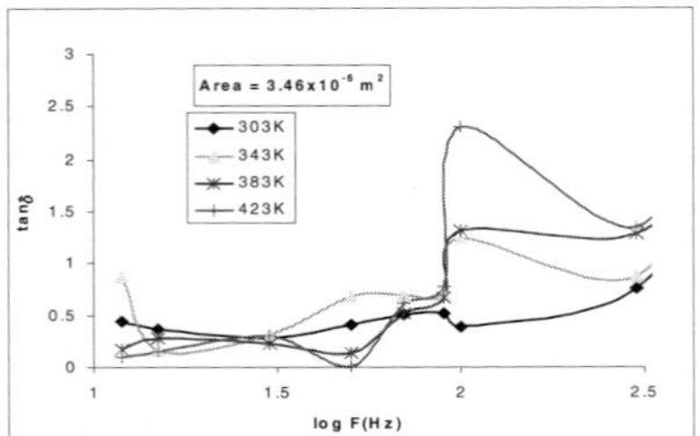

Fig.6. Variation of Dielectric loss with log frequency at different temperatures (d=2565Å)

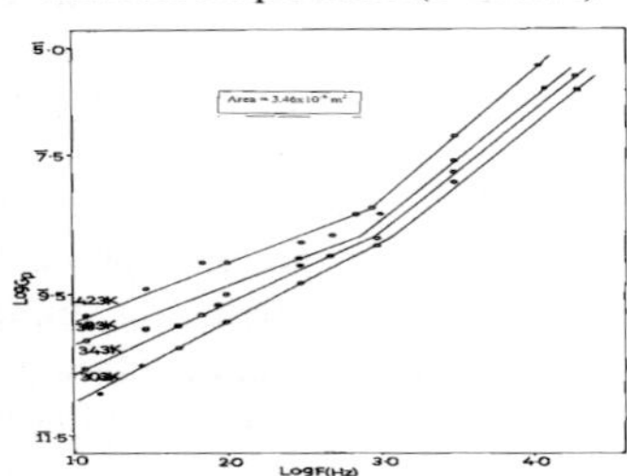

Fig.7 The dependence of a.c conduction on frequency at different temperatures (d=2565Å)

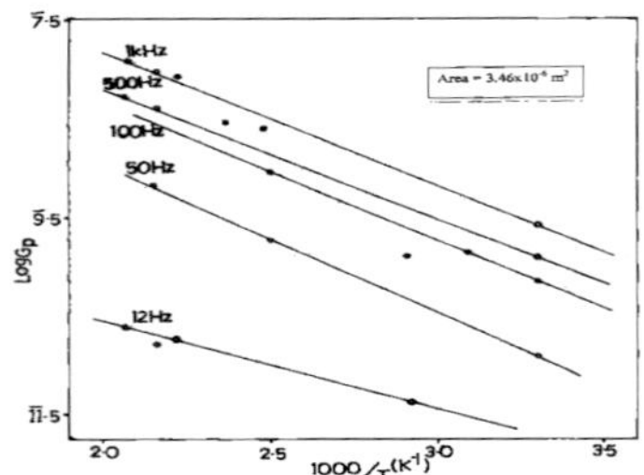

Fig.8 The temperature dependence of conductance at different frequencies (d=2565Å)

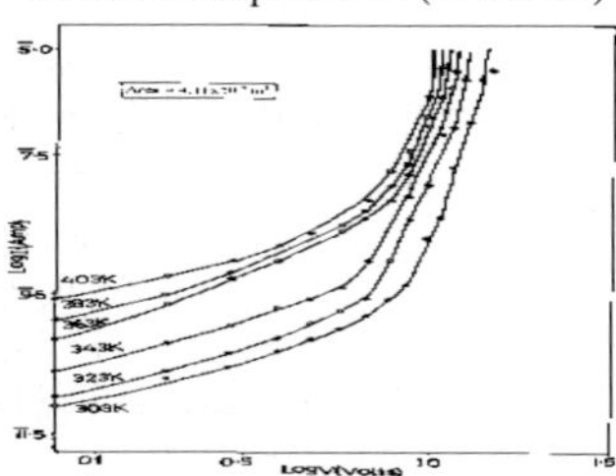

Fig.9. I-V characteristics of for different temperature

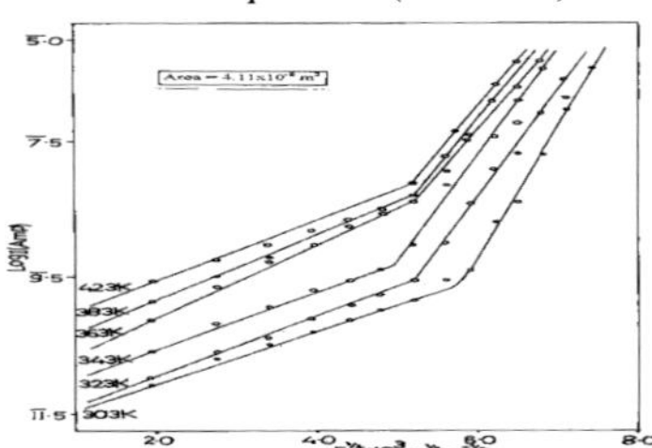

Fig.10. Log I Vs $F^{1/2}$ at different temperatures

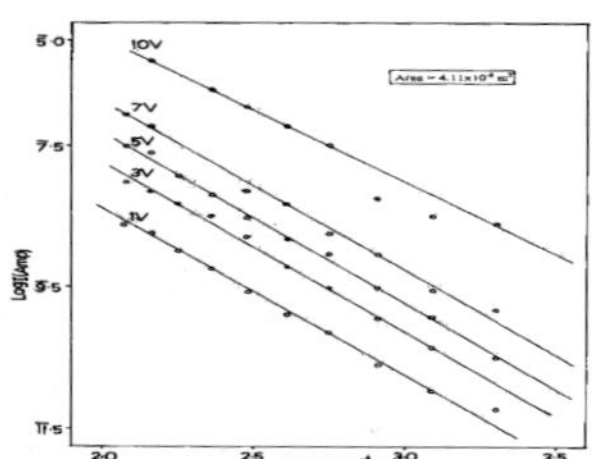

Fig.11. Variation of Log I with inverse temperature at different fields (d=2565Å)

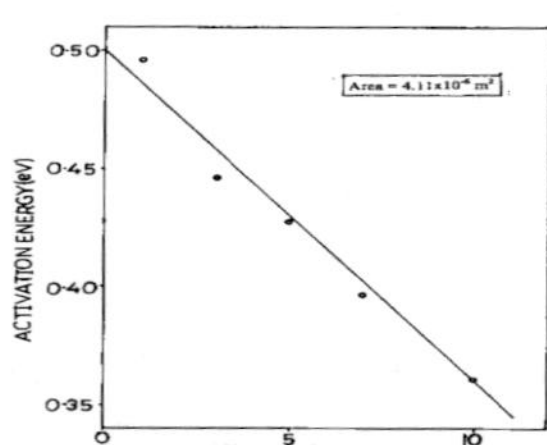

Fig.12. Activation energy dependence on voltage.

Fe-nitride Thin Films Deposited Using Filtered Cathodic Vacuum Arc Technique

W.H Zhong, S P Lau, B.K. Tay, S Li, C.Q. Sun
School of Electrical and Electronic Engineering, Nanyang Technological University, Singapore 639798

Abstract

Nitrogen contained iron films were deposited on Si (100) substrate by using Filtered Cathodic Vacuum Arc (FCVA) technique under various N_2 pressure and at various substrate temperatures. Atomic force microscope, surface profile measuring system, x-ray diffraction, x-ray photoelectron spectroscopy and vibrating sample magnetization were used to characterize the structural and magnetic properties of the films, respectively. The films exhibit a smooth surface and possess a polycrystalline structure. It was found that although the as-deposited films were α'' phase free, within a certain nitrogen concentration, the saturation magnetizations were still higher than that of the bulk pure Fe. This indicates that Fe^+ ion or Fe^{dipole} formation possessing higher magnetic momentum upon nitridation. The study inspires the development of a new technique for deposition of magnetic films.

1. Introduction

The relationship between ferromagnetism and metal – non-metal bonding of iron nitrides was the main subject of ferromagnetic research for both theory and experiment. After Kim and Takahashi reported a giant magnetic moment in the Fe-N film,[1] many works have been carried out to synthesize Fe-N thin films using various deposition methods such as molecular beam epitaxy,[2] multi-shot implantation,[3] facing target sputtering,[4,5,6] reactive radio frequency sputtering[7] and so on. However, very few data have been reported on Fe-N film coatings by the FCVA technique.

The off-plane double-bend (OPDB) FCVA is a promising technique for effectively eliminating micro-particles and producing high quality thin films. The plasma is confined and guided through a toroidal solenoid (a patented [8] steel OPDB torus with a copper coil wound outside) by the curvilinear magnetic field. There is no straight-line path between the cathodic arc source and the substrate; this prevents the neutral particles from reaching the deposition chamber. A set of baffles, acting as a mechanical macroparticle filter, is installed inside of the torus to enhance the macroparticle filtering efficiency. In this way, the system thus can produce films with good controllability and reproducibility.

The deposition of Fe-N films by OPDB-FCVA technique mainly depends on two important factors, namely, the Fe plasma generation rate and the nitrogen pressure in the chamber. If the plasma generation rate is high, it is necessary to input more nitrogen gas into the system to get stoichiometric reaction, but more gas in the system will result in shorter mean free path of the plasma. This means that collisions between the gas molecules and the plasma ions occur more often in the double bend, resulting in a lower deposition yield at the substrate. It is therefore important to establish a balance between the generation rate of the plasma and the amount of gas needed to achieve optimum results. On the other hand, the substrate temperature may affect the microstructure and nitrogen content in the films. In this work, the nitrogen contain iron films were deposited with various N_2 pressure and substrate temperature to investigate the effects on the films structural and magnetic moment.

2. Experimental details

The films were deposited by the OPDB FCVA technique which had been described in details elsewhere[9]. A cylinder iron target (30mm $\times$ ϕ 60mm, 99.9% purity) was mounted on a water-cooled copper block. For the magnetic metal deposition, because the target changes the magnetic field on the surface of target due to its high permeability, the arc is unstable, which can be solved by adding a shield around the target. The shield consists of three main parts: the metal cylinder, the pure iron ring and the BN ring, as showed in figure 1. The metal cylinder covers around the cathode-target and the copper cylinder of the cooling water systems. The wall of the bend is connected to the anode, while the shield is in a floating potential. It separates the electronic field between the cathode and the anode to prevent the plasma formatting between the vertical side of the target and the inner wall of the bend. When the arc forms on the surface of the target, current is generated between the anode and the cathode, which induces a circle magnetic field surrounding the target. On the other hand, the coil entwined around the ectotheca of the bend generates a vertical magnetic field. Because of the high magnetic permeability of the magnetic target, most of the magnetic flux passes through the edge of it, which makes a high magnetic density area in the edge and hence makes the arc spot prefer to being around the edge of the target. In this way, the arc spot is easy to get down to the vertical

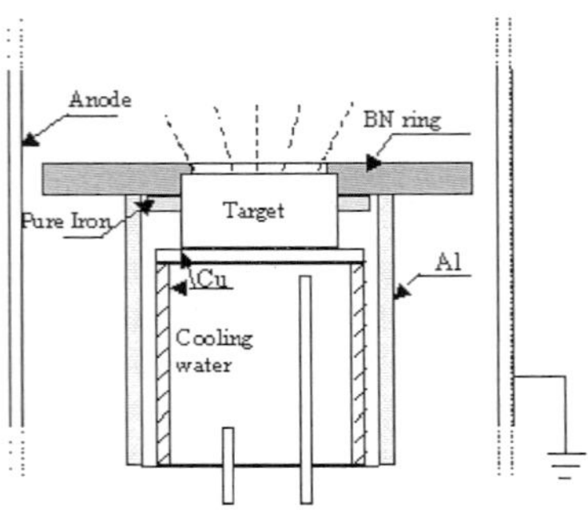

Fig. 1 Schematic illustration of the shield used for magnetic metal target to

side and then extinguishes. Here, after added a shield, the ring of pure iron makes the magnetic filed more uniform and forms a larger virtual target surface. The BN ring is set on the top of the metal cylinder, which shields the target well and prevents the arc burning down to the vertical side.

Two sets of experiments were carried out based on the OPDB-FCVA and the shield described above. The first set is to investigate the effect of nitrogen pressure, which is controlled by adjusting the N_2 gas flow rate, on the deposition rate and the nitrogen content in the films. The gas flow rates range from 5 to 100 sccm, which results in a nitrogen pressure from 2×10^{-5} to 5×10^{-5} torr. The other set is to investigate the effect of the substrate temperature on the properties of the films. The N_2 pressure was kept at 2.1×10^{-5} torr to realize higher deposition rate and the substrate temperatures were changed from 110 to 350 °C with a step of 60 °C. All other parameters were kept constant.

X-ray photoelectron spectroscopy (XPS) (Mg Kα radiation at 1253.6 eV) was used to determine chemical environment of the elements and the N content in the films. All the XPS spectra of the samples were fitted with Gaussian functions and linear background. The microstructures of the films were characterized by the Siemens X-ray diffraction (XRD) with the small (1°) angle of incidence. Vibrating Sample Magnetometer (VSM) was used to measure the magnetic moment of the films. The samples were cut into about 8mm × 8mm square, the thickness and the area of the films were measured by the Tencor P10 surface profile system. The morphology and roughness of the films were characterized by atomic force microscopy and the values of root-mean-square (RMS) surface roughness were evaluated in the area of 1 μm × 1 μm.

3. Results and discussion
3.1 N$_2$ pressure effect:

Fig. 2 shows the dependence of the deposition rate on the N_2 pressure (with arc current 120 A). From the figure it can be seen that the deposition rate decreases with the increase of N_2

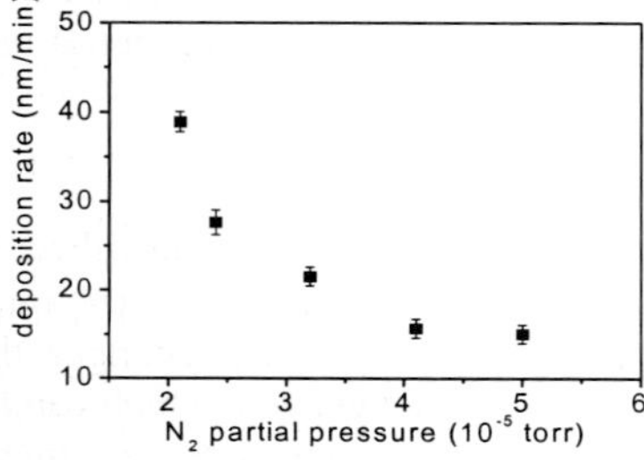

Fig. 2 N$_2$-flow rate effects on the deposition rate

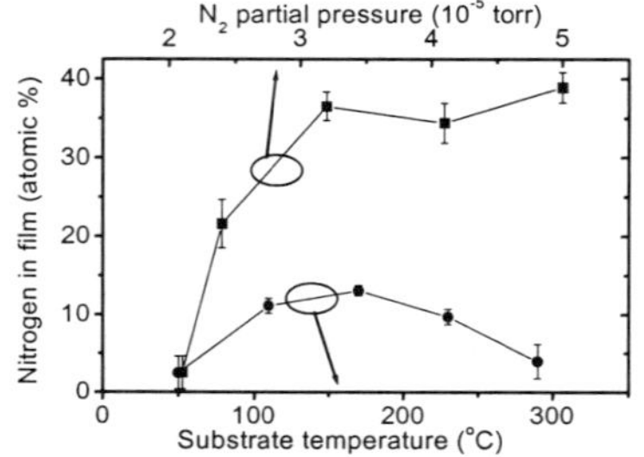

Fig. 3 Gas flow rate and substrate temperature effects on the films nitrogen concentration

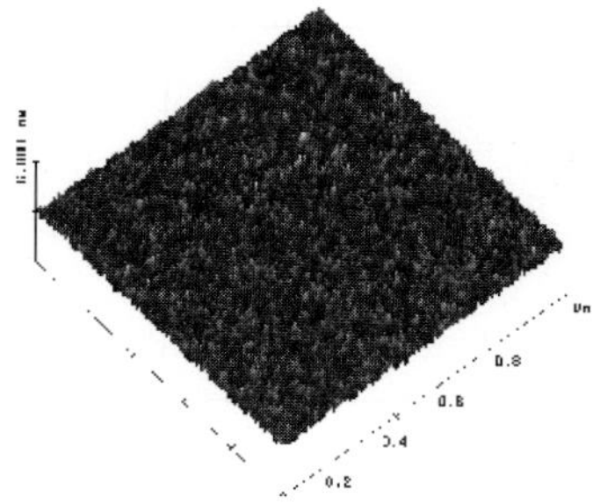

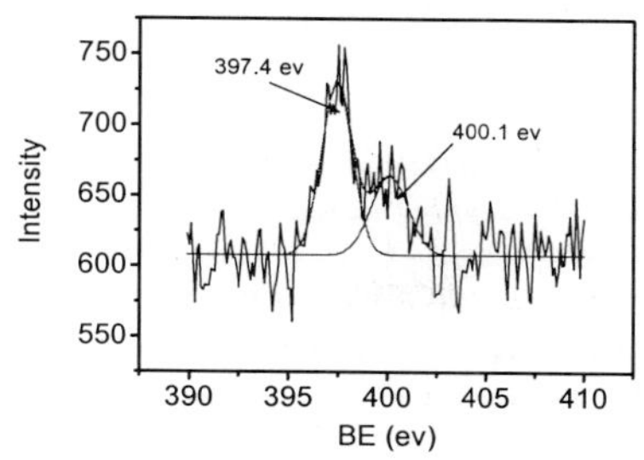

Fig. 4 A typical atomic force microscopy image of the Fe-N films. The RMS ranges from 0.2 to 0.8 nm, which was not apparently affected by the N_2 pressure and substrate temperature.

Fig. 5 XPS spectra of N1s. The nitrogen forms two kinds of bonds with the bond energy at 397.4 and 400.1 ev respectively.

pressure. Increasing N_2 pressure means to increase the frequency of collision between the ions and the N_2 gas. As a result, the mean free path of the ion gets shorter and thus the deposition rate decreases. In another point of view, with the higher frequency collision between the ions and the gas molecules, the ionization of nitrogen gets higher. Subsequently, the nitrogen concentration in the films becomes higher. This tendency can be seen in fig 3, which shows the nitrogen concentration in the films increases with the increase of nitrogen pressure.

Fig. 4 shows the typical surface morphology of the films. Surfaces of all the Fe-N films are clean and smooth. The RMS of the films range from 0.2 to 0.8 nm, which was not affected much by the working pressure. This is mainly because of the effects of the OPDB filter, which can remove almost all the macro-particles and neutral atoms, hence ensures the quality of the films.

XRD data revealed that the films have polycrystalline structures. The crystalline size is around 5 nm. Varying nitrogen pressure affects little of the crystalline size. Although the Fe-N film has been deposited, as can be confirmed by the XPS spectra of N1s in in fig 5, XRD revealed that there is no α'' or γ phase exist. To investigate deeper the nitrogen effect on the iron film deposited by this technique, experiment of substrate temperature effect was hence carried out.

3.2   Substrate temperature effect:

Fig. 3 shows the nitrogen content in the films that vary with the substrate temperature. It was observed that the stoichiometric nitrogen composition in the film has a highest value at a certain substrate temperature with a constant working pressure. Under working pressure 2.1×10^{-5} torr, the nitrogen concentration in the films reaches the highest at substrate temperature of 170 °C. Both at higher or lower temperature than at the critical point, the nitrogen concentration decrease, which suggests that there is an optimised temperature to form Fe-N bond at a certain N_2 pressure. At the lower temperature, the nitrogen ions have not enough energy to form bond with Fe; while at higher temperature, severe vibrational energy may help implanted nitrogen to escape from the film, hence decrease the concentration.

XRD profiles revealed that the films are also polycrystalline, as showed in fig 6, yet no apparent Fe-N phases were identified. With substrate temperature increase from 110   to 350 , the crystalline size increase from 5 to 11 nm, which indicates that higher substrate temperature could enlarge the dimension of particle size, which is a case being the same to melting point,[10] The melting point of nano particles decreases with the solid dimension.

Fig. 7 shows the saturation magnetization of the films deposited under various conditions. No α'' phase is apparent in the film, which was considered as possessing a giant magnetic moment phase[1-7]. However the saturation magnetizations of the films containing nitrogen are still higher than that of the pure iron. This was consistent with the study of Bobo *et al,*[11] who found an enhanced saturation magnetization in the amorphous nitrogen contained iron films. It can be

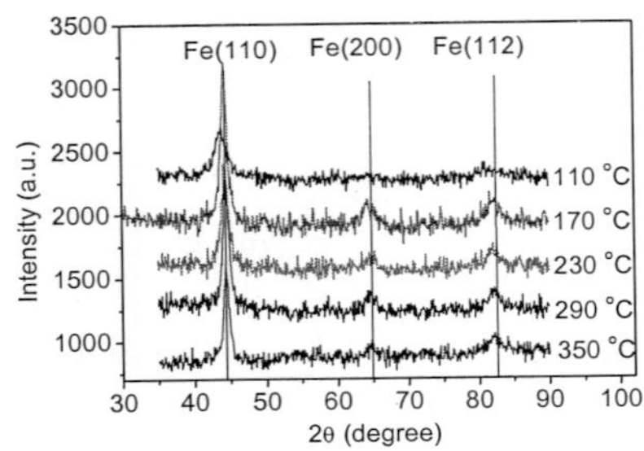

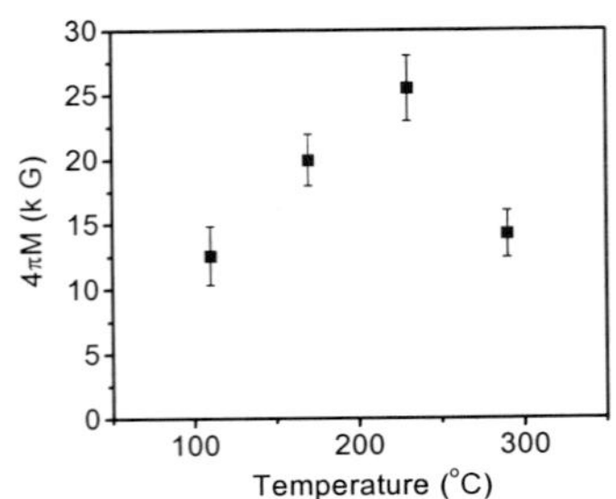

Fig. 6 XRD profiles of the films show polycrystalline structure.

Fig. 7 T_s dependence of the saturation magnetization of the Fe-N films.

explained by that the introduction of nitrogen changes the valence states of Fe atoms. The sp-orbital of a nitrogen atom hybridizes upon interacting with four neighboring Fe atoms. This gives rise to a nitride tetrahedron involving bonding shared electron pairs and nonbonding lone electron pairs.[12] As a result, the valence states of the host Fe atoms are modified with Fe^+, Fe^{dipole}, etc.[13,14] In the former, the Fe donates 3d electrons to the N acceptor. In the latter, the Fe 3d electrons are expelled by the lone pair to an outer shell of itself, 4p or 4d, corresponding to the higher antibonding energy. The N^{3-} and its electrons contribute nothing to the magnetization. The total momentum of the Fe^{n+} varies from 2.5 to 3.0 and the Fe^{dipole} is 4.0 $(3d^54s^24p^1)$ or even 5.0 $(3d^54s^24d^1)$. The average momentum of an isolated tetrahedron $(N^{3-} + 3Fe^+ + Fe^{dipole})$ is then 2.875 or 3.125, being 25 ~ 40% higher than it is a pure Fe (2.22 as measured).[12,15,16]

4. Conclusion

Nitrogen-contained iron films have been deposited using a modified OPDB-FCVA technique under various N_2 particle pressure and varies substrate temperature were found to possess a polycrystalline structure with crystalline size from 5 to 11 nm at different substrate temperature. All the films show very smooth surface with the RMS range from 0.2 to 0.8 nm. No α" phase was found in the as-deposited films. The films exhibit an enhanced saturation magnetization comparing with pure iron films, which indicates that the implanted nitrogen changes the valence state of Fe into Fe^+ or even Fe^{dipole}, and hence improve the magnetic moment of each Fe atom.

*Assistance in the magnetic measurement of Mr Chen Qun and helpful discussions with Professor Ding Jun are gratefully acknowledged. Financial Support by Nanyang Technological University is also acknowledged.

[1] T. K. Kim and M. Takahashi, *Appl, Phys. Lett.* **20A**, 492(1962).

[2] M. Komuro, Y. Kozono, M. Hanazono, and Y. Sugita, *J. Appl. Phys.* **67**, 5126 (1990).

[3] K. Nakajima and S. Okamoto, *Appl. Phys. Lett.* **56 (1)**, 92 (1990).

[4] E. Y. Jiang, C. Q. Sun, J. E. Li, and Y. G. Liu, *J. Appl. Phys.* **65 (4)**, 1659 (1989).

[5] A. Morisako, Mitsunori Matsumoto, Masahiko Naoe, *J. Appl. Phys.* **69 (8)**, 5619 (1991).

[6] D. C. Sun, C. Lin, E. Y. Jiang, S. W. Wu, *Thin Solid Film* **260**, 1 (1995).

[7] J.F. Bobo, H. Chatbi, M. Vergnat, L. Hennet, Lenoble, Ph. Bauer, and M. Piecuch, *J. Appl. Phys.* **77**, 5309 (1995).

[8] X. Shi, D. Flynn, B. K. Tay and H. S. Tan, Filtered Cathodic Arc Source, PCT/GB96/00389, **20**, February 1995.

[9] X. Shi, B. K. Tay, S. P. Lau, Int. *J. Modern Phys. B* **14**, 136 (2000).

[10] Sun CQ, Wang Y, Tay BK, Li S, Huang H, Zhang Y., J. PHYS. CHEM. **B 106(41)**, 10701-10705 (21 SEP 2002).

[11] J. F. Bobo, H. Chatbi, M. Vergnat, L. Hennet, O. Lenoble, Ph. Bauer, and M. Piecuch, *J. Appl. Phys.* **77(10)**, 5309 (1995).

[12] C. Q. Sun, *Appl. Phys. Lett.* **72**, 1706 (1998).

[13] C. Q. Sun, B. K. Tay, S. P. Lau, X. W. Sun, X. T. Zeng, S. Li, H. L. Bai, H. Liu, Z. H. Liu, and E. Y. Jiang. *J. Appl. Phys.* **90(5)**, 2615 (2001).

[14] C. Q. Sun, *Appl. Phys. Lett.* **72(14)**, 1706 (1998).

[15] C. Q. Sun, W. T. Zheng, W. H. Zhong, S. Li, "Bond-band-barrier correlation mechanism for the unusual behavior of a nitride" *Communicated.*

[16] C. Q. Sun, *Surf. Rev. Lett.* **7(3)**, 347 (2000).

Electromagnetic Properties of YBCO Ceramics

Sang Heon LEE
Department of Electronics Information and Communication Engineering,
Sun Moon University, Asan, Ching Nam, Korea, 336-840

We present a fabrication of polarity sensing superconducting magnetometer of Wheatstone bridge type. Superconducting sensor was field cooled in the magnetic field of a 0.15T permanent magnet by placing near the N or S pole. Two variable resistors were used to balance the bridge circuit. Passing a current along the superconductor in a transverse magnetic field, a voltage develops between the ends as the current varied. The magnetic field dependence of the induced output voltage of the HTS magnetometer has been tested in the the 10^{-5} to 10^{-4}T range by changing the distance between the pole of a permanent magnet to the S pole of the superconducting magnetometer. The positive voltage increased, when the S pole of the magnet approached the N pole of the sensor.
Keywords: Electromagnetic, YBCO, magnetic field, flux.

I.   INTRODUCTION

When a sufficiently strong current exceeding the critical current passes through a type II superconductor in the mixed state, the Lorentz force exceeds the pinning force between the flux line and the moving charge and the core lattice is set in motion against viscous drag force inducing a voltage drops in the sample [1], [2], [3].

The early works of Kim et al [4] showed that the voltage drop develops in the type II superconductor of the mixed state, and the flux flow resistivity remains zero up to the critical current. Above the critical current and the slope dV/di increases, the voltage drop increases with current until the liner flux flow region is attained. Yamashita [5] showed that the magnetic sensitivity of the magnetometer was about 10^{-4} T using a thick BiPbSrCaCuO film. We report a new type of superconducting magnetometer which makes use of the characteristic change of flux flow resistivity with external magnetic field and field polarity.

II.   EXPERIMENTAL

The 1-2-3 powder used in this experiment was made by the conventional solid state reaction of Y_2O_3, $BaCO_3$ and CuO powders of 99.9% purity and the powders were mixed and calcined at 950 ℃ for 20h in air and then furnace cooled. The calcined cakes were crushed in an alumina mortar with a pestle, mixed with Ag_2O powder of 99.9% purity up to 2wt%, pressed isostatically into pellets. The pellets were sintered at 950℃ for 10h in air, slowly cooled down to 400℃, held for 24h, and then furnace cooled.

The four-terminal method was used during the current–voltage characteristics measurements. The present authors used a conductive silver paste to attach copper leads to the sample. The magnetic characteristics were evaluated with a magnetometer placed a dc magnetic field, applied perpendicular to the surface of the magnetometer. The superconductors was polarized by the field cooling at 77 K with a samarium cobalt rear earth permanent magnet of B = 0.15 T, 46 mm in length and 10 mm in thickness.

III.   RESULTS AND DISCUSSION

The voltage–current characteristics of a doped Ag_2O YBCO showed in Fig.1. The curves (a) and (c) were obtained at B=0 T with 77 K and the curve (d) at B=0 T with room temperature.

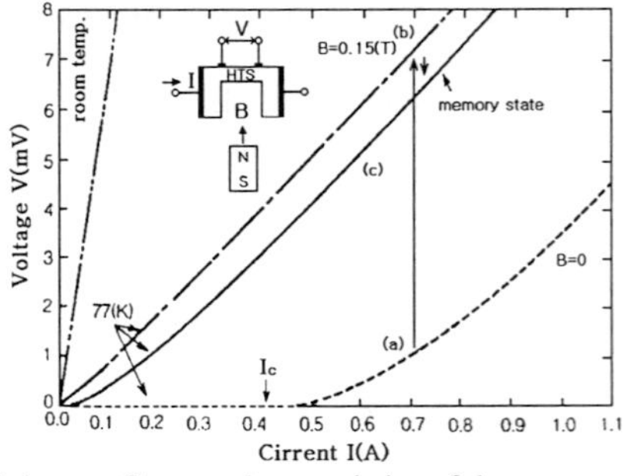

Fig.1 Voltage – Current characteristics of the superconductor.

The curve (b) at 77K gradually approaches the curve(c) after the removal of the external magnetic field. If the voltage is applied again after returning to zero voltage, V–I characteristics are shown in curve(c). This means that the sample is in the memorized state.

The voltage V_{MAG} appeared across the field cooled HTS sample increased with external magnetic field, showing a linear region in the range between 10^{-4} and 3×10^{-3} T, as in Fig.2.

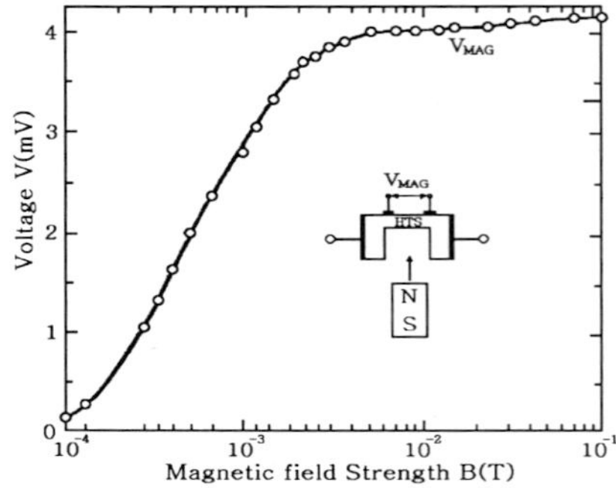

Fig.2 Dependence of the voltage V_{MAG} on externally applied magnetic field at 77K.

The Voltage – Current (V– I) characteristics of the field cooled by the N pole superconducting magnetometer in Fig.3 shows that the Positive $\triangle$ V and negative $\triangle$ V are the increment and the decrement of voltage drop across the HTS sensor with respect to the initial voltage V_0 respectively.

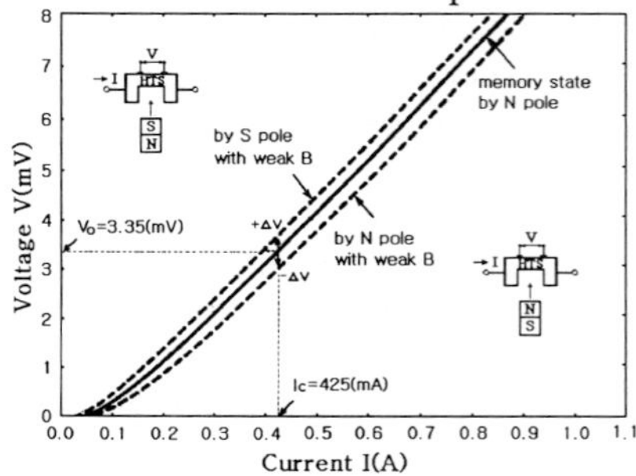

Fig.3 Voltage – Current characteristics of N polarized superconductor.

The decrement of voltage $-\triangle$ V is attributed to the external magnetic flux of N pole, and $+\triangle$ V to the external magnetic flux from S pole. The HTS magnetometer differentiates the polarity of the applied magnetic field and measures the external magnetic field with respect to the initial voltage of the field cooled sample.

The voltage drop across the field cooled superconducting magnetometer varies with external field strength by varying the distance of N and S pole of permanent magnet as shown in Fig.4.

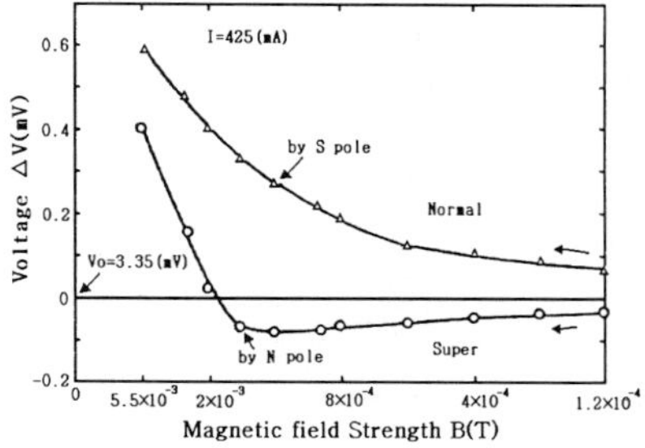

Fig.4 The change in the voltage across the sample memorized by north pole.

The Wheastone Bridge circuit of the field cooled superconducting magnetometer is schematically shown in Fig.5. The output voltage of magnetometer shows approximately linear behaviour with magnetic field in the 5×10^{-5} and 2×10^{-3}T.

Voltage V

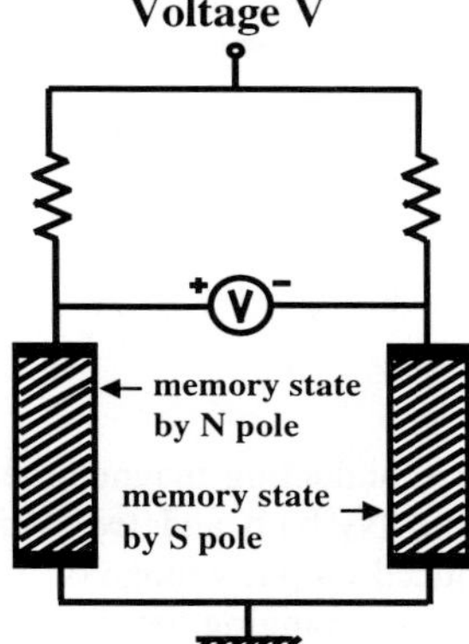

Fig.5 Circuit for measuring magnetic field and polarity.

Superconducting sensor was field cooled in the magnetic field of a 0.15T permanent magnet by placing near the N or S pole and it was assumed that both superconducting sensors had the same magnitude of trapped magnetic field with opposite polarity. The two variable resistors were used to balance the bridge circuit. Passing a current along the superconducting magnetometer a voltage between both ends developed. The voltage developed varied.

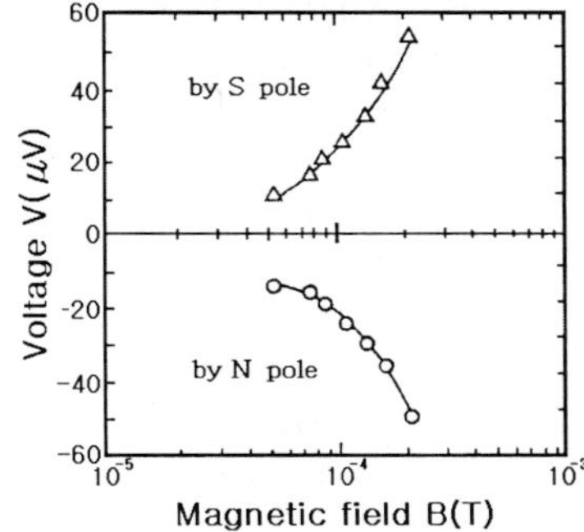

Fig.6 V–B characteristics of superconducting magnetometer.

The magnetic flux trapped in the Ag_2O doped and undoped YBCO sample were measured at 77K in terms of the voltage–current characteristics by using our magnetometer. This magnetometer applying with and without an external magnetic field to the sample is used as superconductor.

In Fig.7, curve (a) is the V–I curve of the magnetometer and the curves (b) and (c) are those for the undoped and Ag_2O doped sample placed to the magnetometer respectively. Under the magnetic field, some superconducting regions of the superconductor are destroyed at the weak link region and the defect.

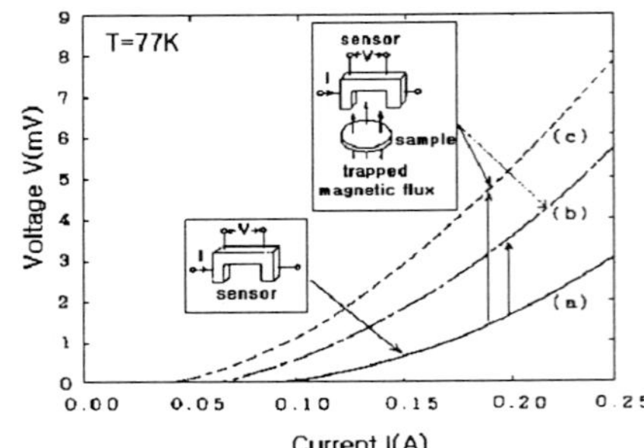

Fig.7 Detection of the trapped magnetic flux in superconducting YBCO specimen by using a superconductor magnetometer.
(a) Current– voltage characteristic of the YBCO magnetometer.
(b) Undoped YBCO specimen is set closely to the magnetometer.
(c) Ag_2O doped YBCO specimen is set closely to the sensor.

As a result, the voltage drop appears across the sample increases with the volume of destroyed regions due to the increase in the magnetic flux trapped in the YBCO superconductor. The larger resistance exhibited in the Ag_2O doped sample in Fig.7(c) is attributed to dopant acting as the pinning center of magnetic flux.

IV.   CONCLUSIONS

We report a fabrication of a superconducting magnetometer of Wheatstone bridge type which consists of a superconductor in liquid N_2 bath and two variable resistors, a null detector. The magnetic field dependence of the induced output voltage of the HTS magnetometer has been tested in the 10^{-5} to 10^{-4}T range and polarity by changing the distance between the pole of the permanent magnet and the S pole of the superconducting sensor. In this case the positive voltage increased, whereas the S pole of the magnet approached the N pole of the sensor.

REFERENCES

[1] K.H.Muller, J.C.MacFarlane,and R.Driver, " Josepson Vortices and Flux Penetration in High Temperature Superconductor", Physica C, vol.158, pp.69-75, 1989.

[2] E.Tjukanov, R.W.Cline, R.Krahn, M.Hayden, M.W.Reynolds, W.N. Hardy, and J.F.Carolan, " Current persistence and magnetic shielding properties of $Y_1Ba_2Cu_3O_X$ tubes', Phys. Rev. B, vol. 36, no.13, pp. 7244-7247, .1987.

[3] J.W.Ekin, "Transport Critical Current in Bulk Sintered $Y_1Ba_2Cu_3O_X$ and Possibilities Enhancement", Adv. Ceram. vol. 2, no.3B, pp. 586–591, 1987.

[4] Y.B.Kim, C.F.Hempstead, and A.R. Strnad, " Flux–Flow Resistance in Type – II Superconductor", Phys. Rev., vol.139, no.4A, pp. A1163-A1172. 1965.

[5] K. Yamashita, and M. Itoh " Fabrication of a magnetic sensor constructed of a thick BPSCCO film" Physica C, vol.372, pp. 1329–1332, 2002.

HYPER-RAYLEIGH SCATTERING CHARACTERIZATION
OF LITHIUM IONS MOVEMENT IN LI$_2$B$_4$O$_7$ CRYSTALS

A.V.Vdovin[1], V.N.Moiseyenko[1], Ya.V.Burak[2]

[1]Physical Department, Dniepropetrovsk National University, 13 Nauchnaya Str., UA-49050, Dniepropetrovsk, Ukraine, e-mail address: mois@ff.dsu.dp.ua

[2]Institute of Physical Optics, 23 Dragomanov Str., UA-79005 L'viv, Ukraine, e-mail address: Vlokh@ifo.lviv.ua

The integral intensity of Hyper-Rayleigh scattering from the bulk of nominally pure ionic conducting crystals Li$_2$B$_4$O$_7$ is studied as a function of temperature in the temperature range, where the break of temperature dependence of ionic conductivity in coordinates ln $(\sigma_{33} \cdot T) = f (1/T)$ is observed. Characteristics of jump-diffusive mobility of lithium ions are determined. The value of the movement activation energy for the temperatures T 400 K estimated from the temperature dependences of hyper-Reyleigh scattering intensity amounts to E$_{am}$ = 0.46 eV, that is close to the value obtained from direct temperature measurements of ionic conductivity.

It is known, that lithium tetraborate crystals Li$_2$B$_4$O$_7$ (LB4) are characterized by quasi-one-dimensional ionic conductivity σ_{33} along the channels of structure in a direction [001], caused by the presence of growth defects (vacancies and interstitial atoms) in lithium sublattice in the temperature range 250 – 700 K. At temperature 400 K the value of conductivity σ_{33}=1.6·10^{-5} Ohm^{-1} m^{-1} [1, 2]. Temperature dependence of ln $(\sigma_{33} \cdot T) = f (1/T)$ shows a break close to T = 390 K where the value of activation energy (E$_a$) decreases from 0.54 eV below 390 K to 0.36 eV above this temperature [2]. In the high temperature region (T 700 K) the ion transport along the directions perpendicular to [001] has been observed [1]. An important role in studying mechanisms of ions temperature migration in ionic conductors plays the methods of elastic and nonelastic light scattering. The method of cooperative scattering on doubled frequency of excitation – the hyper-Rayleigh light scattering (HRS) – are of particular interest in this field, because it possesses high sensitivity to local symmetry changes induced by defects. The HRS line observed for crystals seem to arise unquestionably from defects and static distortions of the surrounding host lattices and hopping mobility of the defects on time scales [3]. The present work is devoted to studies of temperature dependencies of quasi-elastic hyper-Rayleigh light scattering intensity in the range of temperatures where ionic conductivity is observed with the aim of definition of the characteristics of lithium ions mobility in channels of structure along [001] direction.

The samples of nominally pure Li$_2$B$_4$O$_7$ single crystals were grown in the Institute of Physical Optics (L'viv, Ukraine) by Czochralski method. The sample under the study was the rectangular prism of 4x4x10 mm^3 dimensions with planes perpendicular to crystallographic directions [100], [010], [001] accordingly. Hyper-Rayleigh light scattering was excited by radiation of the acousto-optically Q-switched Nd^{3+}: YAG laser (λ = 1.06 µ) with a pulse repetition rate of 8 kHz, duration of pulses 5·10^{-7} sec and with mean excitation power of 200 mW. Measurements were carried out in the Z(YYY)X geometry. The HRS signal was selected by double monochromator of DFS-12 spectrometer and registered by cooled FEU-79 photomultiplier working in a photon counting mode with accumulation. The accuracy of temperature measurements was 0.5 K.

Figure 1 shows the temperature dependence of integral HRS intensity. Starting from T = 400K the sharp increase of HRS intensity with growth of the temperature was observed, correlating with growth of ionic conductivity σ_{33} in this temperature range. It is known that in the range of intrinsic conductivity, the character of temperature dependence of conductivity is determined by the defects formed in a crystal lattice due to thermal energy. In LB4 crystals such defects are Frenkel defects [4] formed in lithium sub-lattice. In this case the conductivity activation energy (E$_a$) includes both the energy of defect formation (E$_F$) and its movement activation energy (E$_{am}$): E$_a$ =

0.5 $E_F + E_{am}$. At further increase of temperature the conductivity is determined by characteristics of disordering of corresponding sub-lattice and depends only on movement activation energy of defects. Assuming that intrinsic conductivity in $Li_2B_4O_7$ crystals takes place at temperatures below 400 K the decrease of an inclination in the dependence of ln $(\sigma_{33}\cdot T) = f\,(1/T)$ with growth of temperature near T = 390K can be treated as transition to conductivity in strongly disordered lithium sub-lattice. Thus it is possible to estimate the energy of Frenkel defect formation in lithium sub-lattice as $E_F = 2\,(E_a - E_{am}) = 0.36$ eV. The mechanism of ionic conductivity is determined by ratio between time of stay of an ion in the lattice site (τ_u) and time of flight between the sites (τ_f). The estimation of specified times within the framework of the diffusion theory of noninteracting particles gives the values of $\tau_u = 4\cdot10^{-8}$ sec and $\tau_f = 4\cdot10^{-13}$ sec for lithium sub-lattice of LB4 crystals, that corresponds to the jump-diffusive mechanism of lithium ions movement [5, 6]. In contrast to the ionic conductivity the HRS intensity at any temperature depends on the contribution of crystal lattice defects of the various nature which are characterized by effective activation energy (E_F*) and jump-diffusive movement of lithium ions [7]. In this case $I_{HRS} \sim \exp\,(-E_a*/kT)$. Then the inclination of the temperature dependence of HRS intensity in the coordinates ln $(I_{HRS}) = f\,(1/T)$ will be determined by the value $E_a* = 0.5E_F* + E_{am}$. The obtained dependence of HRS intensity in the specified coordinates is shown on fig. 2. As is evident from fig. 2 in the range of high temperatures the experimental points can be well approximated by a straight line. The change of an inclination occurs near 400 K. The value of E_a* obtained from the measurements of temperature dependence of HRS intensity amounts to 0.64 eV. Assuming that in the investigated temperature range E_F* $E_F = 0.36$ eV we can estimate the value of $E_{am} = 0.46$ eV. This estimate is consistent with the value $E_a = 0.36$ eV obtained from temperature dependence of ionic conductivity.

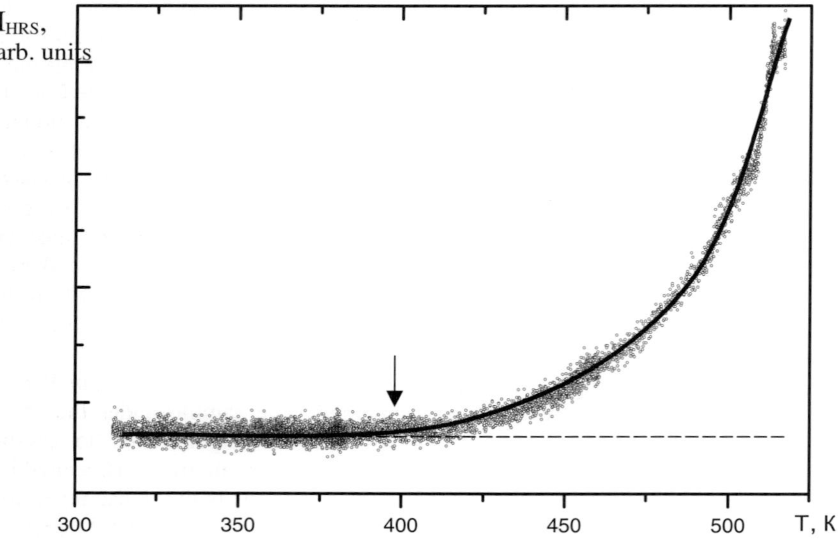

Fig. 1. Temperature dependence of 90^0 hyper-Rayleigh scattering integral intensity.

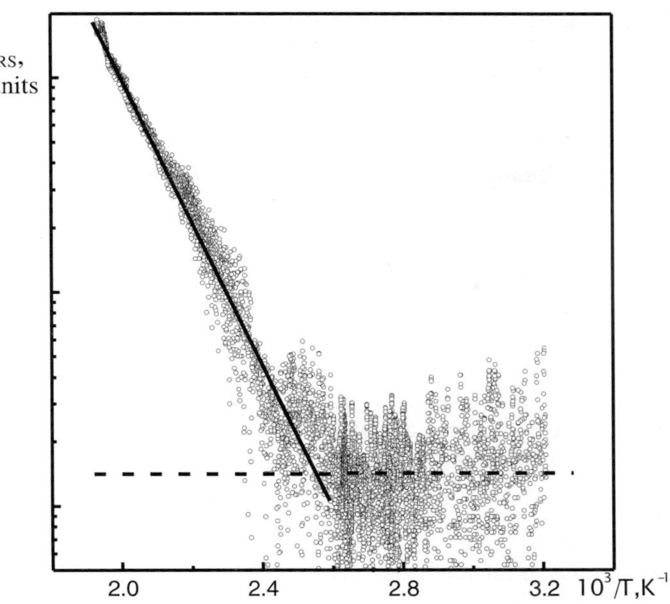

Fig. 2. Temperature dependence of 90^0 hyper-Rayleigh scattering integral intensity in the coordinates *ln I_{HRS} = f(1/T)*.

The sharp increase of HRS intensity with growth of the temperature, correlating with growth of ionic conductivity σ_{33} in the temperature range above T = 400 K, was observed. It is established, that this increase is caused by both growth of Frenkel defects concentration and increase of lithium ions hopping frequency along the channels of structure. The jump-diffusive motion of lithium ions in opposite directions and as consequence fluctuations of their concentration are the reasons of fluctuations of nonlinear susceptibility $\chi^{(2)}$. The activation energy of lithium ions movement determined from the measurements of temperature dependence of HRS intensity amounts to 0.46 eV. This estimate is close to the value E_a = 0.36 eV determined from the temperature measurements of ionic conductivity.

1. Maeda M., Tachi H., Honda K., Suzuki I. Jpn J. Appl. Phys. **33** (1994) 1965-1969
2. Aliev A.E., Burak Ya.V., Lyseyko I.T. Izv. AN. Neorg. mater. **26** (1990) 1991-1993
3. Vogt H. Phys. Rev. B **41** (1990) 1184-1193
4. Button D.P., Mason L.S., Muller H.L., Uhlmann D.R. Solid State Ionics **9&10** (1983) 585-592
5. Physics of Superionic Conductors / M.B. Salamon, ed. / Topics Current Phys., vol.15, Springer,
   Berlin – Heidelberg – New York, 1979
6. Light scattering in solids III. Recent results / Edited by M. Kardona and G. Gyunterodt / Springer-Verlag Berlin-Heidelberg-New York, 1982
7. Dieteriech W., Fulde P., Peschel I. Adv. Phys. **29** (1980) 527

CIRCULAR FLOW OF MR FLUIDS BETWEEN TWO CYLINDERS

J. Huang[1,2], B. Chen[2,*], Y. Ma[3], W. Wang[3]

[1]Chongqing Institute of Technology, Chongqing 400050, P. R. China
[2]Department of Engineering Mechanics, Chongqing University, Chongqing 400044, P. R. China
[3]Department of Applied Physics, Chongqing University, Chongqing 400044, P. R. China

Abstract: Bingham model is used to characterize the constitutive behavior of the magnetorheological (MR) fluids subject to an external magnetic field strength. Based on the momentum equation of continuum mechanics, the effect of applied magnetic field on shear flow within the gap between two cylinders is analyzed theoretically. The expression of the rotating speed is derived to provide the theoretical foundations in the analysis of fluid's circular flow. The results indicate that with the increase of applied magnetic field strength the flow of the fluid changes slow, the yield surface gets close to the inner cylinder.

Keywords: MR fluids; circular flow; yield stress

1. Introduction

MR fluids are suspensions of micron-sized, magnetizable particles in a carrier fluid such as silicon oil, synthetic oil or water. In the absence of a magnetic field applied the suspended particles of the fluids disperse randomly in a carrier fluid. These fluids can flow freely and exhibit Newtonian-like behavior. Upon application of a magnetic field, the particles become magnetized and align themselves, like chains, with the direction of the magnetic field. The formulation of these particle chains restricts the movement of the MR fluids, thereby increasing the yield stress of the fluids. Altering the strength of an applied magnetic field can continuously control the yield stress. The mechanical properties of the MR fluids can be used in the devises controlled by an applied magnetic field[1-4]. The devices that use MR fluids can be classified into direct-shear mode, pressure driven flow mode and squeeze-film mode. In direct-shear mode the fluids are filled in the operational gap between two surfaces, which move in relation to each other. The flow of MR fluids between two cylinders in MR clutches or brakes is typical circular shear flow. In order to understand the flow characteristics of MR fluid, one must analyze the effect of applied magnetic field on shear flow within the operational gap between two cylinders. In this paper, Bingham model is used to characterize the constitutive behavior of the MR fluids subject to an applied magnetic field. Based on the momentum equation of continuum mechanics, the effect of applied magnetic field on shear flow within the gap between two cylinders is analyzed theoretically. The expression of the rotating speed is derived to provide the theoretical foundations in the analysis of MR fluid shear flow.

2. Characteristics of MR fluids

MR fluids can change their rheological characteristics by applying magnetic field. In the absence of an applied magnetic field, MR fluids exhibit a Newtonian–like behavior. However, upon application of a magnetic flied, the behavior of the controllable fluid is often represented as a Bingham fluid having a variable yield strength. In this model, the flow is governed by Bingham's equation:

$$\tau = \tau_y(B) + \eta\dot{\gamma} \qquad \text{at} \quad \tau \geq \tau_y(B) \tag{1}$$

$$\dot{\gamma} = 0 \qquad\qquad \text{at} \quad \tau < \tau_y(B) \tag{2}$$

where τ is the shear stress of fluid, $\tau_y(B)$ the yield stress in response to an applied magnetic field B, η the viscosity of MR fluid with no applied magnetic field, and $\dot{\gamma}$ the shear rate. Equations (1) and (2) indicate that for the shear stress exceeds the yield stress an MR material flows

---

as a Newtonian fluid with a viscosity. Otherwise, it remains viscoelastic property.

3. Circular flow between two cylinders

Fig.1 shows the circular flow of an MR fluid within the operational gap between two cylinders. When the inner cylinder is rotated by the rotating speed ω_1, the MR fluid is sheared and flowed at the rotating speed ω_r in θ-direction. The outer cylinder is rotated by the rotating speed ω_2 because of the shear stress of the fluid. In the cylindrical coordinate system (r,θ,z), z-direction is coincident with the axis of two cylinders, with

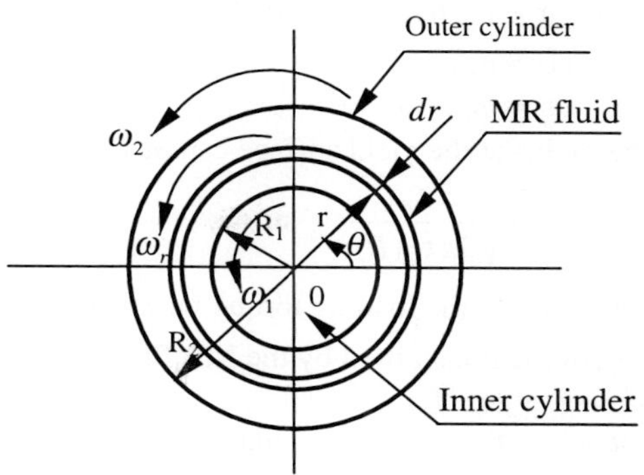

Fig.1 Circular flow between two cylinders

the horizontal direction. In the mode of flow, assume that the MR fluid within the gap is regarded as a cylindrical fluid. Assuming the pressure in r-direction to be no variation, the momentum equation of the shear flow at the narrow gap may be approximated by:

$$\frac{d\tau}{dr}+\frac{2\tau}{r}=0 \tag{3}$$

where $d\tau/dr$ is the shear stress gradient in r-direction, and r the radius of the cylindrical fluid. Integrating the momentum equation (3), shear stress can be indicated as:

$$\tau = c_1/r^2 \tag{4}$$

where c_1 is the integrating constant. The boundary conditions is:

$$\omega_r = \omega_1 \text{ at } r=R_1, \quad \omega_r = \omega_2 \text{ at } r=R_2 \tag{5}$$

where R_1 and R_2 are the radius of the inner and the outer cylinder, respectively. At two surfaces of R_1 and R_2, the shear stresses are respectively:

$$\tau_1 = \frac{c_1}{R_1^2}, \quad \tau_2 = \frac{c_1}{R_2^2} \tag{6}$$

The surface at $\tau_y(B)=c_1/R_y^2$ is defined as yield surface. R_y is the radius of the cylindrical fluid at the yield surface. The field at $\tau > \tau_y(B)$ is yield, and the field at $\tau < \tau_y(B)$ is non-yield. At $\tau_1 > \tau_y(B) > \tau_2$, the fluid of the field between R_1 and R_y is viscous flow. Because the rotating speed decreases with the increases of the radius r, the shear rate $\dot\gamma$ in above equation (1) can be calculated by:

$$\dot\gamma = -r\frac{d\omega_r}{dr} \tag{7}$$

where $d\omega_r/dr$ is the rotating speed gradient in r-direction Applying boundary conditions, the rotational speed can be obtained by equations (1)-(7) as follows:

$$\omega_r = \frac{R_1^2 R_y^2}{R_y^2-R_1^2}\left[\left(\frac{R_y^2-r^2}{R_y^2 r^2}\right)\left(\omega_1-\frac{\tau_y(B)}{\eta}\ln R_1\right)+\left(\frac{r^2-R_1^2}{r^2 R_1^2}\right)\left(\omega_2-\frac{\tau_y(B)}{\eta}\ln R_y\right)\right]+\frac{\tau_y(B)}{\eta}\ln r \tag{8}$$

The fluid of the field between R_y and R_2 is rotated by ω_2 as a solid. According to the above discussion, the flow has the following two particular conditions:

(1) At $\tau > \tau_y(B)$, the fluid within the gap between two cylinders is viscous flow. The rotational speed becomes:

$$\omega_r = \frac{R_1^2 R_2^2}{R_2^2-R_1^2}\left[\left(\frac{R_2^2-r^2}{R_2^2 r^2}\right)\left(\omega_1-\frac{\tau_y(B)}{\eta}\ln R_1\right)+\left(\frac{r^2-R_1^2}{r^2 R_1^2}\right)\left(\omega_2-\frac{\tau_y(B)}{\eta}\ln R_2\right)\right]+\frac{\tau_y(B)}{\eta}\ln r \tag{9}$$

(2) At $\tau < \tau_y(B)$, the fluid within the gap between two cylinders is non-slipping. The rotational speed becomes:

$$\omega_r = \omega_1 = \omega_2 \qquad (10)$$

The radius at the yield surface can be calculated by:

$$r_y = \sqrt{\frac{T_L}{2\pi\tau_y(B)}} \qquad (11)$$

where T_L is the loading torque when MR fluid is non-yield, $2\pi\tau_y(B)R_1^2 \leq T_L \leq 2\pi\tau_y(B)R_2^2$. The torque transmitted by the MR fluid can be calculated by:

$$T = 2\pi r^2 L\tau \qquad (12)$$

where T is the torque transmitted by the MR fluid, L the effective width of the MR effect developed by the MR fluid. Equations (1), (7), (8) and (12) can be manipulated mathematically to yield:

$$T = 2\pi L R_1^2 \tau_y(B) + \frac{2\pi L R_1^3 (\omega_1 - \omega_2)}{h}\eta \qquad (13)$$

where h is the operational gap between two cylinders, $h = R_2 - R_1$.

4. Calculating results and discussion

Fig. 2 shows a calculating example of the rotating speed of the fluid within the gap. For this example, assume the following parameters are given: η =0.94$Pa.s$, R_1=15mm, R_2=16mm, $\omega_1 = 100rad/s$, ω_2=20 rad/s, $r = R_1 + \delta$, $0 \leq \delta \leq h$. Three curves denote the conditions at various magnetic field strengths, respectively. The curve at zero magnetic field is a straight line. The curve at applied magnetic field of 0.3T is a parabola. When applied magnetic field of 0.6T, the curve is a parabola within the field at δ <0.5mm, and $\omega_r = \omega_2$ at $\delta \geq 0.5mm$. This shows the shape of non-yield field. This result indicates that the rotating speed of the fluid changes with the applied magnetic field because of MR effect. Fig.3 shows the place of the yield surface by the calculation at various magnetic field strengths. The calculating parameters are the same as above. The horizontal axis denotes the increment of loading torque ΔT_L. The vertical axis denotes the yield thickness δ_y, $\delta_y = r_y - R_1$. The above of the curves is non-yield field, and the place below the curves is yield field. When ΔT_L=1$N.m$, the yield surfaces are located in the inner cylinder's surface of 0.8mm and 0.1mm at B=0.2T and B=0.8T, respectively. This result indicates that with the increase of applied magnetic field the yield surface gets close to the inner cylinder.

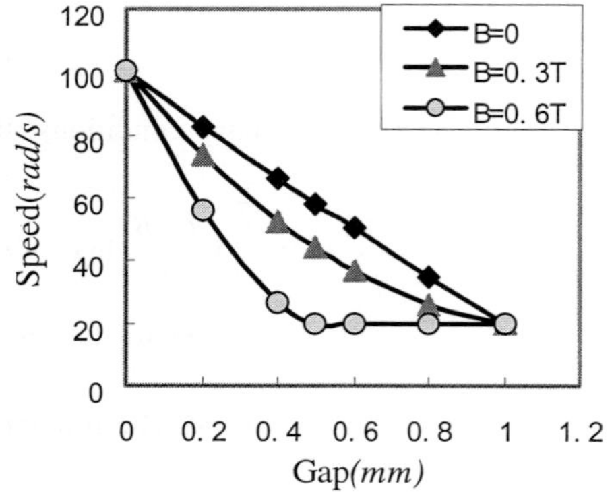

Fig.2 Speed distributions at various magnetic field

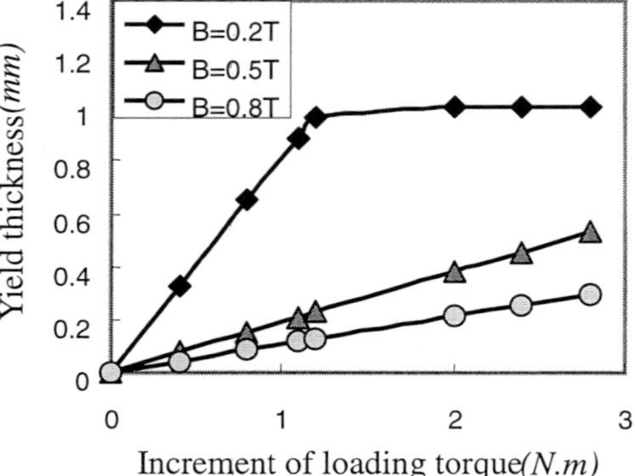

Fig.3 Yield surface place at various magnetic field

5. Conclusions

The expression of the rotating speed is derived to analyze the effect of applied magnetic field on shear flow within the operational gap between two cylinders theoretically. The numerical results show that with the increase of applied magnetic field strength the yield surface gets close to the inner cylinder, and the flow of fluid changes slow.

Acknowledgement

The authors would like to thank the National Natural Science Foundation of China (Grant No. 10272120) for financial support.

References

[1] J.D. Carlson, D.M. Catanzarite and K.A.St. Clair, Commercial magnetorheological fluid devices, Proceeding of the 5th International Conference on Electrorheological Fluids, Magnetorheological Suspensions and Associated Technology, Sheffield. UK, July 10~14 1995, 20-28.

[2] L. Zipser, L. Richter, U. Lange, Magnetorhrologic fluids for actuators, Sensors and Actors A 92(2001) 318-325.

[3] J. Huang, G.H. Deng, Y.Q. Wei and J.Q. Zhang, Application of magnetorheological fluids to variable speed transmission. Proceedings of the International Conference on Mechanical Transmissions (ICMT'2001), Chongqing, Peoples R China, April 5~9 2001: 296-298.

[4] J. Huang, J.Q. Zhang, Y. Yang, Y.Q. Wei, Analysis and design of a cylindrical magneto-rheological fluid brake, Journal of Materials Processing Technology 129(2002) 559-562.

Relation between Dielectric Relaxation Time and viscosity of Heavy Oils

Jong-Ho Park*[1], Chung-Sik Kim[1], Byung-Chun Choi[2], Byung-Kee Moon[2], and Hyo-Jin Seo[2]

[1]Basic Science Research Institute, Pukyong National University, Busan, Korea

[2]Department of Physics, Pukyong National University, Busan, Korea

*Corresponding author e-mail: parkkdp@hanafos.com

Abstract

A Cylindrical–type sample cell has been designed and produced to investigate the dielectric properties in the frequency range from 10^{-2} Hz to 10^{9} Hz of heavy oils using impedance analyzer. The characteristics of the measurement cell are optimized to give high sensitivity. High-sensitivity complex dielectric constant measurements are obtained by calibration with several known fluids. From complex dielectric spectra, we observed the dielectric consisting of two regions for measuring frequency: the low frequency region may be due to diffusion charge transport caused by impurities while the dielectric relaxation mechanism of high frequency region seems to be the non-interacting Debye type. In the high frequency region, it has been observed that the relaxation time was in lineal proportion to viscosity of the heavy oils.

1. Introduction

Dielectric spectroscopy has proved to be a powerful technique in analyzing heterogeneous systems. Dielectric measurements can be used in the oil industry to determine the amount of oil, water and gas in a multiphase flow. The crude oil is a component of different nature. Among these components are dipolar fractions that exist due to different heteroatoms. The concentration of heteroatom are highest in the heavy fractions, resins and asphalts, and play an important role in the forming of permanent dipoles in the heavy petroleum fractions[1,2]. Resins and asphalts are polar and conducting asphalts, which are defined as n-heptance-insolubles, are large polar molecules, which are reported to from aggregates or inverse micelles structures. It is reasonable to believe that these polar compounds are source of the dielectric dispersions measured in asphalts.

However, dielectric spectroscopy implies to measure the complex dielectric constant as function of frequency. Static and high frequency dielectric constant, relaxation time distribution parameter and conductivity are calculated from the spectra. These dielectric parameters depend on the permanent dipole moments, polarisability, dynamics and density of the molecules, ion and free electrons [3,4].

In this paper, we designed cylindrical-type sample cell and calibrated dielectric measurement sample cell. We investigated dielectric relaxation behavior of the refined heavy oil from crude oil and the relationship between viscosity and electrical parameter of heavy oils using impedance analyzer.

2. Experiment

2.1. Measurement and preparation of samples

Table 1 Asphaltenes concentration in the asphalts and viscosity of asphalts each samples.

Samples	Asphaltenes (wt%)	Viscosity (arb.unit)
S1	11.30	2368
S2	12.30	2404
S3	12.40	2639
S4	12.55	2980
S5	12.70	3195

We divided heavy oils samples into 5 classes. The properties of heavy oils are shown in table 1. Dielectric constant of a standard samples (Air; Carbon tetrachloride; Chloroform; n-Heptane; Monochlorobenezene; 1,2-Dichloroethane) and the heavy oils were measured in the frequency regions ($10^{-2} \sim 10^{7}$Hz) using an impedance analyzer (Solatron, SI1260) and the high frequency regions ($10^{6} \sim 10^{9}$ Hz) using an impedance analyzer (HP4191A RF). The experiments were performed at 20 °C on the standard samples and 100 °C on the heavy oils in free atmosphere and the temperature of the sample was measured with a platinum-rhodium thermocouple. The temperature was controlled within ±0.1 °C.

2.2 The design of the impedance cell

In the low and high frequency range, the capacitance and conductance of the cylindrical measurement cell are measured. The admittance, Y, of the cell filled with a fluid of complex dielectric constant $\varepsilon_0\varepsilon^* = \varepsilon_0(\varepsilon' - i\varepsilon'')$ is given by

$$Y = i\omega C + G = i\omega\varepsilon * C_0 = i\omega\varepsilon' C_0 + \omega\varepsilon'' C_0 \tag{1}$$

where C is the capacitance, G the conductance and C_0 the geometric capacitance given by

$$C_0 = \frac{2\pi\varepsilon_0 l_e}{\ln(b/a)}$$

where l_e is the electrical length of the measurement cell. B is the inner radius of the outer conductor and a the radius of the inner conductor. Owing to existing stray capacitances at the end of the cell, the electrical length, l_e, is not equal to the mechanical length, l_m. The measurement cell is designed to maximize the geometric capacitance so that it can be sensitive to small change in the complex dielectric constant. To minimize the influence of impurities in the measured samples, a large sample volume is needed, and mechanical length of 10 cm was designed.

2.3 The estimated complex dielectric constant

Since it is difficult to develop an exact analytical model for the measurement cell, this system is based upon a numerical calibration of the relationship between measured capacitance ($C_{meas.}$) and the static dielectric constant(ε_s') of the fluid. The numerical expression is generally presented as

$$C_{meas.} = C_p + C_0\varepsilon_s' \tag{2}$$

The model parameters C_p (F) and C_0(F) are determined by linear regression on fluids of known dielectric constant, where C_p represents stray capacitances and C_0 the geometric capacitance of the

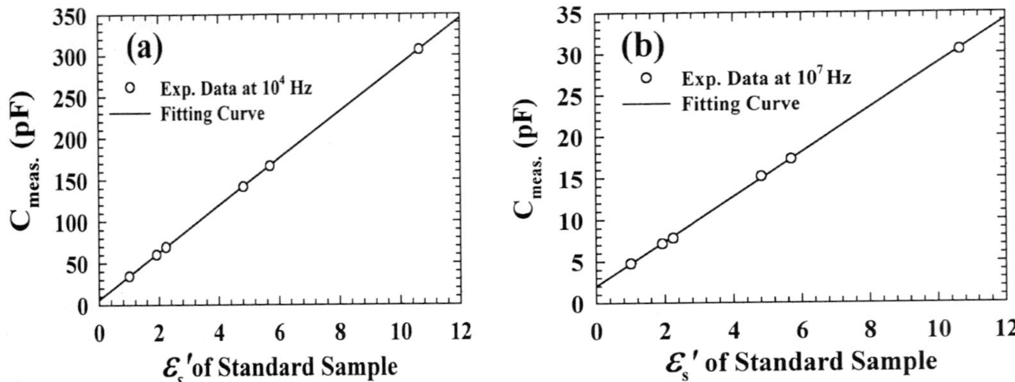

Figures. 1. The relationship between $C_{meas.}$ at 10^4Hz(a), 10^7Hz(b) and ε'_s of standard samples.

measurement cell. Figs. 1 show C_{meas} as a function of ε_s' at 10^4Hz(a) and 10^7Hz(b) in the standard samples. We can see a good linearity between C_{meas} and ε_s'. This implies that when C_{meas} is plotted against ε_s', it should raise upon a straight line with the slope of C_0. Indeed, the present data show reasonable agreement with equation (1) as are seen from Figs. 1. The values obtained $C_0 = 6.186$ ±0.0007 pF and $C_p = 25.223\pm0.0003$ pF at low frequency and $C_0 = 2.675 \pm0.0007$ pF, $C_p = 2.039$ ±0.0003 pF at high frequency.

References in the literature on the imaginary part of the complex dielectric constant of low-loss fluids involve large uncertainties; hence an accurate linear regression model is difficult to obtain. The estimation of the dielectric loss is this system therefore based on equation (3)

$$\varepsilon'' = G_{meas.} / \omega C_0 \qquad (3)$$

where $G_{meas.}$ (S) is the measured conductance containing contribution from both fluid conductivity and dielectric loss.

3 Results and discussion

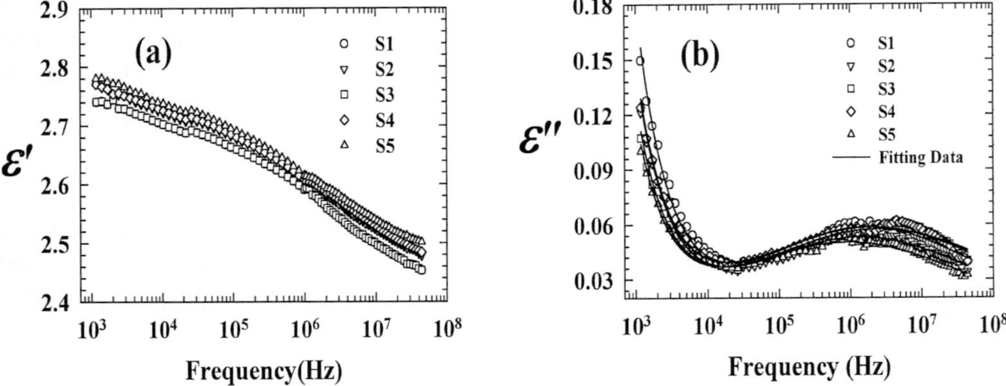

Figures. 2. ε'(a) and ε''(b) vs. frequency characteristics of the heavy oil

Table 2 Fitting parameters, asphaltenes and viscosity for heavy oil

Sample	τ_1 (sec)	σ ($\Omega^{-1}cm^{-1}$)	Asphaltens(wt%)	Viscosity(arb.unit)
S1	6.45×10^{-6}	9.38×10^{-11}	11.30	2368
S2	1.02×10^{-6}	7.11×10^{-11}	12.30	2404
S3	1.23×10^{-6}	5.11×10^{-11}	12.40	2639
S4	1.81×10^{-6}	5.09×10^{-11}	12.55	2980
S5	2.26×10^{-6}	5.33×10^{-11}	12.70	3195

Figs. 2(a) and 2(b) present the variation of ε' and ε' versus frequency using a double logarithmic scale for several heavy oils. ε' and ε'' were obtained by Eqs. 1 and 2. The relaxational peak frequency of ε'' moves to a high-frequency region with increasing viscosity. Moreover, another relaxational behavior appears partially in the low frequency range ($\leq 10^4$ Hz). This implies dielectric behavior of heavy oils cannot be ideally described by the Debye expression. These two different tendencies in the temperature dependence suggest that two dispersion mechanisms should be involved.

The total complex dielectric constant, $\varepsilon*(\omega)$, may be described as follows [5] :

$$\varepsilon*(\omega) = \varepsilon_\infty + \frac{\Delta\varepsilon}{1+(i\omega\tau_1)^m} + \frac{\sigma}{\varepsilon_0\omega}[1+(i\omega\tau_2)^n] \tag{4}$$

The solid lines in Fig. 2(b) are theoretical curves obtained by using Eq. (4). Only the data relative to the imaginary part were used in the fitting. The five parameters, $\Delta\varepsilon$ ($=\varepsilon_s-\varepsilon_\infty$), τ_1,τ_2, m and n were then introduced into the real part of the formula to check whether the calculated values are also in agreement with the measured value of ε'. As shown in Table 2, good correlations were obtained between fitting parameters and viscosity of the heavy oil. The τ_1 was found to increases with increasing viscosity of heavy oil. Also, the τ_1 increases with increasing asphaltene contents The value of τ_1 increased with increasing asphaltene contents, suggesting that viscosity of the heavy oil increased with asphaltenes accumulation

Acknowledgments. This work was supported by Korea Research Foundation Grant (KRF-2002-070-C00042).

References

[1] R.R.F. Kingdom, " An Introduction to the physics and chemistry of petroleum," John Wiley & Sons Ltd., Chichester, New York, Brisbane, Singapore, 1983

[2] H. P. Maruska and B. M. L. Rao, Fuel Csience & Technology Int'l, 5, 119 (1987)

[3] N.E. Hill, W.E. Vaughan, A.H. Price and M. Davies, "Dielectric properties and molecular behavior", Van Reinhold Company Ltd., London, New York, 1969

[4] R. Bartnikas, " Electrical insulating liquids, ASTM, Chelsea, MI, 1994

[5] K. S. Cole and R. H. Cole, J. Chem. Phys. 9, 341 (1941)

AUTHOR INDEX